Weighed in the Balance

George Phillips, Principal of the Laboratory, 1842–1874

Laboratory of the Government Chemist

Weighed in the Balance
A history of the Laboratory of the Government Chemist

P W HAMMOND AND HAROLD EGAN

HMSO

Laboratory of the Government Chemist
is an Executive Agency of the
Department of Trade and Industry

QUEEN'S ROAD, TEDDINGTON, MIDDLESEX, TW11 0LY

THIS BOOK IS DEDICATED TO
ALL STAFF OF THE LABORATORY
PAST AND PRESENT

HMSO publications are available from:

HMSO Publication Centre
(Mail, fax and telephone orders only)
PO Box 276, London SW8 5DT
Telephone orders 071-873 9090
General enquiries 071-873 0011
(queuing system in operation for both numbers)
Fax orders 071-873 8200

HMSO Bookshops
49 High Holborn, London WC1V 6HB
(counter service only)
071-873 0011   Fax 071-873 8200
258 Broad Street, Birmingham B1 2HE
021-643 3740   Fax 021-643 6510
Southey House, 33 Wine Street, Bristol BS1 2BQ
0272 264306   Fax 0272 294515
9–21 Princess Street, Manchester M60 8AS
061-834 7201   Fax 061-833 0634
16 Arthur Street, Belfast BT1 4GD
0232 238451   Fax 0232 235401
71 Lothian Road, Edinburgh EH3 9AZ
031-228 4181   Fax 031-229 2734

HMSO's Accredited Agents
(see Yellow Pages)

and through good booksellers

Contents

page

Preface   vii

Acknowledgements   x

Sources and Abbreviations   xii

List of Illustrations   xiv

I   Introduction: Government Science and
the Revenue pre 1842   1

II   Excise and Inland Revenue Laboratories 1842–1894   9
     1.   The First few years   11
     2.   Growth and Consolidation   33
     3.   Beer and Spirits Work   49
     4.   Staff and Training   68
     5.   The Referee Analysts   86

III   The Customs Laboratory 1861–1894   101
     6.   The Customs Laboratory   103

IV   The Government Laboratory 1894–1911   119
     7.   Formation of the Government Laboratory   121
     8.   A New Home   127
     9.   The Work of the Laboratory   137
     10.   Committee Work   148
     11.   Change   156
     12.   Reorganisation   163

V   The Department of the Government Chemist
1911–1959   169
     13.   The New Department   171
     14.   The First World War and its Aftermath   179

15.   Increasing Versatility 1922—1939   186
16.   Research   200
17.   War Again   211
18.   Recovery   220

**VI   The Laboratory of the Government Chemist
         1959–1991**   241
       19.   A New Beginning: DSIR   243
       20.   The Changing Scene   258
       21.   Agency and After   272

Appendixes   287
       1.   The Laboratory Today   289
       2.   Principals etc.   290
       3.   Organisational Development   292
       4.   Reports of the Government Chemist and
             Laboratory archives   293
       5.   The New Government Laboratory   297
       6.   Examination Papers   305

Bibliography   309

Notes   311

Index   356

Preface

THE LABORATORY OF THE GOVERNMENT Chemist was founded at a period when interest in science was at a high point. The Chemical Society (now the Royal Society of Chemistry) was founded in 1841 for example, the year before the Laboratory came into existence, but at a time when official Government interest in science was small. This lack of interest remained for most of the nineteenth century and meant that the government was very reluctant to support science directly, or indeed scientific and technical education. The Laboratory grew against this background, to a large extent in defiance of the official attitude. As part of the department of Excise, and then of the department of Inland Revenue it grew and flourished, supported indirectly by the Treasury, because the employing departments could see the utility and necessity of having scientific advice and help available to support their other sources of advice and help. The period in which the Laboratory was growing was also one in which the profession of chemistry was growing, and the need for professionalism began to be recognised, i.e. the responsibility of members of the chemistry profession towards the general public, as well as the need for recognised qualifications. The history of the Laboratory helps to illustrate these trends, and demonstrates the way in which two government departments (Excise and Inland Revenue) recognised the need for the staff of their Laboratory to be scientifically educated, in defiance of the general official climate, and indeed paid for them to be so educated, almost from its foundation.

I have tried to give a picture of the development of the Laboratory, and relate it to scientific and social developments outside, in government science and science in general. To a large extent the development of the Laboratory illustrates in practical terms the growth in scientific support of health, environmental and revenue protection in Britain. I have also tried to show the human side of what is inevitably mainly an institutional history. I have not tried to mention everything that happened, such a book would inevitably have become a mere catalogue, but have picked out what seemed to be the most important events and developments. I have dealt with the earlier parts of its history, up to 1900, in slightly greater detail, since this was less well known, and is thus of greatest interest.

Today the Laboratory of the Government Chemist is one of the most modern analytical laboratories in the world. It has a dual origin in the Laboratory set up by the Board of Excise in 1842, and those set up by the Board of Customs in 1860. The development of each laboratory is traced separately until they merged in 1894, and then lost connection with the separate departments with the formation of the Board of Customs and Excise in 1909 and the creation of the separate Department of the Government Chemist in 1911. The title Government Chemist dates from 1911, before which date the head of the Laboratory was known as the Principal. The title was retained when the Laboratory ceased to be a separate Department and became part of the Department of Scientific and Industrial Research in 1959. Those and subsequent changes are summarised in Appendix Two. The present book describes the changes, the reasons for them and the manner in which they developed with, and served, the public need.

No comprehensive history of the work and development of the Laboratory has hitherto existed. A number of brief histories have been written, one for the 100th anniversary, written by the then Deputy Government Chemist, Dr A G Francis, and one by a serving Government Chemist, Dr David Lewis. Another, a small privately printed book by 'John St Clair Cholmondeley formerly Assistant of Excise' entitled *The Government Laboratory* appeared pseudonymously in 1902. This contains a valuable account of some of the early work, and is rather critical of the Laboratory as it was operated in 1902. Several more or less popular articles have appeared at various times, the first by Charles Dickens in his own magazine *All the Year Round* in 1877. From that article the title of this book is derived. These publications, and others, are noted at the appropriate part of the text, and/or the bibliography.

Finally, a few words on the circumstances in which the Preface is signed by only one person although the title page has two authors. Dr Egan, from the outset of his career in the Laboratory, and he joined in 1943 soon after it celebrated its 100th anniversary, intended one day to write its history. He started collecting together records as soon as he could and talking to older members of staff. After he retired he started to sort materials out and put them together, but before he could make progress on a text he died suddenly in June 1984. He had a very distinguished career in the Laboratory, serving in most of the larger departments, and ended with 11 years, between 1970 and 1981, as Government Chemist, a post of which he was particularly proud because of its history. He was only the second person to join the Laboratory in a junior capacity, to serve his whole career there, and to end in the senior post. Dr Egan was widely known and respected internationally for his work in analytical chemistry, and thus particularly well equipped to write the history of a leading analytical laboratory. This was especially so since his interests spread far beyond the scientific. He had studied and published local history material, particularly on his area of London, and collected books avidly, particularly of London and the Home Counties. He also collected much else, what are now known as ephemera, bus and train tickets for example, and coins, of which he had a very large collection. He was very proud of his election as Fellow of the Royal Numismatic Society. He had a great sense of humour, and was a great communicator, interested in people. He had enormous energy. It was particularly sad that a man of so many parts did not live long enough to put these talents to full use in writing this History, a project close to his heart. After his death, having served in the Laboratory for nearly as long a time as Dr Egan I offered to take over the task. I can only hope that he would have approved of the final result.

Writing this history has given me great pleasure. Thanks go to many colleagues, past and present, who have helped in this work (see Acknowledgements). My apologies to anyone whose name has been inadvertently omitted.

PWH
Laboratory of the Government Chemist
(June 1992)

Acknowledgements

IOWE THANKS TO VERY MANY PEOPLE, colleagues at the Laboratory, ex-colleagues, and others, librarians, archivists and staff at other libraries, for their help while writing this book. My thanks must first go to present colleagues for their help. First to Mr Ronald Lees, Head of the Services Division for his very many helpful comments and suggestions and for providing the necessary facilities over the years that this book has taken to write, to David Cottrell for his excellent photographs of divers difficult documents and apparatus, Dr Colin Richards for, *inter alia*, considerable help with the methylated spirit regulations past and present, Mr John Hunt for explaining to me the the intricacies of Agency status, Professor Geoffrey Phillips for reading and commenting on later chapters and for helping with his expertise on the Dangerous Drugs Acts, Mr Paul Beard for help with current Questioned Documents work, Mr Terry Alliston for useful comments on early chapters, Miss Vimy Seth for typing the first few chapters, and latterly Miss Elaine Bradley for typing excellently most of the book, making out my (somewhat) difficult handwriting. Finally my colleagues in the LGC Library who probably wondered if I ever did anything apart from write this History, Ian Neville for getting many obscure books and articles at whose titles he must have occasionally wondered, Andy Wheeler, who spend some time in the PRO at Kew on my behalf, and Euan Bull, and particularly Joy Ruthven, for their help and interest.

Next I must thank Mrs Patricia West, Archivist to DTI for considerable help with LGC archives, both those held by the PRO and those still in her hands, Mr Peter Hoey the Librarian and Mr Ron Hudson of the Royal Society of Chemistry Library for their help and hospitality, Ms Gill Furlong, Archivist to University College London for allowing me to consult records in her care, and Ms Elizabeth Gibson, Records Manager to the College for consulting her registers and solving some difficult problems. The staff of the Inland Revenue Library were also most helpful. Miss Sarah Leonard of the Science Museum helped find a number of the illustrations from their photographic collections.

Thanks are also due to the staff of HM Customs and Excise Library for their hospitality on numerous occasions. To Mr Gilbert Denton, until recently Archivist there I owe particular thanks for his very generous help on all possible occasions, particularly with CUST 121 Treasury Excise files. Had he not been sorting this collection of Excise documents very much important and useful information would not have been unlocked for use.

To Dr John G A Griffiths (member of staff 1932–1947) I owe a great deal of useful information and suggestions, as well as thanks. I have to thank him also for sparing the time for many discussions, for reading my drafts, making many perceptive comments and contributing his memories of LGC, I am sorry that I could not use all his suggestions. To Dr Bob Mesley, until recently a Superintendent at LGC, I owe many thanks for the use of his unpublished work on the Laboratory and the liquor duties, his comments on the chapters dealing with alcholometry and a number of the references in the early history of the Customs laboratories.

Many thanks go to Mrs Daphne Egan and to Mrs Muriel Martin for talking to me about the late Harold Egan, enabling me to give a more rounded picture of him.

Lastly my wife Carolyn helped me in sharing the research at the PRO, Kew and by listening to me talk about this work at great length. As usual she helped immeasurably.

PWH

Sources and Abbreviations

THIS BOOK IS BASED ON PRINTED AND unprinted sources, cited in the footnotes in the usual way. Many of the Laboratory archives (registers, correspondence, work books and files) are either in the Public Record Office at Kew or are about to be deposited there. The first files were deposited in the 1960s and given the access code DSIR 26. They are referred to as:

DSIR 26/   followed by a piece number if at Kew

DSIR 26/–   if the record has been deposited but not yet numbered

CUST   with a number and piece number following are Customs (Excise) records at Kew

CUST 121 TE   are Treasury Excise letters about to be deposited at Kew

Before all of these codes the letters PRO are to be understood.

Much information on work, numbers of samples, and staff numbers comes from the Annual Reports of LGC and its predecessors, (see Appendix 3). If the year is clear these have not necessarily been referenced, unless a direct quotation is used. They are then referred to as:

Report, 1898, which means the *Report* of the Principal Chemist for the year ending 31 March 1898, and so on. The year of publication is not necessarily the same. After 1960 the year of the report ended on 31 December, ie, *Report* 1962 means the calendar year 1962. *Report* 1959 refers to the year April 1958 – December 1959. Customs *Report* refers to the *Report* of HM Commissioners of Customs.

Cholmondeley: John St Clair Cholmondeley, *The Government Laboratory*, London, 1902

Highmore: Nathaniel J Highmore, *The Excise Laws*, 1899 and 1923

Mesley: R J Mesley, *Liquor Duties and the Government Laboratory*, 1974, unpublished manuscript (DSIR 26/–)

Plain Papers: John Owens, *Plain Papers relating to the Excise etc.*, Linlithgow, 1879

Report 1844: *Report from the Select Committee on Tobacco Trade*, 1844, Parliamentary Paper 565

Tanner: A E Tanner, *The Tobacco Laws and their Administration*, Stroud, 1898

Tate: Francis Tate, *Alcoholometry*, London, 1930

The Laboratory, with a capital, always refers to LGC or to its predecessors. The building that it occupied 1897–1964 was usually referred to as 'Clement's Inn' and I have so referred to it in this work.

A list of LGC records held by the Public Record Office is available on request from the Laboratory.

List of Illustrations

Frontispiece George Phillips, Principal of the Laboratory, 1842–1874
1 Clarke's Hydrometer
2 Act establishing the legal use of Sikes' Hydrometer, 1816
3 The Pure Tobacco Act, 1842
4 Excise General Order, 1842
5 Excise Building, Old Broad Street
6 Microscope of 1841
7 Dr Andrew Ure
8 Balance of c.1840
9 Professor Thomas Graham
10 Drawings of microscopic characteristics of tobacco
11 Inland Revenue General Order, 1871
12 Sikes' hydrometer
13 Somerset House, London
14 Ross microscope of c.1862
15 Report on mislabelled coffee and chicory, 1853
16 Sartorius Balance, 1876
17 Appointment of Phillips and Bell as Lime juice inspectors, 1868
18 Thomas Dobson
19 Report of Phillips and Dobson on the use of sugar in brewing and distilling, 1847
20 Bate's saccharometer
21 Sikes' glass hydrometer
22 Advertisement quoting Laboratory results, 1895
23 Patent of George Phillips, 1847
24 Laboratory in University College, London, 1846
25 Professor A W Hofmann
26 Regulations governing behavior of Laboratory students, 1860
27 Note on suitable lodgings, c.1885
28 Dr James Bell, Principal of the Laboratory, 1874–1894
29 Richard Bannister, Deputy Principal of the Laboratory
30 Sale of Food and Drugs Act, 1975

31 Bell, *Analysis and Adulteration of Food*, 1881
32 Senior staff of the Laboratory, c.1894
33 Concert programme, 1890
34 Custom House, London
35 Wine distilling apparatus, 1880
36 Laboratory report on 'factitious' wine, 1864
37 Tea testing cup and bowl
38 Sir Thomas Edward Thorpe, Principal of the Laboratory, 1894–1909
39 'Clement's Inn' building
40 Conversazione programme, 1897
41 Brass plate from Clement's Inn building
42 Laboratory change of address, 1897
43 Steam oven and connected steam bath
44 Fume cupboard
45 Typical Clement's Inn laboratory
46 Refrigeration unit
47 Crown Contracts laboratory, Clement's Inn, 1902
48 Letter to Thorpe from Captain Scott, 1901
49 James Connah, Acting Government Chemist, 1920–1921
50 Main Laboratory, Clement's Inn, 1902
51 Letter to Thorpe from Professor Rutherford, 1907
52 Dr A G Francis, Acting Government Chemist, 1944–1945
53 Apparatus for determining the original gravity of beer, 1913
54 Sir James Dobbie, Principal of the Laboratory and Government Chemist, 1909–1920
55 Specific Gravity bottle
56 Electric heaters for wine and spirit distillation, 1917
57 Staff of the Tobacco Department, 1917
58 Sir Robert Robertson, Government Chemist, 1921–1936
59 Cartoon of c.1930, showing Sir Robert Robertson
60 'Was it Sour', cartoon of c.1928
61 Tobacco section, Clement's Inn, c.1935
62 'The New Captain comes aboard', cartoon of c.1935
63 Main Laboratory, Clement's Inn, 1935
64 'Droll Stories about Paints', cartoon of 1927
65 Hilger infra-red spectrometer, 1922
66 Tate saccarometer
67 Sir John Fox, Government Chemist, 1936–1944
68 Rock salt prism, used late 1930s in infra red work
69 Dr George Bennett, Government Chemist, 1945–1959
70 Polarograph, 1964
71 Ultra violet spectrophotometer, 1966

72   Atomic absorption spectrophotometer, 1973
73   Edwin Nurse, Acting Government Chemist, 1959–1960
74   Dr David Lewis, Government Chemist, 1960–1970
75   Fluoride determination, 1964
76   Automatic beer analyser, 1974
77   Tobacco section, Cornwall House, 1966
78   Smoking machine, 1989
79   'Hey for the Open Road', cartoon of 1948
80   Cornwall House, Stamford Street, London
81   Food laboratory, Cornwall House
82   Scanning electron microscope
83   Modern balance
84   Dr Harold Egan, Government Chemist, 1970–1981
85   Dr Ronald Coleman, Government Chemist, 1980–1987
86   Cornwall House from Waterloo Bridge
87   Alex Williams, Government Chemist, 1987–1991
88   New LGC building at Teddington
89   Dr Richard Worswick, Government Chemist, 1991–
90   Modern optical microscope
91   Entrance to LGC building

Illustration Acknowledgements

I wish to thank the following for permission to use illustrations from their collections: H.M. Commissioners of Customs and Excise for number 15; The Trustees of the Science Museum for numbers 6, 8, 14, 16; The Editor of *Punch* for number 79; the Anne Ronan Picture Library for number 24 and the Government Chemist for all of the others.

Part I
Introduction: Government Science and the Revenue before 1842

Government Science & the Revenue, pre 1842

THE FOUNDATION OF A LABORATORY TO support the Government in its efforts to protect the revenue can be traced back to 1842. This Laboratory later became the Government Laboratory, the Department of the Government Chemist and finally the Laboratory of the Government Chemist. This body is thus the oldest official chemical laboratory in Britain, and perhaps in the world.[1] The need for such a laboratory arose from a desire to protect the revenue, and not from any perception that an official body was needed to protect in any way the population or the environment in which we live. This role came later, as will be seen.

Chemistry had been used on behalf of the government for many years before 1842, although it was largely on an *ad hoc* basis, such as in the work of the Royal Mint on the melting points of metals in the eighteenth century, and in the Royal Arsenal. However the Laboratory of the Government Chemist owes its origin to the duties which had been levied on beer, wine, and tobacco for many centuries. The need for a quantitative basis for the assessment of revenue payable on alcoholic products was of great importance in the eighteenth century, and there may have been small Excise laboratories set up to deal with various problems, the original need for science being felt in the raising of revenue from wine imports. The first monetary duty on imported wine was fixed in 1303 (in the *Carta Mercatoria*[2]). Duties were levied on the volume of spirit imported at this stage however, which only required the application of the old art (or craft) of 'gauging', that is measuring casks to establish the volume of the contents. This applied both to beer and spirits for many years. It was not until 1688 when the concept of 'proof' spirit (a somewhat nebulous concept for many years), as a measure of alcoholic strength was introduced that it was necessary to employ some method of testing the strength of spirits.[3]

The methods used at first to determine the strength of the spirit were very crude. One of the earliest was flammability, by burning off the alcohol from a mixture and weighing the remainder. Another test used was the shaking of the spirit in a phial, when the number of beads which formed at the edges of the surface, together with the speed and duration of formulation, was a measure of the strength of the spirit. The best known method was to mix the spirit with gunpowder and then set light to it. If it burnt steadily it was proof spirit but if there was any degree of explosive violence in the burning it was over proof.[4]

Less crude and more scientific methods were needed by the revenue services the more so as the spirit duties were increased through the years, and differences in strength became of greater financial importance. Methods which would do this had been known for some time. They measured the specific gravity (density), of the spirit, that is the ratio of the weight of a given volume of the spirit to the same volume of water. The volume of spirit will weigh less than the water by an amount proportional to the strength of the spirit. By the early eighteenth century three main methods were known by which density could be measured, the gravity bottle, the hydrostatic balance, and flotation instruments.[5] Only the last of these was sufficiently robust for revenue officers to use in the field on their visits of inspection to brewers and distillers.

Flotation instruments, known as hydrometers and which float at a level corresponding to the density of the liquid, were developed initially at the end of the seventeenth century, but the first practicable instrument was developed by John Clarke of London about 1725. He called it his 'Brandy prover' and it seems to have been used unofficially by Excise officers almost from its invention.[6] Clarke's hydrometer received official recognition in an Act of 1787.[7] As time passed Clarke's hydrometer was refined by the addition of interchangeable weights so that it indicated the strength of all spirits relative to proof.

The results given by Clarke's instrument were not always accurate. As early as 1760, in a letter sent to all Collectors on 30 August of that year, the Board of Customs had directed that a sample of imported spirits about which doubt arose as to the correct charge because of a false hydrometer indication (due to obscuration by added sweetening matter), should be sent to the Board. The letter gave

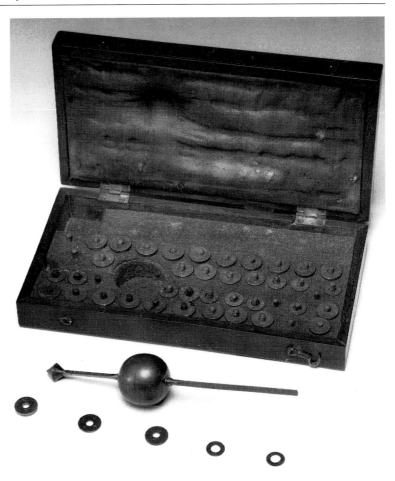

1 Clarke's Hydrometer, late
eighteenth century

detailed directions for examining such spirits in the port of importation, using a crude system of surface tension observations.[8] It seems probable that few if any samples were actually sent to the Board in London; nor was it likely that anything resembling a laboratory was available for testing samples there. The same Act which gave statutory recognition to Clarke's hydrometer also indicated the need for further improvement. The government therefore invited the Royal Society to look into the best means of ascertaining the proper proportion of duty to be paid by any kind of spirituous liquor that came before the Officers of Excise. The work was undertaken by Sir Charles Blagden, secretary of the Society, who concluded in 1790 that the most accurate method would be by measurement of specific gravity. Working with George Gilpin, the Clerk to the Society, he used a

ANNO QUINQUAGESIMO SEXTO

GEORGII III. REGIS.

C A P. CXL.

An Act for establishing the Use of an Hydrometer,
called *Sikes*'s Hydrometer, in ascertaining the
Strength of Spirits, instead of *Clarke*'s Hydrometer,
[2d *July* 1816.]

WHEREAS by an Act made in the Forty-first Year of the Reign
of His present Majesty, among other Things, for making per- 41 G. 3. c. 97.
petual so much of an Act made in the Twenty-seventh Year
of the Reign of His present Majesty, as relates to ascertaining the Strength
of Spirits by *Clarke*'s Hydrometer, it was enacted, that so much of an Act
made in the Twenty-seventh Year of the Reign of His present Majesty,
intituled *An Act for making Allowances to the Dealers in Foreign Wines,* 27 G. 3. c. 31.
for the Stock of certain Foreign Wines in their Possession at a certain Time,
upon which the Duties on Importation have been paid, and for amending
several Laws relative to the Revenue of Excise, as directed that all Spirits
should be deemed and taken to be of the Degree of Strength at which the
Hydrometer, commonly called *Clarke*'s Hydrometer, should, upon Trial by
any Officer or Officers of Excise, denote any such Spirits to be, which was
to continue in force until the Fifth Day of *April* One thousand seven hundred and eighty-eight, and which by several subsequent Acts was continued until the First Day of *June* One thousand eight hundred and one,
should be made perpetual: And whereas, since the passing of the said
Acts, an Hydrometer, called *Sikes*'s Hydrometer, hath with great Care
13 P   been

2   Act establishing the legal use of
Sikes' Hydrometer, 1816

pyknometer to measure this value for various accurately
known alcohol–water mixtures between 30°F and 80°F. This
work was published between 1790 and 1794 and in 1802 the
Government launched a competition for the best instrument
for use in this connexion. This was won by Bartholomew
Sikes, a member of the Excise Department.[9] The Sikes'
hydrometer (and the tables which accompanied it) eventually
replaced the Clarke instrument as the legal basis for measuring spirit strength in 1816.[10]

The introduction of the new (Sikes) hydrometer, which
made it possible to express all spirit strengths as a percentage
of proof spirit, enabled the Government to simplify the spirit
duty. This was done by an Act of 1825.[11] Spirits were for the
first time assessed in terms of proof gallons (obtained by
multiplying the number of gallons by the degrees of proof
spirit).

Imported spirits were assessed for duty in terms of proof
gallons whilst two alternative systems were approved for the
assessment of home produced spirits. The first of these, (and
most important in this context), relied on the attenuation of

specific gravity. Every 100 gallons of wash or wort (wort is the fermentable liquid prior to fermentation, wash is the fermented alcoholic liquor prior to distillation), was charged duty at the rate of a gallon of proof spirit for each five degrees of attenuation (or reduction from pure water), as estimated by Bate's Saccharometer. This instrument had been introduced in 1825, replacing earlier instruments, to ascertain the specific gravity of wort or wash.[12]

In order that Excise officers could determine how much spirit a given wash would produce they were given the power by the 1825 Act to sample and to distil it in any still provided for that purpose by the Commissioners of Excise.[13] It does not appear to be known whether stills were in fact issued to individual officers for this purpose. It does seem possible that this interest in the variations which could occur in practice from the 'official' calculations (by Aeneas Coffee and John Logie, Inspectors General of Excise in 1825) of 1.07 proof gallons of spirit per 5 degrees of gravity attenuation inspired the setting up of a laboratory to investigate this variation and if so that Excise officers sent their samples of wash there to be distilled.[14] It is perhaps more likely that these samples would have been sent by the Commissioners, or even by the officers involved, to one of the many independent chemists who worked for the Excise at this time. As will be seen later, tobacco analysis was carried out then (and indeed later) by 'free lance' analysts, but there is evidence that at least a preliminary examination was frequently carried out by senior Excise officers first, and was sent to private analysts if the Excise could not come to a clear cut decision. Another possibility is that samples were sent to the 'Excise still' at St Katharine's Dock.[15]

The idea of a central laboratory may thus have been in mind at this time and would have made the resort to science in another revenue field, the duty on imported tobacco, that much more likely. Revenue had been raised on tobacco since it was first imported in the late seventeenth century and adulteration had been endemic from almost the same time. Adulteration enabled tobacco on which excise duty had been paid to be eked out to go much further and thus produce greater profit. When smuggling was rife, as in the eighteenth century, adulteration sank into insignificance, but as smuggling was suppressed, it became more prevalent. Adulteration was theoretically restricted by law as part of Pitt's measures in 1783 to suppress smuggling, and this was reinforced in 1821.[16]

It seemed hopeless to try to prevent smuggling and adulteration and the prevailing 'spirit of the age', expressed in the 1833 inquiry into the regulations governing the tobacco trade (part of a general enquiry into the Excise regulations) was against interference in 'diminishing that reliance on themselves which individuals should exercise for their own security'[17] ie *caveat emptor*. The trade had been agitating for freedom from Government control, and argued that they should have the right to do as they liked with tobacco once they had paid the statutory duty on it. This agitation finally resulted in an Act of 1840, known colloquially as the Mixing Act. It forbade the mixing of leaves and plants with tobacco, but did not mention anything else.[18] The intention of the government was to free honest traders from legal restrictions on adulteration and allow them to compete with dishonest merchants in the belief that they would drive the latter out of business while themselves exercising moderation. The Board of Excise expressed grave doubts on this last point.

This Act was received with great satisfaction by the tobacco manufacturers, who received it as a licence to add anything to tobacco except those substances expressly forbidden. Immediately materials, chiefly sweetening compounds (ie sugar, honey, molasses, treacle, liquorice, and nitre) but also including salt, sand and non-tobacco leaves, were added to tobacco, gradually being increased in quantity as prices dropped and profit decreased. It has been said that in some cases shag and roll tobacco sold was more in the nature of confectionery than tobacco, with the addition of adulterants ranging from 5% to 60%. The tobacco trade was delighted with this new freedom, but the government less so since receipts from tobacco began to drop: an average of 6% less was imported in the two years 1840–1842. To add insult to injury, drawback (duty paid back to a trader on re-export), was being claimed (and paid), for a substance which was certainly not all tobacco.[19] Obviously something had to be done about this situation. Cheating the public was one thing, but affecting the revenue was another matter, and despite much opposition from the Trade, on 10 August 1842 another Act came into operation, this one being known as the Pure Tobacco Act.[20] This forbade the addition to tobacco of any

ANNO QUINTO & SEXTO

VICTORIÆ REGINÆ.

****

C A P. XCIII.

An Act to amend an Act of the Fourth Year of Her
present Majesty, to discontinue the Excise Survey
on Tobacco, and to provide other Regulations
in lieu thereof.   [10th *August* 1842.]

WHEREAS an Act was passed in the Session of Parliament
held in the Third and Fourth Years of the Reign of Her
present Majesty, intituled *An Act to discontinue the Excise
Survey on Tobacco, and to provide other Regulations in lieu thereof :* 3 & 4 Vict.
c. 18.
And whereas the Practice has greatly increased of introducing in
the Manufacture of Tobacco and Snuff various Articles other than
Tobacco, either as Substitutes for Tobacco or Snuff, or to increase
the Weight of Tobacco or Snuff, by which Practice the Duties on
Tobacco are greatly injured, and the Revenue further damnified by
Drawbacks being obtained on adulterated Tobacco, and it is therefore
expedient and necessary, in Protection of the Revenue, to make
further Provision than is contained in the said recited Act for preventing such evil Practice, and to amend the said recited Act ; be it
therefore enacted by the Queen's most Excellent Majesty, by and
with the Advice and Consent of the Lords Spiritual and Temporal,
and Commons, in this present Parliament assembled, and by the
12 Z   Authority

3   The Pure Tobacco Act, 1842

substance except water, or oil (usually Olive Oil) in the case
of roll tobacco, and alkaline salts and lime water in the case
of Welsh and Irish snuffs. This Act having been passed the
Board of Excise had to enforce it and this could only be done
with the aid of science, and a laboratory. It is conceivable
that some form of testing of tobacco samples already took
place, since a General Order of the Board of Excise, issued 8
April 1835 forbade the use of unlawful additions to tobacco
and directed officers to send samples of tobacco suspected of
adulteration (with 'molasses etc.') to the Board. It is not
known what, if anything, happened to them there, but it is
possible that whatever procedure was adopted for wine
samples was similar to that adopted for tobacco.[21] Work on
tobacco had undoubtedly been carried out before 1840, since
Andrew Ure, lately Professor of Chemistry at Glasgow, when
giving evidence to the Select Committee on the Tobacco
Trade in 1844, stated that he was employed by the Excise to
analyse tobacco before he left Scotland (which he did in
1830), and subsequently in London.[22] The Board do seem to
have anticipated the need for some kind of in-house laboratory by January 1842, but only in the form of the senior staff,
the Senior General Examiners, becoming acquainted with
processes for analysing adulterated tobacco. This was to save
the expense of using outside analysts.[23] Later considerations
must have caused them to plan a laboratory of their own
because by October 1842 a Laboratory was in operation in
Old Broad Street, at Gresham House, staffed by one George
Phillips, an Excise Officer. October 1842 is generally
accepted as the date of foundation of the body now known as
the Laboratory of the Government Chemist.

Part II
Excise and
Inland Revenue
Laboratories
1842–1894

1

The First Few Years

THE NEW LABORATORY, THE EXCISE LAB-
oratory as it was called at first, was probably founded
on Monday 31 October 1842. The exact date is not
known nor if it was a continuation of the other possible
laboratories described above, but it was certainly treated as
new in all subsequent official communication and returns. A
date at the end of October is probable because in evidence to
a Select Committee in 1855 Phillips gave sample figures
based on a year beginning on 30 October. Confirming this
date of foundation an Excise General Order was issued on 31
October 1842 enjoining 'on all Supervisors and Officers the
strictest vigilance in carrying out the provisions of the Act'
(forbidding tobacco adulteration), encouraging frequent inspections of manufacturers premises and ordering 'if any
Officer shall suspect any Tobacco to be adulterated . . . he
will take a sample . . . and transmit the same to the Board'.[1]

George Phillips, apparently the sole member of staff of
the laboratory was a serving Excise Officer. He had entered
the service on 17 April 1826 and had served in various
stations. By April 1839 he was at Woodbridge, Suffolk, in the
rank of Principal Officer. He seems to have studied chemistry
and microscopy in his spare time, in 1844 he said that he was
self-educated in chemistry, and that he had taught himself the
principles of chemistry to enable him to detect adulterations.
He had certainly examined adulterated samples of tobacco at
previous stations in Gainsborough, Liverpool and Manchester at a series of seizures and trials at those places before
1842.[2] In 1842 he apparently offered to detect tobacco
adulteration under the new Act. His offer was accepted, and
he moved into premises in Gresham House on the east side of
Old Broad Street in the City of London. This was the
headquarters of the Board of Excise. Gresham House (pulled
down in 1853) was a 'very handsome plain stone building of
four stories, with an entrance through the middle of it into a

---

GENERAL ORDER.

Excise Office,
London. }   **31st OCTOBER, 1842.**

In reference to the 5 and 6 VICTORIA, c. 93, "to amend an Act to discontinue the Excise Survey on Tobacco, and to provide other Regulations in lieu thereof," the time granted, under the proviso in clause 15, for allowing the Trade to get rid of their old Stock expiring on the 1st November, THE BOARD enjoin, on all Supervisors and Officers, the strictest vigilance in carrying out the provisions of the Act, and endeavouring to detect such manufacturers as shall use, in manufacturing their Tobacco, any other material than water, except a portion of oil in making up spun or roll Tobacco, and such manufacturers, dealers, or retailers as shall add or mix any other materials or articles to or with Tobacco; and, in order thereto, the entered premises of manufacturers must, as the time will allow, be visited oftener than once in ten days, (as directed by General Order of 21st July, 1840,) and that at irregular times—on every visit the Officer must go through all the entered rooms and ascertain that none of the articles prohibited under the 5th Section, are in possession of the manufacturer, and carefully examine all the Tobacco in the premises, more especially the operations, to endeavour to detect any adulteration. Frequent visits must also be made to dealers and retailers. If an Officer shall suspect any Tobacco to be adulterated, but shall not feel sufficient confidence as to the fact to authorise his seizing it, he will take a sample or samples under the 7th Clause, and transmit the same to the Board. Should it come to the knowledge of a Supervisor or Officer that any manufacturer, dealer or retailer is selling under the ordinary prices of other persons, such trader is to be especially watched, and if, from this and other circumstances or from information which can be relied on, the Supervisor or Officer is sufficiently satisfied that illegal materials are on the private premises so as to be able, CONSCIENTIOUSLY, to make the requisite affidavit, he will obtain a Search Warrant under 7 and 8 Geo. 4, c. 53, s. 34, and endeavour to discover and seize them. It will be observed that, by Section 8 of the Act, all persons are prohibited, under heavy penalties, from preparing leaves or other matters to imitate or to be mixed with Tobacco or Snuff, and as many of the preparations, used in adulterating, are known to be made up for the trade by persons not in it, every exertion is to be made to detect such offenders.

With respect to Snuff, the foregoing directions will equally apply, except that the time is extended, until after the 10th August next, for the trade clearing off their stocks of Snuff and Snuff-work on hand or in operation at the passing of the Act, and Officers will, therefore, until that period, be cautious in making seizures of Snuff or Snuff-work under the 4th Clause, unless they shall have some proof that such Snuff or Snuff-work has been manufactured since the passing of the Act, and not from returns of Tobacco manufactured before the passing of the Act; but the Supervisors and Officers will observe, that the proviso gives no authority to continue the practice of adulteration, and that any manufacturer, dealer, or retailer, detected in adulterating Snuff or Snuff-work, incurs the penalty and forfeiture.

By Order of the Board,

CHAˢ. BROWNE.

4   Excise General Order, 1842

large yard, in which is another building of brick nearly the size of the principal one. ... From the centre of both buildings are long passages and staircases to the galleries, in which are the numerous offices for the commissioners and clerks in the different departments of the Excise.'[3] Despite this, it is not known what room or rooms Phillips occupied.

The tobacco trade seems at first to have treated humorously the idea that Excise would be able to take any effective action over adulteration, doubting that science had any use in this context and believing that adulteration with up to 5% of sugar was impossible to detect. They were surprised when Phillips, who personally visited a large number of manufacturing premises throughout the country, brought many successful prosecutions in the first year of the Act and before the expiration of 1844 some 30,000 lbs of tobacco had been seized in Lancashire and Yorkshire alone.[4] As Phillips himself expressed it in his *Report* to the Commissioners of the Inland Revenue for 1857 'the success of these measures gradually removed the impression . . . that adulteration . . . could not be detected in the manufactured article and a feeling of insecurity in the commission of fraud became general amongst them'. In the years following the Act of 1842 Phillips was not the only person analysing tobacco officially. The Board of Excise continued to use outside chemists too. Thomas Graham, Professor of Chemistry at University College, London claimed to have analysed 100 samples in the fifteen months following the passing of the Act. He gave evidence at trials in Shaftesbury and Gainsborough and was still much involved with the Excise as will be seen.[5]

5   Excise Building, Old Broad Street, home of the Laboratory 1842–1853

A considerable body of knowledge seems to have been built up very quickly, doubtless based on work done before the laboratory was set up. This enabled Phillips to detect the most incredible range of adulterants. Sugar was found in amounts ranging from 1–25%, also molasses, various leaves such as coltsfoot, endive, rhubarb, oak, elm and plane, occasionally various chemicals such as common salt, potassium nitrate, alum, Epsom Salts and Glaubers' Salts, dyes (yellow ochre and green copperas), peat moss, oat meal, malt cummings (the rootlets screened out of germinated and roasted barley), chicory, gum, starch, liquorice, catechu, 'and in some fancy tobaccos, lavender and mugwort'. The most common adulterants were apparently sugar in various forms and potassium nitrate.[6]

The chief means used in the detection of adulteration was the microscope, and a certain amount of what must be called elementary chemistry. Phillips explained his system to a Select Committee on the Tobacco Trade set up in 1844 to survey the state of the trade, and on the part of some of the Committee to press for a reduction in import duty payable. The Chairman of the Committee, one Joseph Hume was of this party, and his questioning of Phillips, Richard Phillips, and Thomas Graham was hostile. A number of samples of adulterated tobacco had been prepared under the direction of the Committee, and these three chemists had attempted to analyse them on behalf of the Board of Excise. Phillips, supported by his colleagues was confident that adulteration could be detected by available means of analysis.[7] In general they did quite well, although the test was an unfair one since the Committee had refused to allow them to have samples of the pure tobacco, upon which they usually relied to confirm their results. Phillips and his colleagues reiterated many times their contention that they should have been allowed pure samples to check their results. In a normal Excise seizure they nearly always did have them. The refusal was apparently due to the personal view of Joseph Hume, the Committee Chairman, expressed in a letter to the Chairman of the Board of Excise. He said this was unnecessary, on the ground that 'no one standard of tobacco can be relied upon', a misunderstanding of why they were needed. Phillips commented bitterly on this attitude 11 years later in 1855 to another Select Committee.[8]

Phillips in his evidence explained that he had carried out

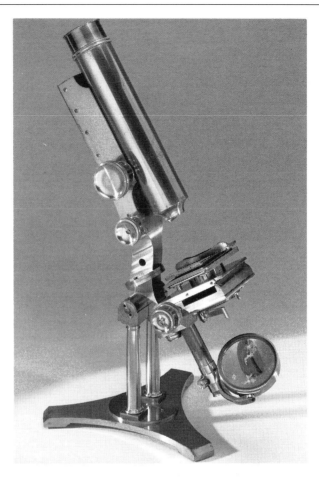

6   Microscope of 1841, by Powell
and Lealand

extensive tests on many samples of tobacco since 1842, in order to draw up a comprehensive picture of what he could expect to find on examining a given type. He explained in great detail his system of examination. He began by drying the tobacco to constant weight, at a low enough temperature not to cause decomposition of the tobacco. For this he used an ingenious steam bath which he described as 'a small boiler connected with two hollow pans by two tubes. The tubes rise from the top of the boiler, and supply the pans with the steam generated in the boiler. In each hollow pan there is a discharge-cock to regulate the pressure of the steam. In the hollow pans water is placed, and on this water float loosely fitted pans containing the tobacco. By this contrivance a regular heat of 176°F is constantly maintained. This heat answers for all the purposes of drying and evaporating, and in no way injures the tobacco'.

After this he relied on a thorough microscopic examination. He showed the Committee coloured drawings of the structure of tobacco and of the common adulterants rhubarb and foxglove and demonstrated the differences. John Lindley, an eminent botanist and Professor of Botany at London University, confirmed to the Committee that microscopic examination enabled an analyst to tell whether or not a tobacco was adulterated in the great majority of cases, although it was not always possible to tell with what material. After his microscopical examination Phillips then determined the ratio of the soluble part of the sample (which he called the 'extract'), to the insoluble part (which he called 'lignin').[9] Where the insoluble part was high he ashed the tobacco at a low temperature to check the quantity (and to look for sand), and also to check for the presence of potassium carbonate (the result he said of ashing his common adulterant, potassium nitrate). If the soluble part of his separation was high he checked it for the presence of sugar by fermenting it at a standard temperature with yeast. Sometimes he could pick out grains of sugar, which he then confirmed was cane sugar by using 'copper oxide'. He and his two collaborators were absolutely certain that there was no naturally occuring fermentable sugar in tobacco – which is not true, (see below). They tended to find sugar in samples to which it had not been added, due to their assumption that if the extract, or the ash, was higher than normal for this type of leaf, then there was probably adulteration. If the extract was high they looked for sugar.[10] Even Andrew Ure, (see above), who had been asked to analyse the same samples as the official analysts, agreed on this point. Ure carried out some very elaborate manipulations on the samples, but came to an even less definite conclusion than the official analysts, who in fact did quite well.

Ure was very hostile to Phillips and the official side in his evidence. He had worked as an analytical chemist for the Boards of Customs from 1834, being appointed consulting analyst and also for the Board of Excise in a more *ad hoc* manner and one wonders if he feared for his work as chemist to the Board.[11] He contended that it was not usually possible to detect adulterations, and that he personally could undertake to submit a sample that could not be detected.[12] He seems not to have doubted his powers. Given the state of analytical chemistry at the time he may have been partially correct, neither he nor Phillips *et al* could find all of the adulterants in the artificial samples, some of which had up to 6 different substances in. Of course as Phillips pointed out, the official side could quite well have determined the presence of many of the common adulterants by simple tests, eg the detection of starch by iodine, but could not possibly look for all of the various salts that might be present, because as he reasonably said 'you might be experimenting for years in one sample'. As Phillips said in 1855, when admitting that they had not been able to detect all of the adulterants, 'We could have detected them then, but we were placed in very difficult circumstances. Mr Hume was the Chairman of that Committee. We protested against the mode in which we were to undertake the examination; we said it was an unfair test of

7   Dr Andrew Ure

chemistry to send us samples of tobacco without our having any standard to fall back on, and then to lay our shortcomings upon the science of chemistry; the tobacco might have been adulterated with the constituents of the tobacco plant itself. Mr Hume said he could see no use in our having standards, and we were refused them, therefore we undertook the examination under those circumstances, and did the best we could'.[13] Something which Phillips and his collaborators could not convince their critics on the Committee of was that if they could prove that there was one substance in the tobacco which should not have been there, then the tobacco was adulterated. It was not legally necessary to find all of the adulterants.[14] The Committee in the end came to no conclusion.

The manufacturers were not content to leave the matter thus. They agreed, in a memorial to the Treasury in February

8   Balance of *c*.1840, once owned by Dr Ure

1845, that reducing the duty thus enabling them to lower the price and not analysis, was necessary to prevent smuggling and adulteration. They further said that the chemical evidence on which prosecutions were being founded was not valid. They believed that the Excise chemists were simply 'asserting a scientific knowledge which they did not possess.[15] The weak point in the official position was of course whether or not there was fermentable sugar naturally occurring in tobacco. Graham had committed himself in 1844 to a statement that even if 0.5% of sugar was detected in a sample, then it was adulterated.[16]

Ure, and three other eminent chemists of the day, (Professors Brande and Cooper and Mr Herepath), were employed to disprove the official position, and they reported that naturally fermentable sugar occurred in tobacco in amounts ranging from 0.75% to 2.06%. The manufacturers went to the Treasury armed with this report, making reference to 'indiscriminate prosecution of the honest and fraudulent, resulting from the adoptions of spurious chemistry'.[17] It does seem possible that this report was part of a concerted effort made by a group (or groups) of manufacturers to show in the courts that the Excise standards and chemists were giving wrong results entirely, and to support their demand for a reduction in the duty on tobacco. Phillips stated in a letter to the Board in 1846 that he had been told by a friendly manufacturer that such was the case and backed his information with comments on two court cases. In one case in Leeds the expert defence witnesses in a case of adulterated tobacco had sworn that 6–9% of sugar had been found in normal tobacco and in another case manufacturers had sworn on oath that tobacco so sweet in taste that the natural tobacco flavour could not be detected was perfectly genuine tobacco. These cases were connected with one involving adulterated snuff in Bristol in which Phillips, Richard Phillips, Lyon Playfair (then assistant at the Royal College of Mines) and a Professor Campbell all agreed that it contained up to 16% sand and earthy matter. The magistrates dismissed the case since the manufacturers argued that this was a natural part of snuff, due to the method of manufacture. Herepath was one of the witnesses for the defence.[18]

At about the same time, and presumably at least partly as a result of what had happened (including the report of Ure *et al*) a considerable effort was being made by the Excise to have the natural constituents of tobacco determined, particularly the saccharine matter. The work seems to have been caused directly by a group of tobacco samples seized originally early in 1845. They were reported by George Phillips, and subsequently by Thomas Graham, as being adulterated with sugar, but at not more than 1.5% (their reports did not coincide). Considering these reports in the light of current events the Excise Board ordered proceedings to be abandoned and the tobacco restored to the manufacturers on payment of expenses. This the manufacturers refused to do. They wanted an admission that no adulteration existed and payment of compensation. They petitioned that 'the Proceedings of the Board of Excise in future be so regulated as to prevent seizures of Tobacco proceeding on fallacious or imperfect Chemical Tests', wording reminiscent of the earlier memorial. They were supported by Messrs Ure *et al.* A considerable amount of correspondence ensued between the Board, their Solicitor and the law officers of the Crown. It was eventually decided that the case should be dropped and after several months it was agreed in June 1846 to pay compensation. Professor Graham strongly maintained that his tests were not 'fallacious', as did the Excise Board, but Graham seems to have felt some doubts and did a considerable amount of work on some British tobacco in the course of which he found a certain amount of sugar. As a result of this it was decided to test a large number of different types of tobacco to determine the natural constituents, particularly the 'saccharine matter'. This would give a standard for the analysts to work to in court.

The Solicitor to the Board originally suggested that Professor Graham carry out the analyses, aided by Richard and George Phillips and John McCulloch, an Assistant in the Laboratory, but in the event the work was carried out by Graham and Joseph Drinkwater, a Senior General Examiner associated with Graham at University College (see below). These two determined moisture, ash, the constituents of the ash (chloride and sulphate) and sand, as well as fermentable matter. As a result of much careful work they found that tobacco usually contained less than 1% of saccharine matter, much less sometimes frequently more, between 1 and 2%, and occasionally up to 7%. Their report was made on 21 August 1846. As a result of the tests it was decided more or less officially not to prosecute on a sugar content of less than

9   Professor Thomas Graham, Professor of Chemistry at University College London 1837–1855

3%. The Board retreated all the way over the trader who had started all this and paid £380 compensation, but refused to admit that the tobacco had not been adulterated, except unofficially, nor indeed that their chemistry was 'fallacious'.[19]

The effect of these events was that there was a definite reduction in the number of prosecutions (35 in 1845, 54 in 1846 and 2 in 1847) and certainly in the adoption of greater care in analysis. Work on checking the method of determining the amount of sugar in tobacco was still going on in 1863 for example and in 1874 a flurry of alarm was caused by the realisation that naturally dried tobacco contained up to 14% of saccharine matter. Bell (Phillips' successor) carried out much analytical work to verify this point, work which was repeated by an outside analyst in 1883 on behalf of the Board of Customs, who were apparently ignorant of the original work. In the original case of 1846 Phillips made no reference to it in his first published Report in 1857, and the numbers analysed fluctuated much as before.[20] The Board of Excise however do not seem to have been entirely happy with Phillips' performance, and together with nine other Excisemen, directed him by a Minute of 2 October 1845 to attend classes at London University and to matriculate in general and analytical chemistry. This attendance, or other methods for training staff in chemistry continued and developed during the rest of the century.[21]

Tobacco was not only tested in the form of leaf, samples of snuff were also received, (as noticed above), since adulteration was equally widespread with this product too. A similar wide range of substances was found, various kinds of ground wood, such as fustic, (a wood used as a yellow dye), logwood and pine, starches from wheat, oats, and maize, also orris root, peat moss, mould, lead chromate, and various colouring materials, ochres and venetian red. Iron oxide was also a very popular adulterant, being found in most of the samples examined. Red lead was also not unknown, with, as might be expected, fairly disastrous results to the user.[22] One problem with snuff was that the Act of 1842 allowed the addition of 'lime water' in the preparation of 'Irish' and 'Welsh' snuffs but laid down no limits to the quantity which could be used. Given the then current habits of the tobacco trade the result was as could have been expected. Some of them (chiefly in Ireland apparently) used what was practically a lime paste.

10   Drawings of microscopic characteristics of tobacco, by Rochfort Connor, 1861, a member of the Laboratory staff

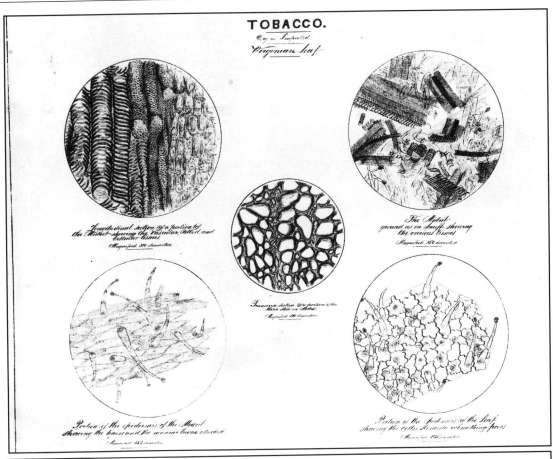

TOBACCO.

Copy as Imported.

Virginian Leaf.

Longitudinal section of a portion of the Midrib shewing the Vascular, Pitted and cellular tissues.
Magnified 300 diameters.

Transverse section of a portion of the Hair Shot in Midrib
Magnified 300 diameters.

The Midrib ground as in Snuff shewing the various tissues
Magnified 150 diameters.

Portion of the epidermis of the Midrib shewing the hairs and the woolen tissue attached.
Magnified 150 diameters.

Portion of the epidermis of the Leaf shewing the cells formula ed nothing pores
Magnified 150 diameters.

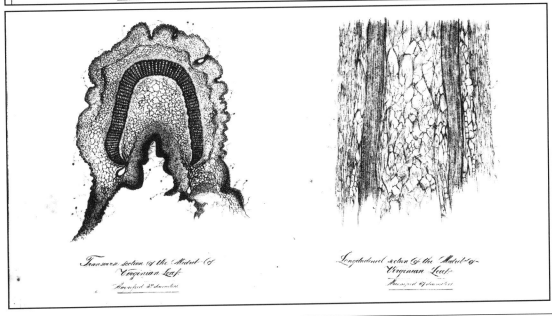

Transverse section of the Midrib of Virginian Leaf
Magnified 87 diameters.

Longitudinal section of the Midrib of Virginian Leaf
Magnified 87 diameters.

Sand was also added, with the argument that this had been imported on the tobacco and had paid duty and to add to the effect of the snuff, alkaline salts were also added. These were also allowed by the Act.[23] To be fair to the manufacturers it does appear that 'white snuff', containing up to 25% lime was popular with mill girls in Ulster. The whitish colour given by excessive lime was usually concealed with red iron oxide, although not always, since one sample received containing 12.6% lime was noted as having a white colour. Phillips complained regularly about this state of affairs, arguing quite reasonably that the unlimited use of sand should be prevented, and that the amount of lime and alkaline salts should be banned or at least limited. Eventually, in 1867 this was (partially) done. The amount of lime, in the form of calcium and magnesium oxides found in the finished snuff, was limited to 13%.[24] In preparation for this change much experimental work was carried out in the Laboratory in March 1867. This was at Phillips' direction, to find out how much calcium and magnesium oxides were contained naturally in tobacco leaf, in order to be able to make allowance for this. Obviously the lesson of sugar in tobacco had been thoroughly learnt. The 13% limit was set as a result of this work. At this point the manufacturers merely began to add more alkaline salts (this term included sodium chloride), these not being mentioned in the 1867 Act, By 1869 Phillips complained in his Report that he was now finding 33–57% of sodium carbonate in some samples.[25] Then, in turn the alkalines, (defined as chlorides, carbonates and sulphates of potassium and sodium, and ammonium carbonate), were limited by an Act of 1878 to 26% of the dried snuff. Adulteration of snuff continued for many years after the 1870s, particularly in Ireland were there were still apparently many secret factories, originally set up in the 1850s and 1860s for the manufacture of 'spurious' snuff. Some of these 'snuffs' contained no tobacco at all.[26]

As a result of the intensive work done by the Laboratory, adulteration of tobacco (and snuff) slowly decreased, although it was still widely practised. In 1857 for example 65% of samples examined were still adulterated, and 20% of snuff samples. Both of them were bought in Ireland. During 1858–1860 (and in many subsequent years), nearly all of the tobacco factories and snuff mills in England and some of those in Ireland were visited by Phillips or his assistants, without any illegal practices being discovered. These visits were largely unexpected by the factories, as were similar visits by Laboratory trained Excise men stationed in towns such as Bristol and Liverpool where much tobacco was manufactured. Such visits had taken place earlier, although not in such a systematic way. In 1852 for example Phillips and Harmon Tarrant (another Senior General Examiner) seized a considerable amount of adulterated tobacco in Birmingham, Bradford and Halifax as a result of visits there and received a reward of £50.10s (£50.50) from the Treasury 'the usual reward at the rate of 1d (0.4p) in respect of 1010lbs of adulterated tobacco seized by them from the manufacturers'. Phillips subsequently claimed expenses for attending court including first class fare to Halifax of £1.15s.8d (£1.79). After his visits in 1860 Phillips said that as a result of what he had seen he had 'no hesitation in in saying that the adulteration of tobacco is now but seldom attempted'. There was certainly a decrease in samples after 1857.[27] Phillips put the improved situation down to the strict supervision exercised over tobacco manufacture by the Inland Revenue. Some of the visits were as a result of anonymous information, usually found to be malicious and untrue, although in one case a 'fraudulently inclined manufacturer' inspired visits to his locality so that he himself could be found to be above suspicion. From his own figures it is apparent that Phillips knew that his assertion that adulteration was then decreasing was valid. It was down from 69% of samples examined in 1844, to 26% in 1858, and 16.6% in 1860. As he reasonably said, with a duty of nearly 700% of the original cost, fraud was bound to be attempted.[28]

Some of the frauds attempted were ingenious, if minor. There was for example the 'Roll' which consisted mostly of cabbage leaves, with only the outside covering being tobacco. Another 'Roll' from Scotland consisted of oakum, with a thin wrapper of leaf tobacco, the whole indistinguishable in appearance from the genuine thing. These were designed more to cheat the public than the revenue. Much of the adulteration actually found in the 1860s was due to liquorice. The method used for the detection of liquorice (based on the separation of the liquorice sugar from the tobacco) had been devised in the Laboratory in 1863 by John Brown, an Excise Assistant then employed in the Laboratory. Because of his work Phillips obtained for him in 1864 early promotion to

Division Officer, (citing the damage to his health caused by the 'poisonous vapours of nicotine') with an increase in salary of £10 *per annum* from £110 to £120.[29] A sign of changing times occurred in 1865 and 1873 when there were seizures of large quantities of fraudulent cigarettes; the

11   Inland Revenue General Order, 1871

GENERAL ORDER.

INLAND REVENUE,
SOMERSET HOUSE,
LONDON, W.C.
}   12th AUGUST, 1871.

An influential deputation of Tobacco Manufacturers having recently waited upon the Board, and stated it to be the earnest desire of the Trade that the law as regards adulteration should be strictly enforced, and that every person found contravening those regulations should be exposed by public prosecution and severely fined

ORDERED,

That every Tobacco Manufacturer be immediately warned that the Board intend to give the fullest effect to the expressed wishes of the Trade ; and reminded that every matter and thing is prohibited in the manufacture of Tobacco, except water, or water and oil in making up Roll Tobacco, or lime water and alkaline salts in Snuff. The penalties being £500 for using any prohibited ingredient—£200 for sending out or having in possession tobacco in any manner adulterated—and £200 if any forbidden material is found on the premises—in addition to the forfeiture of the Goods.

The use of any liquid or other material or thing, (except acetic acid) although used solely for the purpose of dyeing or colouring tobacco, is included in the above prohibition, without regard to the question whether the weight of the tobacco is thereby increased or not. All such practices, besides being illegal, being also unfair and injurious to those Manufacturers who comply strictly with the law ; and being, moreover, frequently resorted to as a means of concealment for grosser forms of adulteration.

It is further ORDERED,

That the several Supervisors and Officers who have Tobacco Manufacturers under survey make every possible exertion to see that those Traders conduct their operations according to law.

With this view Officers who have manufacturers in their residence, must, at least, once a fortnight, visit them twice in the same day ; and their ordinary daily visits should be made at hours varying from early in the morning until late in the evening, so as to be unexpected by the Trader.

On every visit the Officer must very carefully inspect the premises, and also closely examine the tobacco in operation, more especially those portions which are in the earlier stages of manufacture, when soluble ingredients added with the water may be more easily detected before they have been imbibed by the leaf.

Whenever there is reason to believe that illegal ingredients have been used but the evidence appears insufficient to justify the seizure of the tobacco, a Sample should be purchased at some shop where such tobacco is retailed, and sent by post to this Office for analysis before any charge is made against the Manufacturer.

A copy of this Order to be given to every Tobacco Manufacturer.

By the Board,

G [241] 3250 8/71

original Laboratory work on tobacco was on pipe tobacco or cigars. Cigarettes did not start to become popular in the United Kingdom until the 1850s. Those seized in 1863 were made from the sweepings and refuse from a cigar factory, and were found on examination to consist of 40 to 60% of sand, as well as organic matter not tobacco. To add insult to injury, the cigarettes were made to resemble closely expensive Russian cigarettes, the fraud only being discovered when they were smoked.[30]

One sign of the increasingly law abiding nature of the major tobacco manufacturers, is shown in a 'General Order' of the Inland Revenue, issued on 12 August 1871. A deputation from the trade to the Inland Revenue had asked that the law on adulteration be strictly enforced, and said that more frequent inspection of premises by senior staff was desirable. The Order reminded manufacturers that all additions to tobacco and snuff with certain exceptions were forbidden by law and stated the penalties. This request for an increase in inspections was apparently due to the resentment felt by the larger companies towards small manufacturers, who were more inclined to infringe the law. The deputation are said to have referred to the public reliance on the work of the Laboratory, and apparently gave assurances of future cooperation with Excise Officers; the implication being that they had not in the past.[31] As a result of this deputation Phillips submitted to the Board a plan to use experienced analytical staff to carry out inspections in future. In connection with this and to make inspections more systematic, and so that they could all be carried out in two weeks, it was further proposed that the whole of the Kingdom be divided 'into small sections' each of which contained a 'chemical station'. It was apparently thought that a longer period would nullify the effect of the inspections. Manufacturers informed each other 'rapidly, whenever they hear[d] of a special inspection in progress' by means of the telegraph. The Phillips' plan was agreed, although the number of chemical stations eventually set up was considerably less than the 54 in England and Ireland, (plus a variable number in Scotland), originally proposed.

'Chemical stations' were set up at ten places in Belfast, Bristol, Cork, Dublin, Glasgow, Hull, Leith, Liverpool, Newhaven and Southampton, and all other places were probably covered from these as necessary. Thus began the

Laboratory 'out-stations', There were at one time 28 of them, staffed by chemical officers trained at Somerset House. Mr Phillips suggested a general inspection of these stations in the following December, and after obtaining agreement had completed his tour by August 1872. The stations were found to be working satisfactorily, except that in some cases the records kept were in a somewhat unsatisfactory condition. Officers at Glasgow and Liverpool examined large quantities of beer, (see next chapter), as well as making the tobacco inspections, and some stations received many 'miscellaneous' samples, taken by the officer in charge personally, or brought by local Excise men. No records were being kept of these extra samples in many cases, and Phillips recommended that this should be done. Despite his report it was several years before all outstation officers could be persuaded to keep proper records, and in the case of Belfast, to make any returns at all of the miscellaneous samples. In 1872 Belfast had apparently kept no records at all of any samples.

The stations varied very widely in their fittings and suitability as laboratories. The room at Bristol was 'very inferior' and insufficiently supplied with fittings. It was noted that Bristol was likely to become an important station since large bonded warehouses for wines and spirits were being built. In Dublin the room was spacious, the fittings and appointments generally good, and the apparatus well kept, but in Cork the room was both dirty and inadequately fitted out. In Glasgow the room was dark and unhealthy and Phillips proposed that since a large amount of chemical work was done there, the officer concerned should be relieved of the duty of personally drawing samples of bulk beer and carrying them from the docks to the laboratory. The Collectors (officers in charge) of the areas with poor laboratories were ordered to see to the state of the rooms and ascertain what apparatus was needed to equip them properly. Phillips also suggested that the stations continue to be examined each year by a senior member of his Somerset House staff who would check the conduct of local examinations and test the accuracy of the instruments. He had noted that some results from these stations did not always coincide with those from the central Laboratory (where comparison was possible), due as he said to inadequacies in most cases, rather than to doubts about 'the integrity of the officers generally', although apparently the results of the officer at Leith had been recently found wanting in some unspecified way. This problem was eventually met by random duplicate samples being sent for checking to Somerset House. Visits from headquarters to the out-stations continued as long as the establishments existed. The exact location of many of these stations is not recorded, although one, at Newhaven was noted as being in a building owned by the 'Steam Boat Co'.[32]

Ingenious attempts at adulteration went on, many of them from Ireland, a fertile centre of such attempts. Several

12 Sikes' hydrometer, nineteenth century model

times in the 1870s for example gum was added to tobacco, in the belief that the Laboratory could not tell the difference between tobacco gum and other types of vegetable gums. This was certainly not true and in one case in 1876/77 a manufacturer in Ireland was prosecuted for adulterating his tobacco with a small proportion of gum arabic. Two scientific witnesses for the defence declared that they were unable to detect gum arabic in portions of the official sample with which they were provided and the Crown had no opportunity of refuting this evidence. The case was dismissed but on appeal the defence was charged with having adopted a fallacious process of analysis. The prosecution 'demonstrated by actual experiment' that the essential constituent of the gum arabic was thrown away in one of the stages of the process but the magistrates ruled that this evidence (since it was new) was inadmissible, so that the appeal, also, was lost. The case nevertheless appeared to convince the trade that the Laboratory could deal with this type of fraud and that gum arabic could be detected in tobacco, even in very small quantity. This did not prevent this type of fraud being tried in 1881/2, again in Ireland, this time using glycerine, although only on a small scale. In mitigation it was said that the manufacturers had 'only been testing the value of the adulterant, and were not using it in any appreciable quantity'. It is not reported whether or not this statement was put forward in mitigation at the trial.[33]

The last major attempt at tobacco adulteration happened at about the same time. Tobacco had retailed at 3d (just over 1p) per oz, for many years, and when the duty was increased in 1879, there was much consumer resistance to the necessary increase in price of a farthing (¼d) an ounce. To meet this price resistance more water was added to tobacco, after the duty had been paid. The effect was to reduce the amount of tobacco which was needed and so cut the price of manufacture. The quantity of tobacco imported was also reduced of course. No limit for water had been laid down in the *Pure Tobacco Act* of 1842 on the assumption either that it was not necessary or that there were practical limits to the amount that could be added. Samples of tobacco had been examined for moisture in the past, but for the Customs Board not for Excise purposes. In 1863 the *Manufactured Tobacco Act* laid down that drawback could be claimed on tobacco and snuff being exported if duty had been paid on import.

Samples over which a dispute arose as to the moisture content had to be submitted by the Customs officers to the 'officers of Inland Revenue for examination'. Over 2000 samples were received in the first year that the Act operated, a number which remained steady for many years.[34]

After the 1879 Budget change there was a sudden demand by the industry for certain spongy varieties of tobacco that would hold 40 or even 50 per cent of water and excessively damp tobacco began to appear on the retail market. James Bell (who had succeeded Phillips as Laboratory Principal in 1874, see chapter 4 below) advised against action to prevent this addition of water on the grounds that the majority of manufacturers did not indulge in the practice and that the situation would right itself if left alone.[35] This optimism was unfounded and in 1887 the Chancellor restored the duty to its previous level but at the same time ensured that the consumer would get at least some of the benefit of the decrease in duty by inserting a moisture clause in the Finance Act. This clause made it illegal to possess or sell any tobacco containing over 35 per cent of water. It did not actually refer to water, but only to the 'decrease in weight of the tobacco when dried at 212 degrees by Fahrenheit's thermometer' (100°C). This was almost certainly on Bell's advice in order to avoid legal argument about what was actually lost from the tobacco in a laboratory drying oven. The limit on water content became effective at the end of July 1887. A ten-fold increase in the number of samples examined as a result of the new Act occurred in that year from just over 1000 to more than 11000. A year later this increased to over 16000. These were in addition to over 8000 samples examined annually at the chemical stations at the outports. By September 1887 71 manufacturers had been cautioned for exceeding the moisture limit. Some prosecutions were started in earnest but few were necessary, although it was found necessary to clarify the 35 per cent figure. Manfacturers said that it was impossible to produce a tobacco containing uniform moisture at this level. Parts were bound to contain more they said. The Chancellor of the Exchequer then explained that 35 per cent was the maximum allowed in any part of the sample, a clarification not entirely to their liking. Levels of moisture were changed later, but a legal limit remained in operation until the basis of taxation was changed in 1978.[36]

By 1898 the intensive work of the Laboratory on the adulteration of tobacco and snuff had finally borne fruit. By as early as 1860 George Phillips had said, 'the adulteration of tobacco is now but seldom attempted'. This was a slight exaggeration so far as adulteration of home manufactured tobacco was concerned and samples were still received in which non-permitted substances were present. These were largely from imported tobacco, where the manufacturer was unaware of the UK law. In one instance 'Coca' leaves had been added for example, a somewhat odd mixture which would not in fact have any narcotic effect. Another source was smuggled tobacco, usually 'Cavendish' mixture (sweetened tobacco), containing large amounts of sugar or liquorice. The occasional exceptional sample of genuinely adulterated tobacco was received, some fairly extraordinary, as in the case of the manufacturer found guilty of dyeing the outside leaf of cigars with a solution of logwood and alkanet (Dyers Bugloss) root. His defence was that he could not obtain leaf of suitable colour.[37] In fact by the 1890s tobacco adulteration had to all intents and purposes ceased in the United Kingdom.

Growth and consolidation

THE LARGE AND TRAINED STAFF BUILT UP to test tobacco and snuff were not wholly occupied by the analysis of tobacco products. The advantages of having a body of trained chemists (or a trained chemist initially) was discovered very soon after the foundation of the Laboratory and from 1844, two years after the foundation, samples of tea, pepper, soap and beer, were examined. By 1849–1850 a total of 310 samples were analysed for the Board of Inland Revenue, consisting mostly of tobacco and snuff (127 samples), but also 'various liquids' (42), pepper, soap, spirits, coffee, wines and tea. Eleven samples were for the Customs Board, probably mostly to check that they were the substance declared at the port. Just after this date, in 1854, work had increased to such an extent that Phillips was ordered to 'so arrange the business' that he did not usually attend court hearings 'in the country' himself, but that he concentrated on 'subjects of special research and enquiry, and to such cases of adulteration as present peculiar difficulties'. All coffee and pepper samples (added in 1852, see below), should be examined and reported on by junior chemists, 'Mr Phillips adding such remarks as he may deem necessary for the information of the Board'. Beer is in a different category, and is dealt with in the next chapter, but the others were all being adulterated in much the same way as tobacco and snuff, and needed to be examined for the same reason, ie. to protect the revenue, and not primarily to prevent adulteration. Soap was mainly examined up to 1853, when the duty on it was abolished and the interest of the Excise in its purity ceased, (impure soap paid proportionately less duty). It had been examined to see if it had been adulterated with siliceous or earthy matter or with the apparently more usual sodium chloride or sodium sulphate.[1]

The government do not seem to have worried over much about the health of the population in the first part of the nineteenth century. Even Arthur Hassall (a strong opponent of adulteration), noted of snuff, in 1855, that it was 'subject to ..... adulteration, and that of a kind which is not only detrimental to the revenue, but exceedingly injurious to health', in which the order of the drawbacks may be significant. Adulteration meant that smaller quantities were imported and thus less revenue was received. Phillips himself showed a certain concern for the health of the users of his chief source of work, since in 1851 he applied for a patent for the 'preventing the injurious effects arising from smoking tobacco'. This process filtered out, so he claimed, the injurious nicotine, using elaborate filters in the stem of the pipe.[2]

Adulteration in some foodstuffs was as blatant as for tobacco. Pepper for example, on which there was a duty until 1866, was mixed with ground rice, linseed meal, chillies, husks of red and white mustard seeds, and occasionally sago, cereal starches, and rape seeds, also stalks and husks of pepper. Also occurring were powdered slate, gypsum (up to 22%), and ground quartz (which at up to 11%, would be perilously close to ground glass in its effect). There was even a mixture, which could presumably be bought, known as PD (for pepper dust), which was a very common adulterant. This consisted of the above adulterants in varying amounts, usually containing salt, and occasionally pepper.[3]

The adulteration of pepper was usually carried out largely by the wholesalers and was then very skilfully done. According to Phillips 'it is always easy, on analysis, to distinguish between the highly finished article of the wholesale delinquent and the bungled production of the unscrupulous retailer'. At this time apparently most pepper was ground by the wholesaler, (most retailers had ground their own until the middle of the nineteenth century), and it was comparatively easy, and profitable, to add relatively small amounts of adulterant to a large volume of pepper. Not that the adulterants mentioned above were always added in small quantities. In one case in Scotland, in 1864 just before the duty was abolished, a sample of pepper was purchased by an Excise supervisor to whom the retailer expressed the feeling that the pepper was 'not what it ought to be'. On analysis the 'pepper' was found to consist of 25% gypsum, with the rest being ground mustard husks and a little cereal starch. There was no trace of pepper, although Phillips commented that it was a good imitation. A second sample

from another retailer, but from the same wholesaler, consisted of over 60% adulterants, with the remainder being
pepper. The defence against charges of adulteration were
sometimes ingenious. In one case where sago flour had been
used the defendant said that he believed that by mixing his
pepper with such a cheap and innocuous substance 'the
public was served with a better article than could be obtained
for the same money when genuine pepper was purchased'.[4]

As seen in the case of the above sample, the Excise
officers round the country apparently purchased samples of
pepper for examination in the Laboratory, presumably where
adulteration was suspected but possibly also taking samples
at random. The practice of purchasing samples was suggested
by Phillips in 1859 on the ground that the practice of
adulteration was increasing, and it would be advisable to
have regular checks on this commodity. He may also have
been pushed into the suggestion by the remark by Arthur
Hassall in his book on food adulteration, published in 1855,
(repeated in 1861), in which he advocated the purchasing of
food samples to check on adulteration. Hassall was very
scathing about the 'Excise' chemists, whom he accused of
incompetence and of not in fact protecting the revenue. He
even accused them of not even being competent users of the
microscope, (Hassall's statements were often nothing if not
sweeping). This, at least as far as Phillips and his staff were
concerned, was obviously absurd, their competence was
unquestioned, and certainly must have been by 1855, if not
earlier. Phillips noted in 1863 for example that the microscope was 'peculiarly applicable' to the examination of
pepper, (this reference may have been included as a rejoinder
to Hassall of course), and Phillips had been using a microscope to study tobacco since 1842, as Hassall very well
knew.[5] Initially at least Phillips' skills did not go unquestioned however as is shown by the reference by Andrew Ure
in 1844 to a case of adulterated pepper in 1843 when Phillips
and Thomas Graham had claimed to have detected sago as
being present. Ure said that he had proved that it was not
present, a result accepted by the court. Pepper was still being
adulterated in the 1880s, with the occasional sample still
being received.[6]

The tea sold in the early and middle years of the last
century was in a similar state to pepper, being very extensively adulterated, also largely by the wholesalers, although in

this case by the original wholesalers particularly those based in China. Large quantities of adulterated tea were imported from China, a practice which continued for many years. The adulteration of tea has had a long and dishonourable history. In the 1750s, tea imported through Deal was found to be 'dyed and adulterated' and was seized and burnt; a similar order was made in respect of a cargo seized at Exeter in 1761. Adulteration of pure tea in this country was relatively easy to detect apparently.[7] Much of the effort was directed to converting ordinary black tea (produced by fermentation), into green tea, which was more expensive. Much ingenuity was practised in this, and various substances such as Prussian blue, turmeric and gypsum were used in China. Lead chromate and copper arsenite, together with less harmful substances were often used in this country. Genuine tea was frequently mixed with leaves of beech, elm and other trees, or sometimes with the wonderfully named 'Lie tea' manufactured in China, and imported in chests on which that name was inscribed. It consisted of tea dust, foreign leaves, sand and dyes such as indigo made up with starch or gum into small pieces which was coloured to resemble black or green gunpowder tea. Sometimes exhausted tea leaves were dried, mixed with a little fresh tea, and then sold.[8] The number of samples examined each year was very low, about 10, and it is difficult to escape the conclusion that the Inland Revenue department were not very concerned the prevent the adulteration of tea: their interest was not financial but because it was forbidden under various Acts of Parliament. Until 1875 there was little check on imports by the Customs, provided import duty was paid, and the tea licence only brought in just over £76000 for the Revenue before it was abolished in 1869.[9] In 1875 effective statutory control was imposed by the *Sale of Food and Drugs Act*, see chapter 5.

The work of the Laboratory expanded considerably both in variety and in quantity in the years after its foundation, from perhaps not more than one hundred samples in 1842 to over 9500 by 1859.[10] This latter year is chosen as an example because then the Laboratory moved into new rooms in Somerset House. It had become necessary to move in 1853, when the Inland Revenue departments, recently formed from the amalgamation of the Board of Excise and the Board of Stamps and Taxes moved from the old Excise headquarters building in Broad Street to the new Somerset House. For the

13 Somerset House, London, inner quadrangle. The Laboratory rooms were on the right at the top

first few years there was no room for the Laboratory, and so a house at 30 Arundel Street was rented from the Duke of Norfolk on a three year lease. This lease had to be extended until the building of Somerset House was complete. The Laboratory finally moved into its new, and for those days spacious laboratories at the top west end of the building, (the 'New Wing') in June 1859. There are tantalisingly few references to the physical appearance of the Somerset House laboratories. No description of their actual appearance and no photographs exist. One brief glimpse of the rooms and life in them is given in a letter of Phillips written in 1868. This concerns a 'student' (an Excise officer temporarily attached to the Laboratory and attending classes at the Royal College of Science), one D Maclean. He worked in the Laboratory in 1860–62, at no. 10 bench opposite and within a few feet of the sand–bath used to heat inflammable liquids. He had apparently complained of the heat, which he said made him sick and Phillips immediately (he said) enclosed the bath in a 'wooden case lined with metallic plates so as to form air chambers' which obviated the effects complained of. Maclean appears to have been asking for a pension on health grounds. Phillips described his rooms as being 'efficiently arranged and fitted up with special regard to the purposes for which they are now applied'. The rooms were completely full by 1861, with no room left for expansion.[11]

Just before this move and with the agreement of the Treasury the Laboratory was organised as a separate department within the Inland Revenue. Mr Phillips was appointed Principal and he had six assistants, as he had done since at least 1855, but not yet a permanent staff except one George Kay, appointed his Deputy Principal.[12] The organisation of the work carried out by these assistants was described in some detail by George Phillips for the first time in 1860. The samples involved at least 5000 distillations and nearly 16000 weighings. The 'particulars relating to each sample, the results of the analyses and the registration for easy reference of the whole work of the department' were now entered in ruled ledgers 'to avoid the labour of ruling and titling', which improved their neatness and gave the assistants more time to devote to scientific duties. Writing up the results of their work must indeed have taken a considerable amount of time, the existing 'Reference Experiment Books' are written in perfect copperplate hand. In a previous report Phillips had said that registering, recording and reporting occupied enough time to keep any one of the assistants fully occupied. Perhaps this was why by February 1867 a 'Book-keeper', one J O'Loghlan, had been appointed.[13] The increase in staff was necessarily accompanied by an increase in costs, but this was small compared with the increase in work Phillips reported proudly. The numbers of samples had increased five time between 1856 and 1860, but the costs had not even doubled and 'the annual cost of the instruments, apparatus, reagents, and other materials employed in the practical branch of the department' came to about 7d (nearly 3p) for each sample analysed, including 'prolonged investigations and special analyses of a difficult and expensive character.[14] This figure gives the cost of running the Laboratory (excluding staff costs) as just over £300 per year. In 1867 the salaries of the six permanent staff came to £1685 per year and of the fifteen temporary staff and students to £1596, a total of £3281. Of this Phillips received £750 and his deputy £425. In comparison the costs of chemicals and equipment 50 years later in

1911 was £1600, staff costs were then £14,000.[15]

The equipment used by these analysts was of the 'best description'. They included six microscopes, 'three made by Ross, and three by Smith and Beck', and five balances, four of them in almost constant use, and the other 'a valuable instrument by Oertling, being reserved for special investigations, and for such experiments as demand the most extreme accuracy in weighing'. The Laboratory was also well supplied 'with all those appliances which are essential to the proper performance of its duties'. This presumably included glassware and chemicals, but they are unfortunately not specified. What is described is the procedure for dealing with samples when they were received. After saying that the average number of samples received every working day was 36, which was six per 'business hour' (ie a working day of six hours), the account goes on, 'When a sample is received it is at once handed over intact to an assistant, who thereafter becomes responsible for the preservation of the proof of its identity. He then, having endorsed the sample with the date of its receipt and his initials, opens it, being careful not to break the seals and submits it to the inspection of the principal or his chief assistant and, according as its analysis appears easy or difficult, or as the consequences depending on it are unimportant or otherwise, he is directed to examine the sample either alone or in conjunction with another assistant, the results obtained by the two being designed to check each other. The latter precaution is not adopted from a want of confidence in the skill of the assistants, who are well able to determine the fact of adulteration, but in order to prevent any chance of mistake as to the amount of illicit materials which may be present in the sample. The analysis having been completed to the satisfaction of the principal, the assistant after fully recording the results makes his report which, before being sent in, is submitted to the principal or his chief assistant, for approval and counter-signature. Should a prosecution arise from this report, the assistant is required to attend the hearing of the case in whatever part of the Kingdom it may take place and is expected to establish the truth of his analysis'.[16]

The move to new rooms in Somerset House had come just in time to meet a large increase in work and as noted pressure on space was a problem again by 1861. The increase in work was to a large extent due to the large numbers of

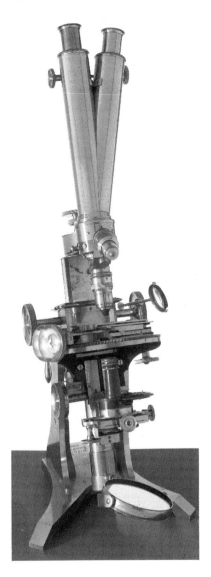

14   Ross microscope of *c*.1862

samples of beer being received for determination of original gravity (see the next chapter), but was partly due to a large and growing number of samples of coffee and coffee and chicory mixture. The law on these commodities was complicated. Up to 1840 the sale of mixtures of coffee and chicory was forbidden by law, but still occurred, so that in 1840 the Treasury allowed the sale of such mixtures. This of course meant that even less pure coffee was sold, and in the late 1840s and the early 1850s a considerable amount of research was carried out on behalf of the Excise to determine whether or not the presence of chicory in coffee could be discovered. Botanical characteristics were tested from the first (a memorandum from Phillips in 1848 discusses these) and by 1849 Adam Young, then attached to the Laboratory, had written to the Board saying that coffee and chicory could be distinguished microscopically. In 1851 Phillips had no doubt that the adulteration of coffee could 'invariably' be detected. Following their usual practice the Board of Inland Revenue submitted the question to various outside chemists for their opinion. Professor Graham of University College and two colleagues examined mixtures chemically, aided partly by James Bell (later Principal of the Laboratory) and then an Excise student at University College studying chemistry under Graham. Professor Lindley and Dr Hooker, eminent botanists and microscopists wrote a report on the botanical distinctions between coffee and chicory and much else such as beans and turnips, which could also be used to adulterate coffee. Doctors Carpenter and Taylor, two eminent medical men, were asked if there was any reason why coffee should not be sold mixed with chicory, and if the consumption of chicory could have any adverse effects. All of the reports produced by these men were published in a limited edition of 100 copies as soon as the last of them had reported, in February 1853.[17]

Perhaps anticipating the reports the sale of mixtures was again forbidden in 1852. This state of affairs only lasted until February 1853 when the sale of a mixture of coffee and chicory was again allowed, provided it was labelled as a mixture. Nothing other than chicory was allowed to be mixed with the coffee. At this point, as Phillips said 'various contrivances were adopted by many dealers to prevent the words required . . . [ie 'Mixture of Coffee and Chicory'] . . . from being observed by the purchasers'. It was then ordered

15 Report on mislabelled coffee and chicory mixture, 1853 (PRO CUST TE 5025)

that the words 'This is sold as a mixture of Coffee and Chicory' should be printed in such a way that they could not be concealed. The immediate response of the fraudulent dealers was to label all their packages with these words, whether they contained a mixture or not. The Inland Revenue Board then (in 1854) issued an order stating that if pure coffee was asked for it should be supplied.

The result of this rather confusing game was that from 1853 onwards large and growing numbers of coffee and coffee mixture samples were received in the Laboratory. These samples were not, unlike most of those being received, taken from warehouses or seized by Excise officers on suspicion, but were purchased from dealers over the United

Kingdom. The various orders obviously did have an effect, since the percentage of samples found not labelled dropped drastically after 1853 (19.5% to 9.5% to 5.8% in two years), and samples found to be a mixture when pure coffee was requested by the purchaser showed a similar fall after the 1854 order (17.4% to 14.7% to 5.8%).[18]

Random samples continued to be bought and examined by the Laboratory throughout the 1860s in quite large numbers. Note books of the period contain groups of several hundred samples at a time being examined. There was apparently a fear that a taste for the mixture was gradually being acquired by part of the community, to the detriment of the revenue (chicory paid less duty than did coffee). It was noted that whilst the consumption of almost every other 'article of comfort and luxury' had kept pace with growing population, that of coffee had decreased contrary to expectations 'when the agreeable and innocent character of the beverage' was considered. Phillips ascribed this to the availability of the mixture of coffee with the cheaper article. In some poorer districts he had found mixtures with from 30–86% of chicory, and in really cheap 'mixtures', up to 100%. He recommended that mixtures should be made illegal. However the government found a much more subtle means of preventing adulteration, and over a period of three years, from 1860 to 1862, increased the duty on chicory to equal that on coffee. The average percentage of chicory found in mixtures dropped in those three years from 39.3 to 22.3.[19]

The large scale sampling of coffee ceased in the following year, and with it the first large scale attempt to check adulteration of a foodstuff, albeit an attempt made for fiscal reasons. In the first year or two after the large increase in duty some traders, inevitably, adulterated the chicory they were selling (often with ground roast peas), but this practice soon ceased. Later coffee was often sold in a largely pure state, although adulteration was still practised, presumably with the roasted beans, peas etc, together with orange pips and iron oxide which had been used before.[20] Before too many years had gone by however coffee was being sold adulterated with burnt sugar, or caramel, a practice which continued for many years. In a batch of 73 samples tested in April 1864, taken in Liverpool and Birmingham, 19 contained caramel, i.e. 26%. Other substances were used as adulterants, for example locust beans (up to 30%) and ground acorns under the name of Pelotas coffee or Coffee surrogate. Roasted date stones appeared in 1878, the by-product of a distillery, which had many tons of them. This mixture rejoiced in the name of Melilotine Coffee. Samples continued to be taken on a small scale for the rest of the century.[21]

As well as increasing in volume the work of the Laboratory was increasing in variety, becoming known outside the Excise service and being used by other departments. In 1870 the Board reported that it felt that 'never was an outlay for public purposes more justified than that which has been applied in the formulation and maintenance of this scientific branch of our establishment'.[22] Work carried out for the Customs department has already been mentioned, other early customers were the Admiralty (in 1871), the India Office, the Post Office and the Stationery Office (all in 1873), the Home Office (1881), and the Commission of Works (1883).

The range of samples dealt with for these departments was very wide indeed. Some of the work was not new, beer for the Navy and the India Office for example, which needed to be 'thoroughly cleansed, well hopped, and properly fermented' in order to keep well. Other samples were bread, cotton drill, (to ascertain whether the bleaching agent had been removed) and linen and silk handkerchiefs. Drug samples were examined, gold lace (for gold content), and a great variety of foodstuffs, much of it destined for navy stores, but some of it for the India Office. In one case a sample of pickles returned from Egypt in a badly discoloured state, was found to have formed ink with traces of iron present combining with tannic acid from the oak cask and pyroligneous acid used in place of vinegar.

Samples from the India Office were a particularly wide ranging group, but 'although the nature of the work was somewhat different from that generally performed' Phillips was confident of the ability of his staff to deal with the challenge. He welcomed the opportunity for them to become 'thoroughly acquainted with the general composition and sophistication of commercial articles which would not in the ordinary course of business pass prominently under their notice, and they are thus able more readily to detect general adulterations, and also to acquire quickness and precision in manipulation' which increased their value as analysts,

16 Sartorius Balance, 1876, weighing to one ten thousandth of a gram

enabling them to carry out a much greater variety of work. Phillips obviously encouraged his staff to spend time in research, to make themselves familiar with all possible scientific techniques, and was proud to say that the addition of new varieties of work had only helped to make the Laboratory more efficient, while saving money by the partial centralisation of the analytical work of several departments.[23] This centralising tendency continued.

Several new areas of work started at this time and these continued (and in some cases continue) to be part of the Laboratory's work for many years. A solitary sample of New River water appears in 1865 (it was examined for density, residue on evaporation and ash), but water samples appear regularly from 1875, possibly for the Home Office or Trinity House for which they were later done on a regular basis. A laboratory was specially fitted up for water analysis in 1879. 'Room 131' in Somerset House was allocated for this purpose, but for some reason was not fitted out properly for at least 12 months after staff had begun to use it, this rendering it 'unbearable' when a combustion was in progress. Given the connection of Sir Edward Frankland with the

Laboratory, (see below) the 'combustion' was probably his method for measuring the carbon and nitrogen content of the organic matter in water. From the ratio of the results was determined whether or not the organic matter was from decaying vegetable matter or highly nitrogenous animal matter and thus potentially containing dangerous germs. The method was complex and difficult and not as good a measure of pollution as Professor Wanklyn's method for determining 'albuminoid' ammonia developed at about the same time as that of Frankland in 1867. What part the Laboratory played in the controversy over what was the best method of measuring water pollution is not known. It was certainly not so public as that over milk and butter samples received under the 1875 *Food and Drugs Act*, (see below), although samples of water were tested for potability under that Act. The bacteriological examination of water began in the 1880s, the earliest reference to the Laboratory's involvement is given in letters of the then Principal in 1894, concerning cultures of typhoid and *B. coli.*, both of them measures of faecal contamination. By 1899 the Laboratory had taken over the analysis of London water supplies when Frankland died.[24]

Other new work was done for the Post Office. This included tests to see if the cancellation marks could be removed from stamps without damaging them, and in approving various formulations of fugitive inks.[25] As early as 1858 the General Post Office had asked for help in the prevention of the fraudulent use of cancelled postage stamps, and Phillips had worked on the problem for over eight months. He explained that the long time he had spent on the problem was due to the conditions laid down by 'Mr Hill', (presumably Rowland Hill, Secretary to the Post Office, and inventor of the penny post), which were that any method for preventing frauds should not involve much trouble and expense. Devising such a method took longer. One of the problems was that the stamps had been covered with a protective film, which meant that the cancellation could be removed with a solvent. The Post Office suggested treatment of the printing or cancellation ink with a chemical substance which would react with the protective film, but this proved to be impracticable. Phillips offered two alternative solutions. In the first the stamps (or 'labels' as he termed them) would be printed on paper slightly tinged in its manufacture with ultramarine blue and the cancellation ink would be weakly acidic, for example with acetic acid, causing the colour to be discharged. Alternatively, a hardened metal cancellation stamp (providing a sharp impression) should ensure that any protective film on the surface of the stamp was sufficiently broken so that the ink used would penetrate to the surface of the stamp and become absorbed. He also drew attention to the greater risk of fraud in the cancellation of bill stamps (as opposed to postage stamps), which had a glazed surface, and suggested the use of paper made from flax or hemp.[26]

The last large area of work to be considered in this chapter is the examination of lime and lemon juice and the spirit intended to fortify it for the Board of Trade (later for the Ministry of Transport). This juice was used to prevent scurvy in seamen. At first Phillips (and James Bell, his deputy) were appointed Inspectors for the Port of London (which included the ports from King's Lynn round to Rye) in October 1868, under the Merchant Shipping Act of 1867, but in the following year from 1 July the Laboratory was given responsibility for the inspection of the juice intended for the whole of the merchant navy. Several hundred samples per year were received from them until the work ceased in 1975. Very few were received in the last few years that the Act operated since other anti-scorbutics were allowed by then. The few samples received were analysed in the Liverpool outstation. As many as 412 had been received in the first twelve months and Phillips estimated that the work was taking the time of two assistants and costing about £100. The cost was waived by the Inland Revenue Board, who agreed not to charge the Board of Trade. The Board of Trade had issued instructions on lime and lemon juice under the *Merchant Shipping Act* in November 1867. These specified that the juice had to come up to a certain standard of quality and must be fortified with a certain quality of palatable proof spirits. These rather vague requirements were supplemented by a more detailed specification which called for fortification of the juice with '15% as proof spirit' of either rum or brandy, with tests for taste, colour, specific gravity at 60°F, odour, consistence, sugar, gum, quantity of citric acid, freedom from certain other acids (not specified although the Laboratory tested for acetic, oxalic and tartaric acids) and freedom from general adulteration. The quality had to correspond to standard juice squeezed in the Laboratory from lemons from known sources.[27] Inevitably, at this date

Received Oct 3rd 1868
G. Phillips

Inspectors' Circular.   No. 301.

No. 3.— ~~February~~ 1868.
October,

LIST OF INSPECTORS

OF

LIME AND LEMON JUICE

APPOINTED BY THE

BOARD OF TRADE.

Merchant Shipping Act, 1867.

The following is a List of the several INSPECTORS OF LIME AND LEMON JUICE appointed by the Board of Trade for the various Ports of the United Kingdom, under Section 4 of "The Merchant Shipping Act, 1867."

The places in England, Scotland, and Wales enumerated in column 3 of this List are assigned to the Inspectors resident at the respective Ports named in column 2, in the order shown. In the case of Ireland, the whole of the Irish Ports are, for the purposes of Lime and Lemon Juice inspection, assigned for the present to one Inspector.

Merchants or Producers who desire the inspection of Lime or Lemon Juice deposited by them in a Bonded Warehouse at any of the places specified in column 3, should therefore address an application to that effect to the Inspector for the District in which the place is classed in the List. A Form [L. J. 1] is provided for this purpose, and can be obtained from the Superintendents of the Mercantile Marine Offices at the Out-ports. In all cases, care should be taken by the Applicants to fill in the particulars required, and to address the application in the manner shown on the form.

21062.

2

ENGLAND, SCOTLAND, AND WALES.

Name of Inspector. Col. 1.	District, and Address of Inspector. Col. 2.	Ports classed in the District named in Col. 2. Col. 3.
J. Falconer King, Esq.	LEITH. *69, Constitution* ~~Street,~~ Street, Leith.	Wick. Inverness. Banff. Peterhead. Aberdeen. Montrose. Arbroath. Dundee. Perth. Kirkcaldy. Alloa. Grangemouth. Borrowstoness. Granton. Leith.
W. Osborne Lambert, Esq., M.D.	SUNDERLAND. 13, Foyle Street, Sunderland.	Berwick. North Shields. Newcastle. South Shields. Sunderland. Hartlepool. Stockton. Middlesborough.
T. Stevenson Usher, Esq., M.D.	HULL. 18, Ocean Place, Hull.	Whitby. Scarborough. Hull. Goole. Gainsborough. Grimsby. Boston.
~~Harry Leach, Esq.~~ *Geo. Phillips, Esq.* *Jas. Bell, Esq.* }	LONDON. ~~41, Great Tower Street Buildings, E.C., London~~ *Inland Revenue Laboratory, Somerset House, London, W.C.*	Wisbeach. Lynn. Wells. Yarmouth. Lowestoft. Woodbridge. Ipswich. Harwich. Colchester. Maldon. London. Rochester. Faversham. Ramsgate. Deal. Dover. Folkestone. Rye.

17   Appointment of Phillips and Bell as Lime juice inspectors, 1868

there could be no requirements for the actual anti-scorbutic agent, vitamin C. Standard samples of lime and lemon juice were to be kept. As might be expected samples sometimes had to be rejected because of adulteration, chiefly dilution, or just because of poor quality, and poor sampling procedure sometimes caused problems. On at least one occasion juice was examined which had been taken from a ship whose crew had suffered from scurvy. It was found 'free from adulteration and in fair condition', but was presumably lacking in vitamin C. In the same year (1876) samples of juice taken from the stores of a returned arctic expedition were examined for the same reason. The tests are not explained, tests for the presence of vitamin C were not available until 1918.[28]

One final comment may be made here on the work of the Laboratory. In the nineteenth century as has been shown, the range of work carried out was wide indeed, but very little has been said about the methods used. Some idea may be obtained from the surviving work books. Most of the methods naturally involved 'wet chemistry', sugar by Fehling's solution or by fermentation and distillation of the alcohol formed but the polariscope was certainly in use by 1872. One sample received in 1867 does demonstrate rather neatly the difference between analytical work then and now. It concerned the seizure of a cask of rum with 'ingredients mixed therewith'. It smelt of oil of almonds, was 'sweet' and tasted of cinnamon. No trace of cyanide (from almonds) was found and from the odour Phillips was of the opinion that the flavouring was a mixture of oil of mirbane (nitrobenzene), oil of cloves and oil of wintergreen. To check this a number of experiments were carried out with mixtures of nitrobenzene, and oils of cinnamon, cloves and wintergreen in different combinations. It was finally decided that taste and smell best corresponded to a mixture of nitrobenzene and oil of cloves, the sweetness being due to the nitrobenzene. This laborious procedure may be compared with the ease and speed with which such an analysis could be carried out with modern equipment, giving an accurately measured answer in a very short time. The impression given by reading the old work books is however of the great care and skill with which the work was carried out by the methods available.[29]

3

Beer and Spirits Work

AS RECORDED IN THE CHAPTER ON 'GOVernment Science and the Revenue', there had long been a government interest in the strength of alcoholic liquor. Such had been the value to the Board of Excise of Mr Phillips' work on tobacco that he was asked to widen his interests to include alcoholic drinks amongst the products analysed soon after the foundation of the Laboratory.

Samples of beer and hops were already being received whenever adulteration was suspected,[1] but other matters were being considered in 1846. Until this year malt and hops were the only permitted brewing materials on which Excise duty had to be paid. Distillers were allowed to use malted or unmalted grain, duty paid sugar, potatoes or mangelwurzels, but mixing of these (except malted and unmalted grain), was forbidden, and this fact, together with the Customs duty on sugar, meant that in practice only grain was used. In 1846 the West Indian sugar planters agitated for sugar and molasses to be permitted in Britain for brewing and distilling. Before this could be agreed it was necessary to establish for duty purposes what quantities of these materials were equivalent to a standard quantity of malt or barley in terms of the amount of alcohol produced.[2]

By an order of 8 September 1846, the Board of Excise asked Mr Phillips and Thomas Dobson (senior to Phillips in the Excise service) to investigate this matter. For this purpose they purchased 10 cwt. of Antiguan molasses (at a cost of £11 14s i.e. £11.70p). They carried out a series of elaborate experiments, fermenting and distilling the molasses and determining the yield of proof spirit per hundredweight. They also tested the use of molasses in brewing beer, concluding here that if the molasses did not taint the beer with their flavour, there was no practical objection to their use. They reported on 14 November 1846, with satisfaction, that they had '18 gallons of good beer, which has no decided molasses

18   Thomas Dobson, Joint Secretary
to the Board of Inland Revenue
1858–1861

flavour', which 'in a few weeks will be fit for use'.[3] They went on to buy three casks of sugar and used them for similar experiments, only this time producing alcohol, and reported in January 1847 that their experiments gave satisfactory evidence that established methods of charging the spirit duty were perfectly applicable to a distiller using sugar and would cause no loss to the revenue.

The Board of Excise reported this conclusion to Parliament, and also said that if molasses were permitted the duty would have to be raised by 3 shillings (15p) or more per gallon of spirit.[4] Quite a large amount of practical work was involved in these experiments, all carried out in just over four months, showing the speed and industry with which Phillips and Dobson had worked. Their results were confirmed by Professor Fownes of University College, who further calculated a table of the equivalents of specific gravity to per cent of alcohol in solution.[5]

Later that year Parliament authorised the use of sugar in breweries and distilleries. Molasses, treacle and mixtures of these with grain were also allowed in the following year, but in distilleries only.[6] By these Acts spirit manufacturers were allowed to use duty paid sugar mixed with their malt. This created a problem with respect to the drawback (repayment of duty paid on materials used) on exported beer. Previously this was allowed provided brewers swore on oath that they had used at least two bushels of duty paid malt per barrel, i.e. that the beer was of a standard strength. There was no check on this, except tasting the beer, a somewhat haphazard procedure. A means of checking the brewers' declaration was needed.

This method was developed by Phillips and Dobson, again working together, and partly at least in their own time.

19 Report of Phillips and Dobson on the use of sugar in brewing and distilling, 1847

BREWERIES AND DISTILLERIES.

RETURN to an Order of the Honourable The House of Commons,
dated 28 January 1847;—*for,*

A "COPY of any REPORT from the Board of the Excise to the Lords of the Treasury, on the use of BARLEY, MALT, SUGAR, and MOLASSES, in BREWERIES and DISTILLERIES."

Ordered, by The House of Commons, *to be Printed,* 29 *January* 1847.

REPORT from the Board of Excise to the Lords of the Treasury, on the use of BARLEY, MALT, SUGAR, and MOLASSES, in BREWERIES and DISTILLERIES.

My Lords,

IN pursuance of instructions received from the Chancellor of the Exchequer in September last, the particular attention of this department has been directed to the comparative value of barley, malt, sugar, and molasses to the distiller and brewer.

Two superior officers, Messrs. Dobson and Phillips, peculiarly qualified by extensive practical and scientific knowledge, were selected for this investigation, and their experiments and calculations have been carefully tested by other officers conversant from long experience with the results obtained on a large scale at breweries and distilleries.

Previous to these experiments, the latest information on the subject is contained in the evidence given before the Select Committee of the House of Commons in 1831, "on the use of Molasses." (*See* Questions 508. 600, 601. 603, and 4152). The equivalent weight of sugar to a quarter of malt in brewing is therein variously calculated, probably according to the quality of the barley and the skill of the brewer. In one very bad year, it was as low as 173 lbs. of sugar to the quarter of malt; in another and more favourable season, as high as 226. The average of these is, 199½ lbs. It has been generally, however, supposed, that 180 lbs. is the equivalent, and this weight has been assumed as the basis of the calculation in a recent publication on cheap beer. In the Bill now before the House of Commons 200 has been taken as a mean, on the authority of the evidence above quoted, and confirmed by our experiments. We have been further induced to adopt this weight from a knowledge that the saccharine produce of barley has been increased of late years by the cultivation of a superior description of grain, and by improved modes of working in breweries.

As to distilleries, the evidence before the same Committee (*see* Question 764) shows that 112 lbs. of sugar produce 11½ imperial gallons of spirits at proof. A calculation varying but little from this was adopted in the Act 52 Geo. 3, c. 3.

This result has been so nearly reached in our recent experiments, that considering the small scale on which they were conducted, and the difficulty of maintaining the equal temperature so essential to perfect fermentation, we have had no hesitation in taking 11½ gallons of proof spirits as the average produce of 1 cwt. of sugar.

In obedience to directions received from the Chancellor of the Exchequer, we have now the honour to forward copies of the reports of the experiments for the information of your Lordships.

26.   A   We

In their report of November 1846 they had hinted that they were confident from their experiments that they could determine the original specific gravities (or original gravity) of the wort (that is the mixture from which the beer was brewed), and thus obtain a measure of the quantity of sugar and malt used to produce the finished beer. Their method was based on one originally proposed by 'Mr Peter Stevenson. Philosophical Instrument Maker of Edinburgh'. This involved a partial evaporation of the beer to obtain a measure of the alcohol content.[7] Phillips and Dobson asked the agreement of the Board of Excise to their carrying out the necessary work. They were finally asked to do so in February 1847 after work carried out under the auspices of the Surveying General Examiners and later Dr Thomas Thomson of Glasgow (Regius Professor at the University), had failed to obtain agreement between the spirit measured by distillation and that calculated from the increase in gravity on evaporation. An evaporation based method was desirable because it would be easier for the Excise officers to carry out in the field.

Working with their usual speed Phillips and Dobson reported on 2 August 1847, although they felt it necessary to apologise for the length of time taken on the grounds that the work was very difficult and that it was 'frequently interrupted by the ordinary business of the Laboratory'. After initial difficulties they had overcome the problem of correlating evaporation and distillation results. They found that the relationship between the two methods was not linear, and produced a 'gravity lost' table to be used with the spirit indication as determined by the increase in specific gravity. Their table was based on evaporation explicitly because Excise Officers would be unlikely to have distillation apparatus available. Subsequent work was also carried out in 1848 by Dobson, Adam Young and Charles Forsey, the latter two temporarily attached to the Laboratory, showing how mixtures of grain and sugar worts could be fermented and the amount of alcohol obtainable from such mixtures. Young and Forsey went on later the same year (December) to confirm their previous results and at the same time and at the request of Thomas Dobson developed a distillation method for the determination of original gravity, in place of the evaporation method of Dobson and Phillips. This had been found to require a 'much nicer Manipulation than could be expected for an ordinary Excise Officer'.[8]

Before the original investigations were finished the Act permitting the use of sugar in brewing had been passed. This required brewers to take an oath to the effect that the worts used before fermentation were of not less original specific gravity than 1.054 or 1.081, according to the rate of drawback claimed. An original gravity of 1.054 was apparently assumed to correspond to two bushels of malt per barrel.[9] It is not known whether Excise Officers used the Phillips and Dobson or the later Young and Forsey method to check the declarations, but by 1851 beer samples were being sent to Phillips' Laboratory to be tested as to the accuracy of the declaration of gravity. It is unlikely that in 1847 Phillips had foreseen that their researches on original gravity would lead to this work, and indeed an ever increasing number of such samples, (from 78 in 1851 to 7544 in 1861).[10] In 1851 Phillips reported that 87.7% of the samples (or 58 of them) were not in fact entitled to the claimed drawbacks. This caused a great outcry from the brewers, disputing Phillips' figures, so that in 1852 the Board of Inland Revenue referred the distillation method to three eminent professors of chemistry, Thomas Graham, Alexander Hoffman and Theophilus Redwood for reinvestigation. Their report said that Phillips' and Dobson's work had been most useful to them, and that a large part of their own work had been carried out by Young and Forsey, under their direction, particularly citing their 'remarkable skill in experimenting'. Graham *et al.* found that the gravity lost tables of Phillips and Dobson were not entirely accurate, due to various factors, including an initial increase in specific gravity of cane sugar worts due to inversion of the sugar. They also found that the evaporation method of Phillips and Dobson did not give results close enough to the distillation methods for use by brewers and produced a gravity lost table based on distillation. It was not very different from the Phillips and Dobson original table, as a leaflet produced by the Inland Revenue in September 1852 triumphantly proclaimed.[11] There is a similar feeling of triumph in a report by Phillips in a case in 1853 where the brewer concerned claimed that the method was 'inadequate'. Phillips said that it was perfectly adequate and that the samples definitely had an original gravity less than that declared. The use of outside chemists in this case, to check the work of the Excise staff was not a particularly rare occurrence. As has been in the first chapter, before the Laboratory was set up the Board employed many consultant chemists to carry out their analytical work and this did not cease in 1842. It was indeed quite a regular event in the first ten years of the existence of the Laboratory both for special investigations and for analysing ordinary samples. Professor Graham was used quite extensively in both instances.[12]

Complaints about the 1847 Act continued to be made, including the claim that the quantity of beer brewed from a bushel of malt was not as great as assumed in the Act. In 1854 it was therefore changed to allow seven rates of drawback on exporting beer, from the two specified in the original Act. The onus of proof of the specific gravity claimed was laid on the exporter, the government apparently feeling that something must be done to check the undoubted frauds being perpetrated but not providing any way of checking the declaration. In 1856 the table and method of Graham, Hoffman and Redwood for determining the original gravity of beer was officially adopted, and given statutory authority for use in checking drawback declarations. The Act of Parliament in which this method was contained was one of the first to embody an official method of analysis.[13] Phillips reported with some satisfaction that many brewers had complained that the method was unfair, but on being given facilities to have special beer brewed and tested by the method, and then have the method explained to them, had expressed their satisfaction that the process did give accurate results. Further, they were 'inclined to ascribe the cause of any disparity between the Laboratory results and their own data, more to inaccuracies in their brewings than to errors arising from the method adopted by the Revenue'. They also had their staff instructed in the method by the Laboratory and adopted it in their own establishments. Brewers were certainly convinced since false declarations dropped from 9.5% in 1856, (the year before the Act was passed), to 2.5% in 1857, the year of the Act, and to 1.4% the year after that. Some brewers required more convincing than others but on investigation it was frequently found that their saccharometers (used to determine the original gravity of the beer), were reading (optimistically) anything up to 5 degrees of gravity higher than they should have done. One brewer apparently never was reconciled, making great efforts to prove the method did not work, doubtless, as Phillips said, due to 'large quantities of his beer having been frequently rejected for drawback on the score of deficiency of strength'.[14] Certainly a great deal of care was taken over the analyses. All samples found to have a gravity equal to that declared were passed as entitled to the drawback, but when a gravity below that declared was found a second estimation was carried out by the same assistant, but under the supervision of Phillips or his chief assistant. If the result was confirmed, a fresh sample was obtained or the duplicate sample was analysed.[15]

As noted above, the number of samples received under the Act of 1856 soon became very large, so large in fact that the ten outstations originally proposed for tobacco inspection (see above p.27) and set up in various ports in 1871 were largely used to test the beer samples. This remained their prime function for many years. The Laboratory had changed from one mainly intended to detect fraudulent adulteration to one also playing an active day to day role in the collection and safeguarding of the Revenue. In recognition of the savings resulting from the checking of beer gravity declaration the Commission of Inland Revenue recommended to the Treasury in 1862 that messrs Dobson and Phillips receive an award of £500 each, a considerable sum of money: just £100 less than Phillips' salary in 1863. The Commission commended the 'considerable chemical talent' and the 'amount of close application and tedious labour which must have tested to the utmost the zeal of the investigators' and which 'must

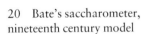

20 Bate's saccharometer, nineteenth century model

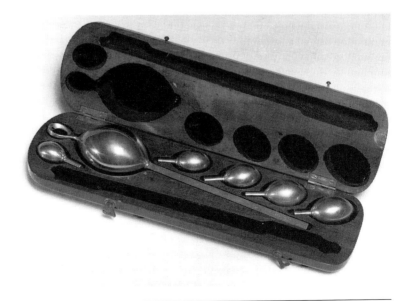

have absorbed nearly the whole of their available time before and after office hours'. This is interesting in revealing that some at least of these investigations were carried out in the investigators' own time.[16]

The work on original gravity and the use of molasses was far from the only piece of research carried out by Phillips. The Board of Inland Revenue seem in fact to have regarded him and his Laboratory as an all purpose enquiry agency.[17] Continuing from his involvement with beer and brewing they asked him in 1854 to investigate 'the distinctive character of raw grain and malt'. Malt was of course the most usual material used in the brewing of beer (until 1847, when sugar was permitted, the only material used). Malt consists of barley grains steeped in water, allowed to germinate and then roasted. Excise duty was paid on this malted barley and not on 'raw' barley. Some brewers were therefore mixing 'raw' barley with malted barley, and using this mixture to brew beer, thus defrauding the revenue. Phillips carried out an extensive number of experiments over twelve months, and concluded that the best way to prevent this fraud was for the official definition of malted grain to include a provision for the grain to have visibly sprouted, i.e. 'all grain in the Malt-Roasters possession shall be deemed un-malted, unless the plumule shall have extended half the length of the grain'.[18] It was thus now possible to prove that a grain should have paid duty. The Revenue immediately (in 1857) took samples of grain from the premises of malt roasters, and Phillips found that out of 53, only two consisted entirely of germinated grain. Samples for this purpose continued to come to the Laboratory until the Malt duty was abolished in 1880.

The numbers of un-germinated grains found dropped steadily, to 38% in 1865, 20% in 1870, and 13% for the last six months when the duty was payable in 1880. There was obviously a small core of obstinate malt roasters. In 1861 another facet was added to the examination of barley. There was a very bad summer and harvest period in 1860, and much of the barley crop would not to germinate. Complaints were made by maltsters, particularly from Ireland, and samples were taken. Their complaints were found to be justified, and a remission of the duty payable was made to these people. This work was also carried on until the malt duty ceased. Never more than a few hundred samples per year were involved, even when samples from malt exporters for drawback were added.[19] Phillips never describes the procedure adopted in his germination testing, but a fascinating picture of his laboratories covered in small pots of sprouting (or non-sprouting) barley comes to mind.

As noted, the malt duty was abolished in 1880. Duty was then charged on the beer itself, rather than the ingredients and brewers were free to use whatever ingredients they chose. Many complaints had been made about the malt duty, and it was certainly a difficult one to enforce. A tax on beer was the obvious answer, related to the amount of malt used, and the method by which this could be done was to determine the original gravity, (i.e. a duty could be placed on beer of a particular specific gravity). This was possible because of the work of Phillips and Dobson on the determination of original gravity, and also by the reduction in the number of brewers and maltsters: there had previously been too many to make the collection of a beer tax feasible.

The new Act charged duty on beer of a specific gravity of 1057 degrees. It was considered that two bushels of malt or its equivalent would produce four barrels of beer of such a gravity. The old tax had been based on the amount of beer two bushels of malt produced and the charge per barrel was now fixed to give the same amount of revenue as the old tax. It was calculated that the two bushels of malt should yield 36 gallons of beer of original gravity of 1055 degrees, but a group of brewers managed to convince the Chancellor of the Exchequer (Mr Gladstone), that a larger amount of beer could be produced from the malt, and wanted an original gravity of 1060 to be fixed. It was finally agreed that the standard gravity should be 1057, thus effectively reducing the amount of duty payable. In 1889 the standard gravity was reduced to 1055, to raise more revenue (the original figure of 1057 produced too little), and because the original calculation had in fact been correct. Inevitably samples of wort were sent to the Inland Revenue Laboratory to check the declaration made by the brewer, or found by the Excise Officer at the Brewery. In 1880, 1332 samples were received, and 11739 by 1890, in addition of course to those beer samples being examined for original gravity on export (7677 and 8621 respectively).[20]

The Act of 1880 had authorised the use of the Bate Saccharometer to measure the gravity of beer worts (together with the Graham, Hoffman and Redwood gravity lost tables), and so at this point the Laboratory was called upon to verify the accuracy of the saccharometers now needed as standard instruments by the Excise Officers. This was officially made part of the work of the Laboratory in 1888, resulting in the need to increase the number of rooms occupied. Hydrometers were also standardised under this arrangement, a standard hydrometer having been held in the Laboratory since 1885 for comparison with the (lesser) standard instruments held by the Customs and the Inland Revenue. Hydrometers had been checked before this, chiefly for manufacturers, but also in one case in 1865 to check an instrument which the Customs thought might read wrongly at high spirit strengths (it did not).[21] In the same year as the Act, 1880, because of the increased use of the Bate Saccharometer, the Laboratory was asked to produce new tables for use with it. The Bate Saccharometer, unlike the other instruments, had a direct reading of specific gravity but this was only correct at 60°F, readings at other temperatures being referred to as 'indications'. The new tables enabled readings at other temperatures to be converted to the gravity of wort as it would be at 60°F. These tables, which covered temperatures from 50° to 150° for apparent gravities between 1000 and 1150 degrees, were prepared and published by 1888 and were supported by other tables which gave for each degree of gravity the proportion of the extract by weight and the excess weight of a barrel of wort compared with one of water, together with 60°F volume correction tables.[22]

Such work had been carried out in the Laboratory more than twenty years earlier, when extending the tables for the Sikes' Hydrometer. This work was to enable the hydrometer to be used for alcoholic solutions stronger than Sikes had envisaged, even up to absolute alcohol, and at higher temperatures (between 40° and 85°F). Phillips was asked to undertake this work in April 1858, and finished it in 1861. He dealt with the problem by producing a second 'Sikes' hydrometer to deal only with very strong spirits. The alternative, which had been considered, was to produce one with a longer stem to increase the range, or to increase the number of poises used to weight the stem.. Either of these alternatives would have meant replacing the entire Sikes' system. The new hydrometer was made of glass for use in the Excise service, while the instrument for use by the Customs was made of

21   Sikes' glass hydrometer, devised
by George Phillips 1858–1861

metal. The tables produced by Phillips did not cover absolute alcohol at temperatures above 50°F, nor were they legalised for use, although they were used unofficially. The length of time Phillips took over the work was explained by him over three annual reports. During the first summer he worked the dewpoint was considerably above 60°F for most of the time, which made obtaining 'correct weighings' at that temperature out of the question, and the succeeding two winters were so mild that he could not obtain 'accurate weighings' below 55°F. It was not until the winter of 1860 that the temperature dropped low enough to enable him to finish his work, and produce the tables.[23]

Another side issue, in a sense, to all the beer work, was the testing of beer for adulterants. In the mid-part of the century, while there had been a tax on the hops used in brewing (abolished in 1862), much work had gone into the examination of spent hops to check for the presence of illicit materials intended to render the hops more bitter, or to make the beer apparently stronger than it was. Grains of paradise (or *cocculus indicus*) was a favourite for the latter use. Samples of spent hops seized from brewers and found to contain 'grains of paradise' were fairly frequently received for examination in the 1860s.[24] *Cocculus Indicus* is in fact a powerful poison. When substitutes for hops, such as quassia, were permitted the problems of the Laboratory were not solved since they still had to determine whether or not the beer had been brewed from non-permitted substances, or had been adulterated by the addition of sugar. Adulteration by the addition of water had long been known but was not forbidden by law until 1885. The *Intoxicating Liquor Act* of 1872 had only forbidden adulteration with 'deleterious substances'. In consequence of the 1885 Act (forbidding dilution), nearly 2000 samples were received in 1886/87 for examination for such dilution. It was apparently prevalent in the London area. In many cases sugar was added to restore the 'fullness' lost by dilution, the whole practice being to the 'serious detriment of the beer revenue'.[25] The numbers of samples examined for adulteration and the proportion found adulterated remained fairly steady for the remainder of Dr Bell's time as Principal. Some brewers (particularly in the Midlands) found that their operations were not conducted efficiently enough to obtain satisfactory yields from the malt. They therefore added sugar to the mixture, without declaring it, to increase their yield, and escaping payment of duty.[26]

Phillips, and his successor Dr Bell, many times stated the near impossibility of telling whether some forms of adulteration had been carried out, and indeed whether or not sugar had been illegally used in brewing the beer in the first place. As time went by methods were developed to cope with these problems, and by 1882 Bell could say that while detection of fraud, (i.e. the use of illicit, non duty paid sugar), was of extreme difficulty, 'on account of the variation of composition of the different kinds of brewing sugars used', 'the chemical and optical results obtained' allowed this determination to be carried out. Reliance was still placed on the discovery of illicit materials on brewing premises by the Laboratory trained Excise officers however. By 1881 though Bell could say that beer then virtually never contained poisonous substances.[27]

As part of the beer work, the Laboratory also examined various light, non-intoxicating beverages to see if they contained more proof spirit than was allowed duty free by law, (2%), and thus should pay duty. There was a steady increase in these products around 1890, possibly due to the influence of the temperance movement. They had names such as tonic stout, tonicine, and botanic beer, and despite the claims on their labels and temperance use often contained 4 to 6% of proof spirit, derived from alcohol added with flavouring ingredients or from a deliberate, slight fermentation to provide some aeration. The label of one non-alcoholic beverage examined in 1886, and found to contain 23% proof spirit, had stated that total abstainers who consumed it were both pleased and surprised at its comforting and exhilarating effects, as well they might have been.[28] The number of samples of beer and brewers material continued to expand, so that by the time of the merger with the Customs Laboratory in 1894, the examination of such samples represented a major feature of the work of the Laboratory. In 1893 nearly 23000 were examined, just over 47% of the total.[29]

In 1872 the Laboratory took part in an enquiry by the Home Secretary concerning the amount of salt naturally present in beer. The *Intoxicating Liquors Act* had forbidden, *inter alia*, the addition of sodium chloride to beer, one unforeseen result of which was that after the Act had been passed beer brewed without salt had in some cases been returned to the brewers as unpalatable. The Brewers' Association pointed out that some water used in brewing contained less salt than others, so that it was difficult to prove addition to any particular beer. In addition, since salt increased the keeping properties of beer, brewers using water which naturally contained less salt were at a disadvantage. Samples of beer (110 of them) were taken in different parts of the country to check the varying amounts of salt naturally

22   Advertisement quoting Laboratory results, from the Farnham *Weekly Herald*, 1895

THE FARNHAM UNITED BREWERIES,

LIMITED.

BREWERS OF

PALE & MILD ALES & STOUT.

(Declared by the Chief Officials at the Laboratory, Somerset House, to be 'well-hopped,' of good quality,' 'all free from adulteration,' *vide* Report issued to Military authorities who had taken samples at random from the stock in a Canteen, and had sent them up for analysis.)

Supplied direct from the Brewery (in any sized casks.

Extra Quality, Condition, and Flavour.

WINE AND SPIRITS OF THE BEST QUALITY.

LION BREWERY (LATE **TRIMMER'S**),

FARNHAM, SURREY.

present. Some 47 samples of brewing water were also taken. The brewers' assertions were found to be true, and the Home Office suggested that where the amount of salt present did not exceed 50 grains per gallon, the Inland Revenue officers should not enquire as to whether it had been artificially added or not. The variation of salt in beer was taken up again by Dr Bell, Phillips successor, in 1877, in a case submitted to him under the *Sale of Food and Drugs Act*, 1875 (see chapter 5 below). The prosecution had alleged that the beer had been adulterated with common salt, and Bell's result confirmed the local analyst's figure. Subsequent enquiry showed however that the materials used by the Brewery concerned would account for that amount of salt. Bell commented on the difficulty of telling how much salt beer should contain because of the variability of the natural products used in brewing. This he had shown in an 'external enquiry' into the amount of 'salt and other chlorides present in different samples of water, malt, sugar, hops, and genuine beers' obtained from different parts of the country. This desire to fully check his facts, and make allowance for natural variability, before accepting that adulteration had occurred was something Bell followed in all his work under the 1875 Act.[30]

Before 1855 all spirits, distilled or imported into the United Kingdom, were liable to duty. This was the case no matter what their purpose, be it for consumption or use in manufacture. There was some resentment in industry at this duty, which caused British products to be relatively uncompetitive compared with imported goods which paid no spirit duty. In 1853 a lubricant manufacturer applied to the Board for permission to use spirit free of duty and Mr Phillips was asked to look into the possibility of allowing duty-free spirit for industrial use generally. His experiments led him to suggest the addition of wood naphtha (crude methanol) as a denaturant, rendering the spirit too nauseous for use as a drink. The advice was again sought of Professors Graham, Hofmann and Redwood, to whom Mr Phillips' proposal was referred. They were asked whether, and by what means, ordinary spirit might be rendered so offensive to the taste or smell as to make it unfit for human consumption while at the same time not impairing its use for the manufacture of esters or for solvent or other industrial purposes, to what branches of industry or the arts should the use of such a product be limited, and what controls would be necessary in its preparation. They were also asked to show that the mixed spirit could not easily be purified or made palatable. Graham *et al* made extensive and practical trials, using amongst other things the liquid obtained by the pyrolysis of rubber as a denaturant. This they had to admit had such a revolting smell that a product containing it would never be admitted to shop or home, as they expressed it. They were unable to improve on Phillips suggestion, and their report indicated in some detail the current usage of alcohol ('spirit of wine'), in 'arts and manufactures', and where freedom from duty would be welcome.

They also confirmed Phillips' finding that it would be very difficult indeed to purify or make the denatured spirit usable as a palatable drink. They finally recommended that as much as 10% of wood naphtha be added to the alcohol, due to the difficulty of determining the presence of less, and that the 'methylated spirit' should not be freely available, but only to licensed manufacturers. This was to prevent it falling easily into the hands of 'individuals of perverted tastes who, in extreme cases, may use it for producing intoxication'. There were widespread worries over the considerable amount of drunkenness at this time. When the authorising Act, allowing the use of methylated spirits free of duty, had been in operation for about 12 months, in 1856, Thomas Dobson, the Secretary of the Board of Inland Revenue, recently Phillips' collaborator, wrote to a number of major users to confirm that they found the new 'methylated spirit' useful. All replied that they did. Frederick Abel, the director of the War Department Chemical Establishment at Woolwich Arsenal also found a noticeable improvement in the workmen there using alcohol to moisten explosives, the few who continued to indulge in the alcohol being readily identified by the peculiar odour imparted to their breath. Sadly, Mr George Hutchinson of Glasgow, the man who originally asked if something of the kind could be done, replied that he could not say how useful the concession would be since his works had been destroyed by fire one month after the Act came into operation.[31]

The new arrangement by which alcohol was denatured by crude methanol was supported from 1855 by the examination of samples of naphtha for their suitability for this purpose.[32] Much of Phillips' report for 1858 was devoted to methylated spirits. It had been claimed in the *Pharmaceutical*

Journal that good pure drinkable spirit could be recovered from sweet spirit of nitre (ethyl nitrate in alcohol) made from methylated spirit, and Phillips showed that whilst nitrosation did not separate methanol from ethanol, and that the product was comparatively free from disagreeable flavour, the nauseous character of the wood spirit still remained. He was of the opinion that even if this latter could be eliminated the process would be as costly as was the duty that was being evaded. Wood naphthas were tested from this time for their suitability (i.e. being sufficiently nauseous) for protecting the revenue, about 50 samples being examined each year.[33]

Despite further claims to the contrary Phillips continued to regard illicit distillation as being of much more danger to the revenue than the rectification of methylated spirits, and said so most forcefully in his 1863 report. As he said such rectification must always cost very much more and be more difficult to conceal in operation than ordinary illicit production of spirit. In the same report he said that he had carried out a series of experiments, in accordance with the conditions laid down in two patents, and in neither case could he make either methylated spirit or wood spirit potable. Examples of such spirits exhibited as having been purified by these processes had been purified by distillation he believed. The samples were also still not drinkable. Further work was carried out on the purification of methylated spirit in 1865 and various patent methods were tested during this period. Phillips had a considerable interest in distillation, going as far as obtaining a patent in 1847 for the 'Purification of certain oils and spirits'. It is not known if his method was ever put to practical use. It involved an ingenious double worm condenser.[34]

In the following few years Phillips frequently referred in his *Reports* to samples of 'Indian brandee' and 'Indian whiskee', neither of which resembled the spirits suggested, being highly impure hypronitrous ether made from methylated spirits, the 'brandee' being coloured and the 'whiskee' not, and both very nauseous.[35] One solitary 'Ginee' appears. There was obviously a thriving trade in these, skilfully marketed. One 'Whiskee' gloried in the name of 'The Pure Islay Mountain Medicated Whiskee'. The 'Indian brandee' was said to be a specific for nearly every disease, and to be composed of 'the most costly and rare products of India, which had by great skill been so combined and applied as to

A.D. 1847 . . . . . . Nº 11,965.

Purification of Oils and Spirits.

PHILLIPS' SPECIFICATION.

TO ALL TO WHOM THESE PRESENTS SHALL COME, I, GEORGE PHILLIPS, of Park Street, Islington, in the County of Middlesex, Chemist, send greeting.
WHEREAS Her present most Excellent Majesty Queen Victoria, by Her
5 Royal Letters Patent under the Great Seal of Great Britain, bearing date at Westminster, the Sixteenth day of November, in the eleventh year of Her reign, did, for Herself, Her heirs and successors, give and grant unto me, the said George Phillips, Her especial licence, full power, sole privilege and authority, that I, the said George Phillips, my executors, administrators, and
10 assigns, and such others as I, the said George Phillips, my executors, administrators, or assigns, should at any time agree with, and no others, from time to time and at all times during the term of years therein mentioned, should and lawfully might make, use, exercise, and vend, within England, Wales, the Town of Berwick-upon-Tweed, and in all Her Majesty's Colonies and
15 Plantations abroad, and in the Islands of Jersey, Guernsey, Alderney, Sark, and Man, my Invention of "CERTAIN IMPROVEMENTS IN THE PURIFICATION OF CERTAIN OILS AND SPIRITS;" in which said Letters Patent is contained a proviso obliging me, the said George Phillips, by an instrument in writing under my hand and seal, particularly to describe and ascertain the nature of
20 my said Invention, and in what manner the same is to be performed, and to cause the same to be inrolled in Her Majesty's High Court of Chancery within six calendar months next and immediately after the date of the said in part recited Letters Patent, as in and by the same, reference being thereunto had, will more fully and at large appear.

23   Patent of George Phillips, 1847

become a perfect boon to the human race'. Phillips goes on to remark, 'it is sad to reflect on the unblushing audacity of such statements, made by persons who deem themselves honest'. The 'brandee' was usually partly purified methylated spirit but more usually not purified at all, still containing hyponitrous ether, sweetened and coloured, occasionally with such additions as rhubarb and fennugreek. These and similar preparations, which were many, were sold as medicines, but were consumed by poorer people as stimulant beverages. Phillips remarks that the sale of such compounds appeared to be limited chiefly to certain areas in Lancashire and Cumberland. Another source of difficulty was the importation of pure vinous alcohol to which had been added a small amount of wood alcohol with intention to evade duty, following

which the product, partially rectified and mixed with unde-
natured alcohol, might escape detection. Phillips then
advised the examination of all imported wood spirit.[36]

Throughout the rest of the century, and into the next
several hundred samples of wood spirit (or naphthas) were
received. Samples were occasionally rejected as being too
pure for the purpose i.e. they would not render the mixture
nauseous enough. Methylated spirit being so much cheaper
than pure 'spirit of wine', the former was occasionally found
to be substituted for the latter in such medicinal preparations
as tincture of rhubarb, sweet spirits of nitre and paregoric.
Methylated spirit sold through retail outlets as 'finish' was
required to contain a minimum of 3 oz per gallon of gum
residue (seed-lac, shellac, sandarac or, later, common resin).
Some manufacturers added the requisite amount of gum but
failed to dissolve it in order for it to be used as a drink.
Attempts continued to be made to sell methylated spirit as a
drink. One instance of the sale of methylated spirit,
sweetened, coloured and reduced in strength, (similar to
'brandee'), as a beverage, was found in East London in 1873.
It was 'difficult' to understand how tastes of persons could
become so depressed as to enable them to drink a liquid
having such a disagreeable smell and flavour', but the victims
were 'persons in very humble circumstances who were not
able to obtain spirit of better and more wholesome quality'.[37]

The tests for the presence of methanol in a case in which
methylated spirit was alleged to have been added to two large
vats of duty-paid spirit were challenged in court in 1885.
Some of these were used with success to provide an 'ocular
demonstration' in the court proceedings, (obviously col-
ourimetric tests) the samples were estimated to contain 3% of
methylated spirit, equivalent to 0.3% of wood spirit. These
tests were probably Gunning's or the Riche and Bardy test,
frequently mentioned in the contemporary workbooks.[38] In
another case in the same year the owner of an illicit still in
Ireland, unable to produce a product of the strength of
ordinary whisky, had fortified it with methylated spirit. In
December 1885 Dr Bell discussed the tests and the regulations for the use of duty-free spirits in the arts and manufacture with official chemists from France and the Netherlands
in Amsterdam. In this early example of an international
scientific co-operation it was agreed that the tests available
for the detection of methanol in various compounds were

completely adequate for the purpose. It was first agreed that those attending the discussions would continue to correspond with each other, to inform each other of new types of fraud, and of any improved methods for the detection of wood spirit.[39]

Over 800 samples of wood naphtha for methylation were received annually by 1892, when at Dr Bell's suggestion a small quantity of petroleum (or mineral naptha) was also added to methylated spirit for retail sale. This was to discourage the use of the product for beverage purpose which had developed in some rural areas in Northern Ireland. The addition not only made the spirit more offensive in character but also caused it to become cloudy when diluted with water. 'Ordinary' methylated spirit (frequently now called Industrial methylated spirit) was still available, by permission of the Board of Inland Revenue, for industrial use.[40] Samples of both 'wood naphtha' and 'mineral naphtha' are listed in the tabular summaries from 1892 and continued thus into the next century. Regulations as to the quality of wood naphtha to be used were made by the Commissioners of Inland Revenue in 1899. These described the method used by the Laboratory in testing the naphtha. The level of addition of wood naphtha was reduced from 10% to 5% in 1905, as had originally been envisaged by Graham, Hoffmann and Redwood as soon as methods for the determination of methanol had improved, and a similar reduction (together with the addition of methyl violet dye) for 'mineralised' methylated spirit took place in 1918, due to a shortage of wood naphtha caused by the first world war. In 1925, as the result of a conference in which the head of the Laboratory participated, pyridine was added to produce the now familiar retail methylated spirit. Other varieties of denatured alcohol for specialised purposes are now available and the wood naphtha introduced by Phillips is at last being phased out, but the soundness of his original choice has been justified many times.[41]

4 Staff and Training

GEORGE PHILLIPS WORKED ALONE FOR most of the time immediately after the establishment of the Laboratory in 1842, although one extra assistant was employed for at least part of the time in 1844 and 1847, after which his success and the consequent increase in work forced the employment of three extra chemists. It is usually believed that one additional chemist, Thomas Dobson, was employed from 1844[1], but while Dobson undoubtedly worked with Phillips, it seems certain that he only did so on short term projects and apparently not in official hours. He certainly could not have been Phillips' assistant both because he was senior in grade, and because the staff was officially stated to consist of one person in 1845 and 1846, i.e. Phillips himself when it is known that Dobson was working with Phillips on the use of molasses and sugar in brewing. In a Laboratory report of 1848 Thomas Dobson is said to have 'suggested' some work. This suggests that Dobson was in some sense in overall charge of the Laboratory as might be expected, given his senior rank.[2] It seems from the signatures on sample reports in the years before 1847 and the mid 1850s when it is known that the staff fluctuated between two and six that those working with Phillips were Excise students of University College (see below). Such names as Charles Forsey, John McCulloch and Adam Young appear, particularly Young who signs reports between 1847 and 1853. It seems therefore that while there was a shifting staff some of them shifted less than others.[3] In 1845, when the staff was officially reported as two, the salaries paid totalled £250. In 1845 and 1846 when it was only one, the salary paid was £200, i.e. Phillips' own salary.

On 31 January 1858 the Laboratory was constituted as a separate department in the Inland Revenue. The Commissioners had obviously decided that it was there to stay. They appointed George Phillips as Principal of the Laboratory, with George Kay as his deputy, officially 'Assistant'. Kay had joined the Excise service in 1843 and had been ordered to attend University College in 1851. He continued to do so very successfully until 1854, and probably went directly from University to the Laboratory. Phillips' salary was set at £500 *per annum* rising to £600. This would today represent something like £23,000 rising to £27,000. Kay's salary was set at £200, rising to £300, both salaries taking effect from 31 January 1858. Kay died in harness as 'Chief Assistant' in 1867. Phillips expressed great regret, noting Kay's great ability, versatility and energy. When necessary Phillips had always left the department in his charge 'with the most perfect confidence'.[4] A number of other assistants, eight in 1859, worked in the Laboratory on a temporary basis, chosen from young Excise officers who showed promise during their training. In 1859 one at least was still a trainee who had served in the Laboratory from 1852. These officers usually returned to the Surveying branch of the Inland Revenue service after one or two years in the Laboratory, but this was obviously not very satisfactory. As Phillips said, to become a good analyst required several years training, particularly in view of the increasing complexity of the work. This required continuity of service, not a continually changing staff.

The Inland Revenue Board agreed in February 1867 that this should be done and obtained the necessary permission from the Treasury. The Laboratory was then reconstituted with three assistants plus a Book Keeper joining Phillips and Kay as permanent members of staff. The three assistants chosen by Phillips from his current temporary staff were Richard Bannister, William Harkness and Henry James Helm and the Book Keeper was John O'Loghlen. All of these stayed in the Laboratory for the rest of their careers, all reaching high rank. The salaries of the Assistants were £130 rising to £400 after 27 years. Mr O'Loghlen received £120 rising to £250 after ten years. George Kay was given a rise to £450 rising to £500 and Phillips now received £600 rising to £700. Since by an agreement of 1863 Phillips already had a maximum salary of £750 this was retained as personal to him. The salaries of both Phillips and Kay had been raised by the Treasury in 1863 at the request of the Board of Inland Revenue, on the grounds of extra work and responsibility, but again as personal to them and not changing the salaries fixed in 1858. All of the Assistants appointed in 1867 had been analysts in the Laboratory since at least 1860 and O'Loghlen certainly had not always been restricted to book-keeping in the past whatever he did in the future. On at least one ocassion in 1863 he carried out scientific duties examining barley corns to check that they had germinated properly. By 1866 each 'officer' was allowed the generous leave of 28 days continuously and twelve others, plus statutory holidays, and sometimes special leave for such events as marriage, although 'not as a matter of course'.[5]

In the early annual reports frequent reference is made to the training functions undetaken by the Laboratory for revenue officers, both the general training for a limited period of excise examiners due for promotion to supervisor in the detection of adulterants, and students who received extended training in laboratory work on detachment from their normal duties for a period of two to three years. The latter also received instruction in qualitative and quantitative inorganic analysis firstly at University College during the period 1845 to 1859 and later at what became Imperial College. As mentioned already, there had been a close relationship between the Laboratory through the Board of Excise and the University of London from the early days of the Laboratory, commencing with the direction of the Board in 1844 to George Phillips and nine other officers of Excise to matriculate in general and analytical chemistry. This was after Phillips' evidence to the Royal Commission on the Tobacco Trade, as described in chapter two.

The ten Excise students, together with five others not mentioned in the Minute registered at University College on 12 January 1845; they were obviously too late to register at the beginning of the autumn term. Another student from the Excise registerd on 28 February. George Phillips certainly matriculated as ordered but equally certainly did not attend any classes. Most of his colleagues did attend classes. Phillips felt perhaps that he was above such things. Another of his colleagues, Joseph Drinkwater, also failed to attend any. Drinkwater was a Senior General Examiner, a very senior grade, but it was not for this reason that he did not attend classes because he appears to have spent most of his time in the 1845 academic year aiding most of his colleagues in the general analysis of tobacco, snuff and soap as well as helping them examine samples for added adulterants. In doing this they were following his suggestions as he noted in a report he wrote for the Excise Board on 5 October 1846. In the early part of 1846 Drinkwater spent most of his time carrying out extensive experiments on the constitution of tobacco with Professor Graham, as already described, and at the same time worked extensively on the composition of proof spirit at the University laboratories. Drinkwater was a chemist of some attainments.[6]

The University of London had opened in 1828. The Professor of Chemistry at the time the Excise chemists started was Thomas Graham and the Professor of Analytical and Practical Chemistry was George Fownes, (from Cambridge), appointed in 1845. Fownes died in 1849, and was succeeded by Alexander Williamson, who in turn succeeded Graham in 1855, and it was Williamson under whom most of the subsequent Excise students studied. The connection between the Excise and the University was undoubtedly the fact that that John Wood, Chairman of the Board of Excise, was also a

24   University College, London; new laboratory for the School of Practical Chemistry, 1846 (Illustrated London News)

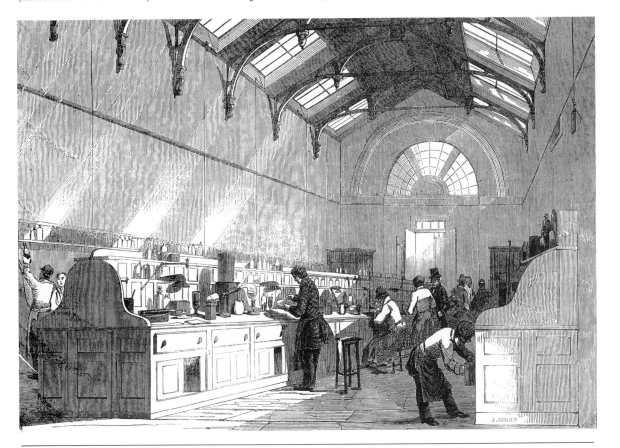

member of the Council of University College from 1846. Wood obviously saw an opportunity to help the College (it was going through a difficult period at that time) and at the same time help the Excise. There can be no doubt that a knowledge of chemistry was useful in the circumstances in which Excise officers found themselves in the 1840s and 1850s and that the knowledge of the few (particularly in senior positions) would percolate down through the service. This at any rate was the reason given in December 1844 by the Board of Excise to the Treasury for sending students to college. The Treasury had queried the new expense of £366 in the first year. A further reason was to reduce the amount paid to outside expert witnesses, to produce their own home grown expert witnesses as it were. The Treasury duly gave permission, and Excise students continued to attend classes in chemistry, practical chemistry, mathematics, natural philosophy and occasionally French and German, (not all students attended all classes), at University College for the next twelve years, until 1858. The numbers sent varied from year to year, and each year the various professors involved sent to John Wood, on his instructions, a report on all the students. These were invariably laudatory, truthfully so judging by the number of prizes which they won. In most years at least one or two names of Excise students appeared, and in some years they won most of the prizes given in the chemistry faculty. Winners of prizes included George Kay, already mentioned as Phillips' first deputy and James Bell his second deputy and successor. The reports were always shown to the students, who signed a paper to say that they had seen it. Some of the students carried out research projects for Graham or Williamson, some of which resulted in papers published in the *Transactions* of the Chemical Society.[7]

In addition to the annual reports from each professor a monthly report on the attendance (or otherwise) of each student at all lectures was sent to the Board. If the attendance was unsatisfactory the student was asked to give the reason. This was given either on the report itself or in a separate letter. The excuses were mostly sickness, sometimes due to accidents in the laboratory, or official business, or 'engaged in experiments for the Honourable Board'. In one or two cases the reason was such that since the student felt that most of his French class were well in advance of him he did not 'consider it was necessary to attend' the examination, (this was George Kay in 1854). The students were allowed to have the whole of August off in the years when they were studying, and they usually attended classes for two years and occasionally three or four. All of their expenses were paid and they received their Excise salary.[8]

Many of the Excise men who studied at University College went on to attain high office in the Excise service. Adam Young, an 'Officer' (strictly a Division Officer) and aged 29 in 1845 when he first attended college and Edward Dodd, Senior General Examiner in 1845, both became Assistant Secretary to the Board, and Charles Forsey became Inspector of the Inland Revenue. Some students achieved high office in the Laboratory. These achievements may partly reflect the high calibre of the staff selected, but partly because having successfully attended College the expectation in the Excise would be that these were men who would achieve high grade. It certainly was the expectation among the students, and some of them solicited promotion on the grounds of their success at College. There is no doubt that promotion was quicker amongst the University students than in the Excise service in general.[9]

This situation naturally caused adverse comment amongst Excise officers not chosen to attend College, and practically as soon as a suitable vehicle for it was established, in the form of the *Civil Service Gazette* in 1853, correspondence on the subject appeared at intervals. Much of it was on the lines of how the 'University Pets' were the beneficiaries of 'barefaced favouritism', of the iniquity of educating men at the public expense, and of how the 'whole system' was 'intended for patronage and jobbing rather than utility'. All of these letters were anonymous, signed by such authors as 'An Exciseman' and 'Anti Hocus-Pocus'. Most of their letters were fairly predictable, but those signed by the latter were much more vicious and written by someone who obviously had a certain amout of inside knowledge, and may indeed have been in a fairly senior position in the Inland Revenue judging by what he knew of the situation. He had obviously seen some at least of the relevant documents and did not seem to have liked Phillips at all. He was very much against the whole system. He came very close indeed to saying that John Wood was guilty of deliberate fraud on the revenue. Certainly he, and others, were quite sure that promotion was too rapid for the 'University Pets' and that this was wrong.[10] It may be that others agreed with this, because soon after John Wood died in 1857 the experiment was brought to an end, in October 1858.

The reason give by the Board for ending this education of some of their staff was that since their chemical department was now so efficient and was housed in a 'spacious and complete Laboratory' they considered that the training of their staff in chemistry would be better and 'more economically conducted under the superintendence of Mr Phillips'. The cost in the last years was about £800 *per annum* (or possibly £600, the estimates vary), which was a considerable sum. It had risen from the original £360 at the start to a maximum in 1853 of £1038. The average cost of each student in 1854 was £96 each 'exclusive of maintenance'. As noted, the decision to end the placing of students at University College was taken in 1858. The new Laboratory was not then ready and Phillips was still in temporary accommodation in Arundel Street. His comments on the change were not therefore as sanguine as those of his employers. He expressed his 'deep sense of the responsibility' which had fallen on him, a feeling not lessened by 'the fact the apartments in which the business of my department was then temporarily carried on were totally unfit for educational purposes'. After he had left Arundel Street in June 1859 and moved to Somerset House the progress of the nine officers he was then instructing was considerably improved. As part of the new system they were examined in December 1859 (later students were examined in June) by Dr Augustus Hofmann of the Royal School of Mines (later the Normal School of Science, then the Royal College of Chemistry and finally Imperial College) and did extremely well. In 1861, following the same procedure the results were so good that Dr Hofmann said in his report that 'during my professional career I do not remember a similar examination. Out of ten candidates he had placed nine in the first class (those with marks of more than 80%), one candidate indeed received 100% marks, and one was just placed in the second class. In addition to their chemical studies the students were also trained in the use of the microscope and its application to the detection of adulteration. The examination was both written and oral and included a practical test. It was regarded as an extremely searching examination and the oral part was said to be more severe than that to which regular students at the College were

25   Professor A W Hofmann, Royal School of Mines

REGULATIONS to be observed by the Students in the
Inland Revenue Laboratory, Somerset House.

1st.—The hours of attendance are from 10 in the morning to 4 in the afternoon.

2nd.—The " Attendance Book " is to be delivered to the principal or his assistant within five minutes after 10 o'clock in the morning; and each Student, in succession, is to be weekly responsible for the proper performance of this duty.

3rd.—No Student is to leave the department during business hours, or introduce strangers, without the consent of the principal or his assistant.

4th.—Any Student unable to attend to his duties through illness, must immediately, by letter, acquaint the principal or his assistant of the fact; and, should his absence continue more than one day, he must forward a medical certificate to the Laboratory.

5th.—When engaged upon operations which give rise to offensive or poisonous vapours, the Students are most strictly enjoined, not only for the sake of their own health, but for that of other persons employed in the building, fully to avail themselves of the appliances furnished for the purpose of carrying off and rendering innoxious such vapours. Any wilful or careless infraction of this important regulation will be severely reprehended.

6th.—Whilst each Student will be amply supplied with necessary materials and apparatus, he will bear in mind their expensive nature, and avoid, as much as possible, their injury and waste. He is also enjoined to keep his bench and apparatus clean and orderly, as a slovenly and immethodical Student will never make a good analyst.

7th.—The Balances and other instruments and apparatus of a delicate or costly description must be used with great care, and kept clean and perfect; and any injury which they may sustain must be reported to the principal or his assistant immediately after such injury occurs or is discovered.

G. PHILLIPS.

Laboratory, 2nd April, 1860.

G 100   3 | 60

26   Regulations governing behavior of Laboratory students, 1860

subjected a few days later. The most successful candidate was presented with a first class microscope by the Inland Revenue Commissioners and the two next with books to the value of £5. In 1859 the cost of the new system was £50 *per annum*, being the expenses of the students in attending the examination. It also included two guineas (£2.10) per year for the travel expenses of each student and a 'small sum for the prizes'.[11]

This system of educating the Laboratory assistants continued for many years. Most, if not all of those who took part were placed in the Laboratory to work for at least a year or two, some staying permanently. Dr Hofmann was succeeded as examiner in 1866 by Dr Edward Frankland and he in turn by Professor, later Sir Edward, Thorpe. With a trace of satisfaction Phillips in 1866 remarked that Frankland's system of examination was different from that of Hofmann but that the examination results were equally good. This result confirmed Phillips opinion that 'the method of instruction pursued in . . . [his] . . . department is one well adapted to impart a sound knowledge of chemistry'. At some time after 1859 (and before 1875) the students began to attend the regular theoretical lectures at the Royal College of Chemistry, apparently on a part time basis. This was in addition to being instructed in practical chemistry and microscopy at the Laboratory. They still took a separate examination in theoretical and practical chemistry from the regular College students together with a rigorous oral examination. By 1875 92 students had taken the course and of these 75 had obtained first certificates. Four of these had obtained full marks, and 26 of them 98% or more. In some years all students obtained full marks. The 'competition' by which the students were selected after 1858 was a obviously a rigorous one. In 1885 the students were required to enter a 20 year bond for £54. This was because one student had left the Laboratory immediately after he had finished his course, presumably to get a beter paid job. The bond was estimated as being made up £35 for practical training, £5 for lecture fees, £10 for books and £4 for fares to and from South Kensington. The fares were second class only. The bond was repayable and one twentieth of the sum still to be paid by the ex-student was deducted for each year subsequently spent with the Laboratory. The bond was adjusted in 1889 to run from the time the studentship commenced.[12]

Most of these students were presumably happy with their situation, they had after all made some effort to get there, but some at least found that living and travelling in London was much too expensive. In 1872 fifteen of them signed a petition to the Board. They pointed out that one of them when in his country station had been able to rent a house of seven rooms with a stable and garden for £14 a year as against £25 a year for two rooms in London. The health hazards of working in the Laboratory were mentioned also. Phillips supported this request, but the most the Board would do was to pay the Assistants an extra £10 a year for their books and any equipment which they had to buy. The Assistants were still protesting (and leaving the service) over the cost of living in London in 1895. An allowance of £25 was then granted to some of them. Senior staff apparently did what they could to help.[13]

Another new training scheme had been instituted a short while before the University College one was ended, following a request in 1856 from the Examiners of Excise to attend the Laboratory 'to receive instruction in the more useful scientific checks'. As a result, the Board had arranged from 1856 for six of the younger officers to be selected each to receive laboratory instruction for a period of 28 days. They were subsequently examined in the 'elements of chemistry' and mathematics, the best candidates then forming a source of

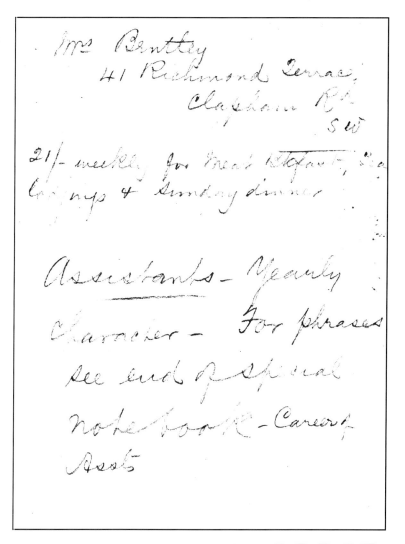

27   Note on suitable lodgings for young Laboratory staff, *c.*1885 (from PRO DSIR 26/205, a note book of H J Helm)

recruitment for new assistants. By 1862 141 Examiners had completed the short course, of whom 120 had become supervisors, representing some 30% of these officers within the service. This number had increased to 170 by 1864, by which date 59 of the longer term students had also passed through the Laboratory's chemical course. By then only six students (together with 14 Revenue examiners) could be accommodated, owing to space limitations and the increase of business of the Laboratory.[14]

By 1876 these 'expectant supervisors' received instruction in the determination of the original gravity of beer, the alcoholic strength of wine and 'the best modes of the detection of adulteration' in dutiable commodities. They had to rely on their own judgement of the latter when they were subsequently placed in charge of districts. They were not always able to complete the training, even though the full month was regarded as a very short period for this. Phillips complained in 1864 about the case of a longer term assistant who had been transferred prematurely. Owing to an increase in the amount and importance of the work of the Laboratory, Dr Bell (by then Principal) suggested a reorganisation in November 1876, with the addition to the staff of Assistants (Second Class) and improved salaries all round. He pointed out that the salaries had last been increased in 1867, since when the number of samples had risen by 54%, in addition to which the staff now had the onerous duties of being referee analysts for the *Sale of Food and Drugs Act* of 1875 (see below). In later correspondence it was suggested that the term 'analyst' was more appropriate than 'assistant' to make their status less ambiguous. Many of his proposals were approved, after he had argued very strongly against Treasury suggestions that the salaries of laboratory assistants be compared with his 'Assistants'. He pointed out that his 'Assistants' were expert analysts. The new salary improvements were dated from 1 April 1877.[15]

A measure of the increasing scientific sophistication of the permanent staff of the Laboratory is shown in their request in November 1872 that a Library be set up. To quote a draft memorandum drawn up by James Bell, then Deputy Principal, 'in consequence of new and diversified work having been added to the Laboratory, great difficulty is experienced in not having standard publications for reference, and in many instances when information is required with respect to any article, books have to be borrowed from outside chemists'. One wonders which 'outside chemists' lent them books. Some at least of the staff bought their own reference books, particularly H J Helm, a senior analyst, who left his books to the Laboratory when he retired. Part of the memorandum consisted of a modest list of 11 books and 14 journals. By the time Phillips came to draft his formal letter to the Board of Inland Revenue not long after the proposal had become much more elaborate. He stated quite reasonably that it was necessary for staff to be up to date in their knowledge in order that their results should bear scrutiny, and that since the books were mostly expensive 'few of the gentlemen are in a position to purchase them'. Persuasively Phillips said that money would be saved by their having a Library in that no time would be spent in repeating the work of other chemists. He further said that 'the small room connecting with the general laboratory could be fitted up as a Library' at a cost of £30, and that the cost of the 'annexed list' of books and journals would only cost £60, and that a grant of the same amount each year would suffice for subsequent purchases and subscriptions. The list of books with Phillips' letter was much longer than the first draft. It included books not only in chemistry but in microscopy, botany and physics as well as a list of foreign language dictionaries including one Latin dictionary. The journals included *Nature* and the *Journal of the Chemical Society*, to both of which the Laboratory has subscribed ever since. The Board of Inland Revenue duly agreed with this request, and a note on the first memorandum noted that periodicals were to be applied for through the Store Keeper in March of each year.[16]

The Store Keeper referred to had been appointed in the preceding year. This was the culmination of a long drawn out process which had begun in 1868. Phillips had applied to the Board, requesting that someone be appointed to look after the chemicals and equipment. One Robert Alexander, a Porter, had been doing the job but Phillips felt that it would be better if he could be appointed officially to the post and be properly paid for it. This was agreed, provided he passed a test in 'reading. writing and the first three rules of arithmetic' and Alexander was appointed Keeper of Chemicals and Scientific Instruments on 10 August 1868, at £70 per year rising to £120. His duties were to keep a list of all equipment held, together with a daily record of what was used and see that it was returned and locked up at night. He was also to keep a catalogue of all the official and other books and to be the last to leave each night' and before leaving to see that the water and gas are turned off and that the outer door is secured'.

By 1871 Phillips had decided that an Assistant Keeper was needed. He managed to convince the Treasury that he was not asking for two people to do one person's job by persuading them that much of what was needed was not being done. The person appointed was one John Terry, nominally a Messenger but in fact carrying out much more 'onerous' duties (according to a minute written by him). Terry had been in the Laboratory since it had been made a separate department, having been appointed as a porter at £1 per week in May 1859. Terry's duties were to look after lime and lemon juice samples, the chemical and botanical specimens and the scientific books recently bought for the new Library. To cope with this responsibility he was placed part way up his salary scale of £70 rising to £120 at £80 *per annum*. Terry resigned in 1875 to take up a better paid job as Tobacco Inspector in the Isle of Man.[17]

Just before Terry's resignation George Phillips retired at the age of 68 on 31 March 1874 having been in the Revenue service since 1826, nearly 48 years.[18] He died on 16 January 1877. In recognition of his outstanding services to the Inland Revenue and before that to the Excise during his long service he was awarded by the Treasury, as an exceptional gesture, a pension equal to his salary of £750 per year. According to a later account he retired reluctantly from his 'interesting post', but was compelled to do so because of impaired eyesight. Phillips was obviously a man of great administrative ability, perhaps not a great chemist but certainly a more than competent one and undoubtedly a practical man, as seen from his two patents. He was interested enough in chemistry to join the new Chemical Society in April 1845 as 'Chemist to the Excise'. He was probably a fairly formidable man. The Inland Revenue Board usually took his advice, frequently writing to the Treasury in the words suggested by Phillips. His abilities were amply recognised by the Board in 1863 when writing to the Treasury to request permission to raise his and George Kay's salary. They observed that in recent discussions the Chairman of Excise 'could not fail to have been impressed . . . with the intimate knowledge evinced by Mr Phillips in all matters connected with that important [tobacco] trade and the confidence invariably reposed by the Traders in any statement made by Mr Phillips bearing upon the subject'. They further observed that he was not adequately recompensed by his present salary for his services and had no doubt that he had forgone considerable chances of promotion by staying in the Laboratory. They had no hesitation in asking him to undertake unusual chemical investigations. In 1858 for example because the Commissioners had recently had their room covered with paper coloured with copper arsenite they asked Phillips to investigate its 'alleged poisonous influence'. This he did, proving to his own (and presumably to the Commissioners') satisfaction that it had no effect whatever. He further stated at the end of his report that he and his family had occupied a sitting room papered with such paper for three years with no ill effects.

No personal letters of Phillips appear to have survived, and we have little information as to his personality. There are some signs that he was of a kindly disposition. When reporting the death of a 'labourer' in the Laboratory in 1859 for example he went out of his way to say that he took the opportunity 'to pay a tribute to his worth by mentioning that he was an industrious and faithful servant and entirely devoted to the interests of the service', a kindly gesture because not necessary. He seems to have been on friendly terms with his colleagues, two of them Richard Bannister and Henry Helm were his executors and described in his will as his 'friends'. He was undoubtedly a very hard worker and the Laboratory was very much his Laboratory, his creation. He took an active part in the work throughout his whole career. Examples occur of his suggesting new ways of carrying out analyses when he was not satisfied with results. Ten years after he had retired his 'wonderful personality and ability' were still remembered. Bannister (a man not given to lightly praising someone) believed Phillips to be 'worth himself and Bell put together'.[19]

Phillips was succeeded on 1 April 1874 by his Deputy, James Bell. Bell had served as Chief Assistant or Deputy Principal since 27 April 1867. He was born at Altnamaghan, Newtownhamilton in Co. Armagh in 1825 and joined the Excise in 1846. He was not immediately appointed to the Laboratory but was chosen to attend University College

28   Dr James Bell, Principal of the
Laboratory, 1874–1894

between 1851 and 1852 and served in the Laboratory for at
least a year after this. He seems then to have served as an
ordinary Excise officer until recalled to the Laboratory on the
death of Kay in 1867.[20]

As described the staff of the Laboratory slowly increased after it had been made a separate department in
1858. Increases in staff were authorised in 1870, 1873 and
1878, at which point there were eleven permanent staff
including the Principal and his Deputy. The analytical staff
were divided into First and Second class Analysts by 1878,
the previous division, from 1873, was Upper and Lower class
Analysts, a slightly invidious classification. In 1884 the work
of the Laboratory had increased to such an extent since 1878
that in February Dr Bell applied again for additional permanent staff. The increase had been caused partly by the

substitution of beer duty for malt tax and partly by the many calls being made on the Laboratory by other departments of state. Some 14000 samples had been examined in 1872–73 when the current number of staff had been fixed and had risen to over 24000 by 1884. One perennial problem in the Laboratory was that at any given time analysts were out of the Laboratory attending court hearings giving evidence or on regular surveys of manufacturers' premises. This situation was exacerbated after 1875 by the position under the *Sale of Foods and Drugs Act* (see below). Principals regularly commented on this in Reports and in submissions for more staff. A standing grievance of analysts was that travel and subsistence while attending hearings was not allowed at first class rates, despite their efforts to obtain them.

The number of rooms occupied at Somerset House had also increased, from four to eight during the same period. Dr Bell proposed that the number of Second Class Analysts should be increased from four to six, with a similar increase in the number of First Class Analysts. In addition he asked for the appointment of two Superintendent Analysts to check and control the work done by the Temporary Analysts. These Superintendent Analysts were to receive a duty allowance of £50 a year, and did not have to sign on when entering the building in the mornings, as did more junior staff. The Temporary Analysts were at the same time to be reduced by three. The request was only made because the Laboratory, principally 'established for the service of the Excise Department, has now practically become a General Laboratory for all Government Departments'. The proposals were agreed by the Treasury, who asked that some check be kept on the requisition of the Laboratory's services by other government departments. Dr Bell replied that with the exception of the lemon juice samples from the Board of Trade and tobacco and snuff samples from the Board of Customs, no sample was analysed except on the written request of the Secretary or other head of the department concerned and that he had no reason whatever to believe that the privilege was being exercised unduly.[21]

By 1885 the permanent staff had therefore been increased to six First Class Analysts, six Second Class Analysts and two Keepers of Chemicals, in addition to whom there were 18 Temporary Assistants and eight Students. The total cost of the permanent staff was £3925, perhaps £175,000

now. Two of the First Class Analysts, William Harkness and
Henry James Helm, both of whom had served in the Laboratory for many years, acted as Superintending Analysts as
agreed. They received £550 a year. The other staff had not
received an increase since 1877 and they wrote in 1887 to Dr
Bell ('the Chief' as the copy was annotated in the margin),
pointing this out. This was despite the very large increase in
work and pressure on them. The overcrowded state of the
laboratories, 'in an atmosphere much vitiated by necessary
gas flames and deleterious vapours of various kinds' are
emphasised. No response is recorded, although Bell took up
some of their points in an appeal to the Board for yet more
staff in 1889. The analysts tried again in 1889, and the result
of all the activity was the appointment of two further analysts
in 1890. One of them left almost immediately to take up the
post of Chemical Adviser to the government of Trinidad. By
now the work of the Laboratory had been organised into four
'branches', each under a 'head' and a 'sub-head' assisted by
one or two analysts and temporary assistants. The branches
were Beer, Spirits, Tobacco and Miscellaneous (that is non
revenue work).[22]

The salaries received by the staff, starting in 1859,
compared favourably with those of similar grade officers in
the Civil Service generally, so far as this can be determined.
The staff were in a rather anomalous position of course since
the work they carried out was not like that of any others in
the Inland Revenue nor indeed in the Government service.
The Treasury had argued in 1877 that the salaries of the
Laboratory Assistants should be compared with those of
similar staff in private or University laboratories. Bell argued
strongly that his staff had more onerous and demanding
duties than such people. He won this argument, and his views
generally prevailed in the future. Comparing the higher
grades of staff, the Deputy Principal for example with
Assistant Secretaries, a reasonable comparison, the salaries
were approximately equal. Analysts did much better than the
probably comparable clerks of the Second Division, receiving
in 1875 up to £400 a year as against a maximum of £184 for
the clerks. Both of these figures may be compared with those
for bank clerks at about £190.[23] The salary of the Principal
of the Laboratory was always about half of that of a
Permanent Head of Department (in the nineteenth century at
least), but in 1872 when it was £700 *per annum* it was

29   Richard Bannister, Deputy
Principal of the Laboratory
1874–1889

calculated that not more than 60,000 British families out of a population of about 4.6 million were in the wealthiest or 'comfortable' class, which required about £800 per year'. It has been calculated that in 1880 the average semi-skilled/skilled worker, with an average wage of £83 a year and with a wife and three children could feed clothe and house himself quite adequately with this amount. In 1880 most of the Temporary Assistants in the Laboratory earned £96 a year. Fewer than a million wage earners received this amount. By 1914 the same family as above would have required £117 *per annum*, a Temporary Assistant then earned £130 maximum. The income tax limit was £200 per year, at that level the salary of most Second Class Analysts, a family was well off indeed.[24]

CHAPTER 5

The Referee Analysts

IN HIS REPORT FOR 1861 MR PHILLIPS DWELT on the extent of adulteration as disclosed by the samples analysed in his laboratory. His work was undertaken mainly for revenue protection purposes, but he concluded that the difficulty in dealing comprehensively with the adulteration of food and beverages in many other areas had been much under-rated. The extent of fraud would not be greatly reduced whilst the practice continued to be profitable to the producers and dealers, or until the consumer realised that the value of a commodity was made up of its price and quality.[1]

The question of the purity of food, and its deliberate fraudulent adulteration had been widely discussed in the middle years of the last century. As has been seen already, it was something of a golden age for adulteration, and the most poisonous things were added to food, as well as to tobacco, and drink. This had been known for many years, the book by Frederick Accum, *Treatise on Adulteration of Food*, published in 1820, caused a great stir, with its discussion of the colouration of Gloucester cheese with red lead, and green tea made from black and white thorn leaves dyed with verdigris and 'Dutch pink'.[2]

The *Bread Act* of 1836 forbade the addition of anything extraneous to flour and bread, so that Accum's book, (and those of his plagiarists) seems to have had some effect – if not a great deal in a practical sense, but by the 1850s there seems to have come a general awakening to the amount of adulteration which was going on, and indeed to its dangers. Dr Arthur Hassall (a gentleman who had no doubt as to his abilities) became chief analyst of an Analytical Committee set up by the *Lancet* in 1851, and as a result of his work for this committee published the first edition of his book *Food and Its Adulterations* in 1855. Hassall's book described a serious state of affairs, and at this point even the government seems to have become alarmed. A Select Committee was set up in

1855 to hear evidence on the adulteration of food and drink. Much alarm at the 'unrestricted sale of poisons' was expressed and it was suggested in the Report that adulteration could be prevented by every municipal authority adopting a system similar to the Inland Revenue authority, that is by appointing analysts. Hassall himself gave evidence as to the very large amount of adulteration occurring in foods and drugs and how they could be detected. His chosen method of detection was by using the microscope. He repeated a charge already made in his book that the 'Excise' had not availed themselves of all the resources of modern science in determining adulteration. The Excise in the person of George Phillips also gave evidence. Phillips styled himself 'chief officer of the Chemical Department' (of the Inland Revenue that is), and he described the training of his analysts, (at that point there were six of them), who received two or three years formal training at University College London and who subsequently continued their training under more senior colleagues. Simultaneously they visited traders' premises to see for themselves the appearance and characteristics of adulterated and genuine articles. They were then placed at Inland Revenue stations around the country where their knowledge was put to practical use. Phillips did not think that the amount of adulteration was as great as Hassall thought. In answer to a question he said that he had not read Hassall's book, but had read the reports on which it was based.[3]

The Select Committee resulted in the so-called Adulteration Act, ie the *Adulteration of Food and Drink Act*, 1860. This Act allowed the appointment of Public Analysts but made it a discretionary matter. It was not compulsory, so that in fact few were appointed and those few often medically rather than chemically trained. There does not seem to have been much effort to enforce the Act. The standards for composition and purity of even basic food commodities was little understood. The Act thus became rather a dead letter, as did a second, the *Adulteration of Food and Drugs Act* of 1872 although there was more effort to enforce it in some areas. Following this Act, yet another Select Committee was appointed in 1874. All of the major analysts of the day gave evidence to this Committee, including James Bell, now Principal of the Inland Revenue Laboratory, and Public Analysts appointed under the 1872 Act. These included James Wanklyn, Public Analyst for Buckinghamshire, of whom more below. Bell, in his evidence, showed that his Laboratory had experience of analysing a wide range of food stuffs, (as well as the staples of tea, coffee and beer), as noted in the last chapter.

Opinion in the Committee was in favour of setting up a system of public analysts with greater powers, and with a central body acting as referees in cases of disputed analyses. Witnesses were asked if they thought that the 'Somerset House' Laboratory (as it was usually called) could take this role. Most of them, particularly the trade representatives, according to Bell, but including at least one Public Analyst, Dr Augustus Voelcker, agreed that this would probably be acceptable, and seemed to agree that the Inland Revenue Laboratory had a good reputation.[4] Professor James Wanklyn however violently disagreed. He was of the opinion that staff of the Excise Laboratory did not enjoy a high reputation amongst chemists; he certainly had a very low opinion of them.[5] It is interesting to speculate if Wanklyn's extreme hostility may have been due to his even more extreme hostility to Professor Frankland, of the 'South Kensington College' (later Imperial College), in a dispute over the best method for the determination of organic nitrogen in drinking water. Frankland worked closely with the Inland Revenue Laboratory, who may thus have been guilty by association. Wanklyn was apparently a man much given to quarrelling, and resigned from the Society of Public Analysts over a slightly absurd editorial matter (he was on the Publications Committee of the *Analyst*) in 1876.[6] This attitude of Wanklyn was taken up in the *Chemical News* edited by William Crookes (the eminent spectroscopist). Crookes seems initially not to have had any connection with the Public Analysts although he was later elected to membership of the Society of Public Analysts in 1875, but he whole-heartedly took their part, repeating in an editorial in July 1874 the remark of Wanklyn mentioned above.[7] He adds that he is sure that Dr Voelcker would not wish to have an analysis over-ruled on 'such authority', presumably a dig at Voelcker's apparent acceptance of the Inland Revenue Laboratory acting as referees.

This editorial provoked a letter from Phillips, by now retired, in which he defended the Laboratory, pointing out that its expertise was in fact wide and did include foodstuffs, a point Bell had already made. The same editorial also predictably produced a letter from Wanklyn praising its views. Phillips' letter was, equally predictably, ridiculed in the next issue of *Chemical News*.[8] The chief objection seemed to be to an official body sitting in judgment over professional chemists — ignoring of course that Bell and his colleagues were equally professional chemists. All permanent members of staff had been trained at University College or the Royal College of Science, as well as receiving internal Laboratory training. The ten permanent menbers of staff, with an average of over seven years experience each (and in some cases considerably more) were certainly trained chemists. The agitation against the proposal continued, being the subject of a motion at the founding meeting of the new Society of Public Analysts in August 1874. This motion was in fact couched in moderate terms, the objection now being to the recommendation of the Select Committee that the 'Somerset House Laboratory' decision would not have to be tested in the courts by cross-examination on oath.[9]

A counter proposal was made by the Public Analysts for referees to be appointed from amongst the Public Analysts themselves, which seemed rather like acting as referees on their own work (a point admitted, and defended by Crookes in *Chemical News* in August 1874). The lobbying which undoubtedly went on seems to have had effect, since the first draft of the new legislation omitted any official referee, merely appointing the Public Analyst of an adjacent district to analyse a disputed sample. A clause referring to 'Somerset House' (in this version, 'chemical officers in the employment of the Inland Revenue'), as referees was inserted into the Bill at a revision in May, and received without comment by *Chemical News*.[10]

Opposition at this stage centred on the proposal that the referees certificate would be final. Assurance must have been received on this point, and guidance issued after the Bill became law said that defendants could call the analyst, from the context of the remark the referee analyst, as a witness, although nothing actually appeared in the Act to this effect.[11] Somerset House chemists did subsequently give evidence in court. The Act (the *Sale of Food and Drugs Act*, 1875) finally became law on August 11, with the 'Chemical Officers' of 'Somerset House' being appointed referees. It made the appointment of public analysts a duty on local authorities, and gave the courts the power for the first time to challenge the findings of either the prosecution or the defence by referring to the Inland Revenue Laboratory for an independent opinion. This required the inspector taking the original sample to divide it into three parts, one to be analysed by the prosecution analyst, one by the defence analyst, and the third to be kept by the inspector until required for analysis by the 'Somerset House' analysts. Regulations had later to be issued concerning the size and division of samples since some of them were too small to be analysed properly. The passing of the Act provoked the comment in *Chemical News* that it had been 'strongly (though unsuccessfully) resisted by the Analysts . . . because the standing and experience of the Inland Revenue chemists is altogether unknown'. The hope was expressed that their staff would be strong enough for the work.[12]

After the Act was passed, both sides had to settle down and make it work. Methods of food analysis in the early 1870s were for the most part in a very unsatisfactory state. Very little attention had been paid to the systematic examination of food with a view to determining its quality, although the Inland Revenue Laboratory had experience in detecting adulterants in most of the foods likely to come to them, in some cases considerable experience. The first public analysts and the 'Somerset House' chemists had to devise their own methods, and at first little agreement or uniformity was evident. Another problem was the question of training and/or qualifications for those appointed to be Public Analysts. Many of the witnesses to the Select Committee in 1874 agreed that many of those who had been appointed under the 1872 Act were not analytical chemists.

The Laboratory staff were less acquainted with milk, butter and flour than with some other foodstuffs although Bell stated in his evidence in 1874 that he had analysed milk samples out of interest to discover for himself the natural variation in constitution. However as soon as the Select Committee recommended that the Laboratory be made referees, work was started immediately on a systematic investigation of milk butter and flour as well as other substances. A very large number of milk and butter samples were analysed. Of the latter, in July 1875 three are marked in the register as being received from Ewell in Surrey, 'per Mr Bell', perhaps informal samples purchased personally by Bell from his local dairy, (he lived in Ewell). The results of the milk analyses were submitted to the Inland Revenue Board. These were not published but the results of the butter analyses were published in 1876 as a Parliamentary Paper together with a method for the amount of non butter-fat (if any) which had been added. This method was based on the method of Hehner and Angell originally announced by Dr Hassall in 1874 and denounced by Professor Wanklyn.[13]

30   Sale of Food and Drugs Act, 1875

[38 & 39 VICT.]   *Sale of Food and Drugs.*   [CH. **63.**]

CHAPTER 63.

An Act to repeal the Adulteration of Food Acts, and to make better provision for the Sale of Food and Drugs in a pure state.   [11th August 1875.]

A.D. 1875.

WHEREAS it is desirable that the Acts now in force relating to the adulteration of food should be repealed, and that the law regarding the sale of food and drugs in a pure and genuine condition should be amended :

Be it therefore enacted by the Queen's most Excellent Majesty, by and with the advice and consent of the Lords Spiritual and Temporal, and Commons, in this present Parliament assembled, and by the authority of the same, as follows :

1. From the commencement of this Act the statutes of the twenty-third and twenty-fourth of Victoria, chapter eighty-four, of the thirty-first and thirty-second of Victoria, chapter one hundred and twenty-one, section twenty-four, of the thirty-third and thirty-fourth of Victoria, chapter twenty-six, section three, and of the thirty-fifth and thirty-sixth of Victoria, chapter seventy-four, shall be repealed, except in regard to any appointment made under them and not then determined, and in regard to any offence committed against them or any prosecution or other act commenced and not concluded or completed, and any payment of money then due in respect of any provision thereof.

Repeal of statutes.

2. The term "food" shall include every article used for food or drink by man, other than drugs or water :

Interpretation of words.

The term "drug" shall include medicine for internal or external use :

The term "county" shall include every county, riding, and division, as well as every county of a city or town not being a borough :

The term "justices" shall include any police and stipendiary magistrate invested with the powers of a justice of the peace in England, and any divisional justices in Ireland.

[*Public.–63.*]   A 2   1

In the operation of the Act many problems were caused by the lack of recognised methods of food analysis and of the Laboratory and the Public Analysts working to different standards. It is useful here when considering the disagreements which arose in the interpretation of the Act to note that the equipment available to the analysts of that time were the balance, the polarimeter, the microscope, hydrometers and calibrated glassware. The work which they did with these and their frequent close agreement are a measure of the high level of skill which most of them showed. One of the problems was the rigid minimum standards adopted by the Society of Public Analysts. These were published in the *Proceedings* of the Society in 1876. The Laboratory was well aware of them. They subscribed to the *Proceedings* and at least two of the senior analysts copied them into their work books, but disputed their accuracy. Not all public analysts worked to these standards, at least one refused to accept any standard but his own. The other difficulty was that no standards were laid down in the Act and in its certificate the Laboratory felt obliged to say that if a sample could have been genuine, for example that a naturally occurring milk could contain a certain (low) percentage of fat without being adulterated with water, then they had no choice but to say so. They had no legal authority to set standards, nor had the limits of the Society of Public Analysts any legal status. The figures which the Laboratory used as a guide in the frequently disputed milk and butter samples were the analytical results of the several hundred samples analysed by themselves. The Public Analysts accepted this in a theoretical sense, but mostly could not apparently see why the Laboratory did not accept their standards. These were usually based on averages. That the Laboratory staff were not allowed to join the Society lest their judicial status was impaired also caused some annoyance.[14]

There was a considerable demand from 1875 for standards in all of the foodstuffs which could be adulterated and many analysts and magistrates' courts wrote to Somerset House to enquire what standards they had adopted for a particular type of food, usually milk or butter. The reply was invariably that the Laboratory did not adopt particular standards because of the wide natural variation in for example milk fat and 'solids not fat' and the relationship between them. The standards which the Laboratory did adopt in practice were revealed in letters, (which were usually very helpful), in discussions, and in the evidence given in court. The level of fat in milk to which they then worked (2.5%) was revealed by a case in 1879 for example and their empirical allowance for decomposition of milk in 1883. In this case the certificate of analysis did not state how the allowance was made, merely that 'after making allowance for the decomposition of the milk' the result was such and such. This allowance was important in the days before widespread refrigeration. No regulations had been laid down to say that samples should be packed in ice, and milks were frequently received in a state of advanced decomposition. The allowance was at first an empirical one based on the state of fermentation, but one with more scientific basis was later developed, based on the the ash and chloride content. Later still a method was developed by Bell, and published by him in his book on the analysis of foods and drugs.[15]

The Laboratory was at fault to some extent in being reluctant to admit publicly to the standards which it did in practice work to, or to be more open as to the methods which it used. The lack of publications on these matters by the Laboratory staff was used as the reason for the claim by some of the Public Analysts that the Laboratory staff were only routine chemists who never contributed to the advance of scientific knowledge. It was later admitted that not all Public Analysts were of very high attainments. Nor were some of their methods as reliable as they believed.[16] The method used by them for the determination of non fatty solids and fat in milk for example was found in 1882 to give erroneous results. It gave a half per cent or more too high in the case of extracted fat. This alone accounted for much misunderstanding, and its discovery brought results of milk analyses nearer to those of the Laboratory which had never used their method. Watering of milk accounted for a very large number of court cases in the years after 1875.[17] Other problems were caused by the exreme rigidity with which some Public Analysts interpreted their standards, as if no samples were allowed to be below them. Professor Wanklyn himself stated categorically in his book on milk analysis that 'milk is a secretion of constant or only slightly varying composition' a statement totally at variance with the facts. One analyst stated publicly in 1882 that the Society of Public Analysts had laid it down that genuine butter should contain not less than 80% of butter fat, anything less showed adulteration. This was regardless of evidence that it was perfectly possible for a genuine milk to contain less than this. This analyst took the same view about the Society's milk standards and charged the Laboratory with ignorance in not following these standards. These cases became something of *causes célèbre* and Bell protested vigorously.[18]

In the great majority of cases the Laboratory agreed with the results of the Public Analysts, it was the interpretation on which they occasionally differed. In some cases the Laboratory may well have been wrong in their interpretation or in their results, it is not really possible to tell now. In the case of salt in beer, referred to in chapter three, it appears that they may certainly have been wrong in one particular case. This was a beer sample taken in Bath. The Laboratory result confirmed that of the local Public Analyst (over 60 grains of salt per gallon), but said that such an amount could occur naturally. Results published subsequently seemed to show that at this particular time this was probably not so.[19] One of the difficulties was that neither side seemed to make any great effort to understand the position of the other. After the attitudes taken by the Public Analysts before the 1875 Act it is perhaps not surprising that neither party felt able to conciliate the other. For many years the *Analyst*, the journal of the Society, spared no opportunity to show the Laboratory or its staff in a bad light. This attitude extended to the rank and file on both sides, some members of the Laboratory staff took a less than helpful attitude to requests for information from the Society of Public Analysts. There is for example a very terse correspondence between Richard Bannister, Deputy Principal and Mr Allen, Sheffield Analyst in 1894. The Society asked Bell to speak to them in 1878 on the standards he adopted for genuine milk, butter etc and on the methods he used. Bell politely declined, but offered to meet the Secretaries of the Society to discuss matters. He further said that all Laboratory results were open to inspection by any of their members who felt 'disposed to favour him with a visit'. Many apparently did so including the joint Secretaries and methods and results were discussed. Private meetings were apparently usually amicable: perhaps because neither side needed to strike public attitudes. A further request for a meeting was declined by the Board of Inland Revenue on Bell's behalf. They said that they did not consider it expe-

THE

ANALYSIS AND ADULTERATION

OF FOODS.

BY

JAMES BELL,

Principal of the Somerset House Laboratory, Vice-President of the Institute of Chemistry, etc.

Part I.

TEA, COFFEE, COCOA, SUGAR, &c.

Published for the Committee of Council on Education
BY
CHAPMAN AND HALL, LIMITED,
11, HENRIETTA STREET, COVENT GARDEN.
1881.

31 Bell, *Analysis and Adulteration of Food*, 1881

dient. An offer to meet privately was not the same as a public meeting nor was it equivalent to publishing the methods used by the Laboratory. Eventually though in 1881 and 1883 Bell published a two volume work *The Analysis and Adulteration of Food*. This contained the results of much important new work and descriptions of methods used. It included the results of the Laboratory's investigations into the variation in the composition of milk and the important laboratory method for the analysis of sour milk.[20]

As late as 1893 the Public Analysts complained in a memorandum to a Committee considering the future of the Customs and Inland Revenue laboratories that despite efforts they could not find out what methods of analysis were in use at the Laboratory. They also implied that discussions with them had been refused and that even such methods as Bell had included in his book were unknown to them. It was a very bitter letter in some ways, and Bell in his reply described it as 'erroneous and unfair' and 'misleading'. It is difficult not to agree. Bell refuted the charges in detail and pointed out with some satisfaction that new standards being adopted by the Society were the same or similar to those being used by the Laboratory for the last 18 years. This attack by the Society is strange in view of the fact that in the same year Bell was invited to speak at the first annual dinner of the Society to which anyone not a member was invited. In fact bad relations had usually not extended to the heads of the respective organisations. He was received courteously and made a speech prophesying an era of peace and goodwill between the Public Analysts and his department. He was a little premature, but relations did improve within a few years.[21] .

In the same year as the memorandum of the Public Analysts (1893), Bell issued a report on a further investigation into the composition of milk. This was in order to discover whether or not any change had occurred since his last such investigation in 1876. The investigation gave the results of analyses on the milk of 273 individual cows and the mixed milk from 55 dairies, all from a wide range of country farms and town dairies. Bell concluded that there had been a definite improvement in the quality of milk, with the fat content on average higher than in previous investigations.[22] This particular milk investigation illustrates well the change in the nature of the work of the Laboratory which had occurred with the introduction of the 1875 *Sale of Food and Drugs Act*. Initially it had been entirely related to protecting the revenue, henceforth it was also concerned with protecting the public health, and later animal health.

Having weathered this eventful period, Bell, Dr Bell as he had become, retired on 4 March 1894. He had postponed his retirement for two years to enable new arrangements for the Laboratory to come into being. These are related below. Bell was then 69, and special arrangements had been made for him to stay on after the new Civil Service retirement age of 65. As related above, he had been educated at University College London in 1851–52 as one of the Excise students, and indeed had a highly successful career there, but had left the Laboratory after about 1853. Between then and his transfer back to the Laboratory in 1867 he had presumably made use of his chemical knowledge as had been envisaged when the University scheme had been started, but he had not entirely neglected his chemical studies either. In 1865 he published a new and revised edition of *The Excise Officers' Manual* by Joseph Bateman. This contained much scientific information, showing both that he had not forgotten what he had learnt and that he had kept in touch with his old colleagues some of whom contributed to the book. Bell also published important work on fermentation in 1870.[23] His important work on the analysis of food was recognised in 1882 by the award of a PhD from the University of Erlangen and in 1886 of a DSc from the Royal University of Ireland. In 1887 he published a book on the chemistry of tobacco.[24] He was elected a Fellow of the Royal Society in 1884 and made a CB in 1889. He died in 1908.

Bell's judgement had been respected by the Inland Revenue Board, who regarded him as a shrewd adviser, of strong commonsense and evident fairmindedness in resolving disputes between the Board and various trade interests. In person he was apparently genial and tactful, described by one who knew him as the personification of kindness. His personality seems to have endeared him to his staff, a number of whom attended his funeral, and evidently regarded him as a friend. Some of his staff had helped him with his book on food adulteration. He obviously took an interest in staff welfare in a professional sense. In 1877/78 for example he suggested to the Board that the subscriptions to the Institute of Chemistry of the whole of his professional staff should be paid by the Board since membership would benefit them professionally but not personally and it was thus not equitable that they should pay it. He was greatly interested in professional matters and played a large part in the foundation of the Institute of Chemistry (later the Royal Institute of Chemistry), serving as its President 1888–91. He did have from the beginning a high opinion of his own status. A little over a year after his appointment as Deputy for example Phillips applied on his behalf for him to be allowed a daily

32   Senior staff of the Laboratory, c.1894. From left to right: J G O'Loghlen, J Cameron, E G Hooper, H W Davis, G Lewin, C Procter, G N Stoker; second line, W Harkness, R Bannister, D R Bell, H J Helm, J Holmes; lying down, C H Burge, J Woodward

subsistence rate when travelling of 20 shillings (£1) per day, on the grounds of his status and that he should not be embarrassed for lack of money and not undertake official journeys willingly. This was allowed, as it obviously had not been to his predecessor George Kay.[25]

The Laboratory over which Bell had presided was much the same as that left to him by Phillips, with the exception of the food and drugs work. It was still very much rooted in the past as an Inland Revenue, indeed an Excise laboratory, with by far the greater volume of work being carried out for revenue purposes. Phillips until a very late stage in his career had personally made seizures of tobacco and given evidence in court and it is apparent that the staff regarded themselves as very much Excise officers, on guard wherever they went to look out for contraventions of the Excise laws. Thus in 1867 Richard Bannister, at that time a senior assistant in the Laboratory and John Lewis a junior assistant seized a 'suspect sample' from a tobacconist which they 'noticed in the window when passing the shop'. Samples were also legitimately received from the public as possibly contravening the Revenue laws, one such as late as 1883.[26] However the range of samples continued to expand during Bell's tenure of office as Principal and the Laboratory was not entirely parochial or lacking contacts with the wider world. The Board of Inland Revenue were sufficiently proud of their Laboratory and it was sufficiently well known to the world at large for at least once an official party to ask to be allowed to visit it. This was the 'Chemical Examiner' to the Government of Bengal who requested a meeting with the 'Superintendent of the Excise Laboratory'.[27]

The Laboratory staff had continued to grow in numbers with 14 Analysts (First and Second Class), two Keepers of Chemicals and 41 Assistants and Students in 1893. Most of the staff enjoyed at least 36 days leave, 42 days after ten years service, or 48 if on a salary of more than £500 *per annum*. The latter included most of the First Class Analysts and above. All of them had to sign an attendance book except for the Principal and his Deputy.[28]

Despite these 'class' differences there seems to have been a community spirit, one which the Laboratory long retained. By 1887 there was an Inland Revenue Laboratory Assistants Annual Dinner, and by 1890 the first of a series of annual reunions for the Laboratory staff 'Past and Present,' That for 1894 was described as a 'Concert and Cinderella Dance'. These continued at least until 1895. They were elaborate affairs, usually held at the Holborn Restaurant or Anderton's Hotel in Fleet Street. They took the form of a concert and dance, with refreshments, rather than a dinner dance. They were organised by senior members of staff with Dr Bell as 'President' when he was Principal. The concerts were given by members of staff, with Mr Connah, later Deputy, playing (literally) a very important part in helping to organise things and playing piano and cello. A 'Christmas Annual' of somewhat laboured humour ('Lab-Gas') was also produced later, a few times at least, by the unattached officers seconded to the Laboratory.[29]

This community, a state supported laboratory, had thus changed and grown in 52 years. It had expanded considerably to meet changes in need perceived by the Boards of Excise and Inland Revenue. It had not in a true sense been supported by the Government, although the Treasury had regularly sanctioned expenditure on it albeit sometimes

33   Concert programme, 1890

rather reluctantly. This is consistent with the *laissez faire* policy of governments in the nineteenth century, a policy which did not change until the 1914–18 war.

The Laboratory thus went into a new era, ready for a new Principal and in a good shape for a change of pace and more contact with the outside world of science.

Part III
The Customs
Laboratories
1861–1894

6

The Customs Laboratories 1860–1894

THE LABORATORY AT SOMERSET HOUSE had been accepting samples from Government Departments other than the Inland Revenue for some years by 1860. Samples for the Customs Department had appeared regularly from 1856. Given this experience it is difficult to understand why in 1861 the Commission of Customs decided to set up their own 'wine testing stations', except perhaps for a desire to control their own work. As will be seen, the two laboratories worked quite closely together. These tests had become necessary because up to this time, from at least 1303 (in the *Carta Mercatoria*), the customs duty on wine was charged by volume (the imperial gallon from 1825), irrespective of strength. In 1860 the Customs Laws were amended, changing the basis of the duty to a scale based on proof strength. Facilities for testing strength of wines and somewhere to do it were now required. The Act came into operation on 1 January 1861 and in preparation for this day the Customs Commissions set up a suitable laboratory in London, together with 'a set of new stills and other instruments for the purpose'.

The man in charge of the Custom House Laboratory was Mr James Johnstone, Principal Inspector of Gaugers in London. He had been directed to get into touch with George Phillips to discuss a method by which the strength of wine could be 'rapidly and accurately tried'.[1] The method actually employed to determine proof strength was to distil a standard amount of wine, collecting the alcohol into a calibrated receiver, and to restore its volume to the original volume with water (see plate 35).[2] The method was tested during 1859 on nearly 1500 samples of wine, and found to work satisfactorily. Each determination was said to take only 25 minutes.

During 1860 a further three laboratories were fitted up, so that as well as the main one in the Customs House in London, there were others at the London, St Katherine and

34   Custom House, London

Victoria Docks, also in London, in order 'to enable the mercantile community to present their samples for testing at the most convenient station for themselves . . . , and to secure dispatch in the operation . . .'. Staff were also selected to undergo a course of instruction in London, (possibly at the Inland Revenue Laboratory, although Phillips makes no reference to them) 'to qualify them to perform the operation of testing wine, both in London, and at the outports'. The outports were ports outside London designated (by permission of the Treasury) as 'testing ports'. There were eleven of these (twelve including London). Laboratories had been equipped in the outports in the same way as the London ones, and by the beginning of December 1860 preparations were so far advanced that notice was given to importers that their wines could be tested in advance of the legal first day of testing, 1 January, and their wines delivered from bond on that day to avoid delay.[3]

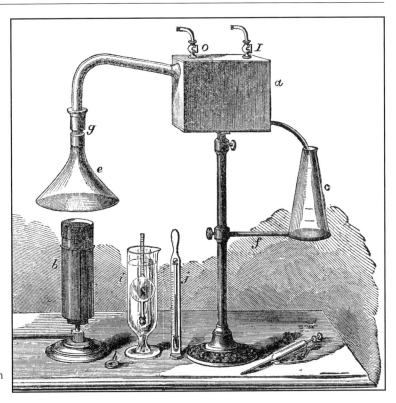

35 Wine distilling apparatus, 1880,
as modified by James Johnstone, with
a 'Phillips' condenser

The situation shows a very considerable degree of
forward planning and organisation, as indeed does the fact
that in the first year of operation the laboratories successfully
handled no less than 115769 samples, 70519 in London
alone. The new system was apparently introduced without
considerable delay to the importers and the staff were praised
for their application, despite 'the demand upon their time and
attention' being 'constant and severe for many weeks'. The
Commissioners were obviously very pleased and proud of
what had been accomplished. Twenty complaints only were
received and of these only four mistakes were found to have
been made. The numbers of staff involved were small, nine
gaugers in the three London Laboratories, who evidently did
the analytical work, and six weighers. The large numbers of
samples were handled expeditiously despite many of them
having to be sent to the nearest Laboratory from the point of
importation: the Act of 1860 allowed the Commissioners of
Customs to designate ports into which (and only into which)
wine in casks could be imported. There were forty of these
approved ports initially, with five others being added soon

after. Samples (in a bottle) were directed to be placed in a box, closed and sealed with the 'seals of office' and sent by rail or 'other conveyance to the nearest or most convenient testing Port'.[4] Sample bottles had to be capable of containing 'about 3 gills' (425 mls), and a form of test note was approved by 19 December, and ordered to be printed on 31 January 1861. These test notes were printed on blue paper, and were rather smaller in size than a post-card.[5]

The cost of setting up the Customs House Laboratory was £326 13s 5d, and the three docks laboratories £477. The former was £98 10s over the expenditure sanctioned by the Treasury, the work was more extensive and cost more than expected. Finance for the docks laboratories had not been asked for at all since the need for them had not been anticipated. The Treasury sanctioned the extra expenditure.[6]

The operations of the Customs House Laboratory, and probably the whole new organisation, was controlled by the Mr James Johnstone referred to above. In 1864 he received a special payment from the Treasury of £200, in recognition of his work for 'the improvement he has effected in the instruments and apparatus used in the Laboratories in that [ie the Customs] Department and for his services generally in carrying out the new system of assessing the new duties on wine'. Johnstone was presumably one of the two Gaugers employed at the Custom House, (although it is difficult to see such a senior man carrying out actual distillations), the other could have been Mr James Keene. In 1863 Keene, a gauger, tested samples of most of the wines exhibited at the International Exhibition of 1862, mostly from Europe but a few from Australia. He did not think very much of the quality of these latter wines.[7]

Very shortly after the introduction of the new wine duties in 1862, they were modified in the light of experience gained. Four bands of duty had been set, corresponding to wine of different strengths and these were reduced to two, corresponding to table wines (under 26° proof) and fortified wines. Two months after this a category of strengths 'unknown' was allowed to be declared, to avoid under or over declaring wines near the 26° thus causing the importer to be penalised. This latter category needed sampling at the old rate, but the other new rates allowed a much reduced rate of sampling, to approximately two thirds of that before the change. There was also some reduction due to less wine being imported. Due to this, and a simplification of the testing procedure, the number of staff was reduced to less than half its original strength.[8] Such a drastic reduction may have been regretted by the following year since by then there had been much work necessary to check a growing practice in the import of 'spurious wine', mostly alcohol with (sometimes) sufficient genuine wine added 'to disguise the compound by the imitation of colour and flavour'. The Annual Report does not say what was added to 'disguise the compound' if genuine wine was not, but it was probably usually cider as seen from the Somerset House results. The samples of these 'wines' were examined in the first instance by the Customs officers (presumably in their Laboratories). In cases where the composition of the sample was in any doubt they were sent on to the Inland Revenue Laboratory. In 1864 Phillips comments that 'within the past year, the Customs have adopted very stringent measures for the purpose of suppressing the obnoxious practice ... of importing fabricated wines'. He had analysed 94 in the past year, and commented that many such samples from the Customs were difficult and tedious to analyse, and it was 'not too much to say that one of such samples requires for its analysis more time than did fifty samples of coffee when examined for the presence of chicory'. This is borne out by the rather unusual length of time taken to report the samples, an average of ten days. The problems involved is shown by a comment by H J Helm in November 1863 that 'I am of the opinion that they are factitious wines [alcohol and cider in this case] mixed with a portion of genuine wine, but it is doubtful whether this opinion could be maintained by the results of the analysis only'. The Commissioners of Customs noted in the same year their gratitude to Phillips for his work on samples of tobacco and snuff presented to Customs for drawback, saying that the advantage of being able to obtain a really scientific opinion such as that of Mr Phillips was very great, the number of such samples had just risen from 21 in 1862 to 2254 in 1863, due to the provisions of the *Manufactured Tobacco Act* that year (see p.30 above).[9]

In 1866 new regulations as to the number of samples needed under the 1860 Act were introduced, simplifying them and reducing the numbers. In 1867 the Victoria Dock testing station was closed due to the few samples being received, and in 1872 Mr Johnstone, until then in charge of

36 Laboratory report on 'factitious' wine, 1864 (PRO DSIR 26/206, a workbook of Richard Bannister)

the Customs testing stations, was superannuated at the age of 58, due to decline in the amount of work. He was succeeded as Principal Inspector of the Gauging Department and head of the testing stations by Alfred Baker, a member of the gauging department since 1860.[10] In 1874 further work was carried out by James Keene, by then an acting Inspector of Gaugers, on wine submitted to the International Exhibition of that year. Many of the samples were from Spain and Portugal and were of interest as part of an ongoing survey of the strength of natural wines, as opposed to fortified wines, such as port and sherry. Keene decided that the Australian wines present were this time of better quality. From Keene's results the Commissioners decided that the figure of 26° proof at which wines were regarded by the law as natural (ie not fortified) was justified.[11]

A few years after the reduction in work in the wine testing stations noted above the situation was entirely reversed, and samples analysed increased again. This was caused by a further change in the law. In 1876 the law concerning the Customs duty on imported beer was changed to be similar to that concerning domestically produced beer. Instead of a flat rate per barrel, three rates were introduced for original gravities between 1065 and 1090 (see above p.54). The Customs, as well as the Inland Revenue, were then required to verify the actual gravities declared by traders instead of checking that they fell in a particular band. The distillation method introduced in 1856 was specified for checking declarations, and samples were tested in the Custom House testing station (and only there) since the 'delicate and special character' of the work, needing 'the highest attainable accuracy and uniformity in the results' required 'officers specially skilled in the work'. Over 1000 samples were being examined per year by 1879, rising to more than 7000 by 1894.[12] The Commissioners did wonder, rather plaintively, if the method of charging duty on the beer could be changed to a tax on the alcoholic strength of the beer (rather than on the original gravity), carefully suggesting a rate to bring in the same amount of revenue as the current system, on the ground of 'greater simplicity of method and a large economy of time'.

The reason for this preoccupation with the amount of time taken up by the beer work may well have been because this was not the only new area of work in operation, or about to be given, to the wine testing stations at this time. One set of new samples had been received from 1863, although these were not referred to in the Annual Reports. These were of fusel oil. This 'oil' is the last fraction, or tailings from the distillation of potable spirits. It was widely used in the preparation of other chemicals or varnishes and lubricating oils and fruit essences, although the initial impetus for importing it appears to have been as a substitute for turpentine, the price of which had increased dramatically due to the American Civil War and consequent reduction in supply. It consisted largely of higher alcohols, particularly of amyl alcohol, but also of a certain amount of ethyl alcohol. In 1863 it was allowed to be imported free of duty, provided it did not contain more ethanol than the home produced article. The examination of these imports was entrusted to the laboratory under James Johnstone, at the Custom House, and George Phillips was asked to supply a method to determine the alcohol present. The method he developed consisted of shaking the 'oil' with an equal volume of water and determining the spirit content of the separated aqueous layer with Sikes hydrometer. By this method he obtained a figure of nearly 15% ethanol in a sample of home produced fusel oil. The experiments at Somerset House are described in a workbook of 1859–1864, Phillips himself, his Deputy George Kay and H J Helm carrying out the work. The Treasury thereupon ordered that fusel oil of greater than 15% ethanol should pay duty. This method remained in use until 1897, (usually only 2–300 samples were involved), despite the fact that a large proportion of the 'alcohol' determined by it was not ethanol, and occasionally disputes arose. One of these involved a sample which by Phillips' method contained 44% of spirit, but which an independent analyst said contained 10%. A new method was adopted in 1895, when Benzine was used as a solvent.[13] Considerable interest was shown in fusel oil in 1890/91 in the investigations of the Select Committee on British and Foreign Spirits. The object of this committee was to enquire whether, on the ground of public health, some spirits should be kept in bond, to mature, before being sold for consumption. Fusel oil, a component of most spirits, tends to disappear as the spirit matures and its possible toxicity was therefore of interest to the committee. There had been some concern on the health effects of fusel oil in 1873, as possibly being the chief cause of intoxication amongst the lower orders. No conclusion was then reached. Dr Bell gave evidence in 1890 that he had done a considerable amount of work, having a German report on the physiological effects of fusel oil translated and carrying out many tests on spirits of different modes of manufacture and maturity in order to determine the higher alcohol content. Mr Cobden Samuel, then head of the Customs Laboratory, (see below), described some experiments he had conducted on himself. On five days he consumed 0.24 fluid ounces of fusel oil contained in 4 fl. oz. of 'a very fine fifteen year old brandy' without any ill-effect; it did have an unpleasant taste though. Samuel also had 'three penny worth' (apparently less than a liqueur glass full) of ordinary brandy with his lunch every day. After ceasing this experiment he repeated it for four days with pure alcohol and water of the same strength. This gave him a persistent headache and indigestion. He repeated the brandy experiment (with a cheaper brandy) for 19 days in the following year, without any ill-effects, but an assistant who tried the control with pure alcohol and water had to give it up after a few days.[14]

The work on fusel oil was joined in 1881 by a considerable amount of work on imports of spirits. It had long been known (since at least 1760) that measuring the alcohol content of a wine or spirit containing sweetening or colouring matter with a hydrometer gave an inaccurate result, that is the additional matter obscured the true alcohol content. Some wine samples had been distilled in the testing stations if it was suspected that this was happening (263 in 1879 for example) as were spirits too but the inaccurate results on spirits were ignored until 1887. In 1881 provision was made 'that in any case where by reason of the presence of colouring, sweetening, or other matter, the correct strength of any spirit cannot be immediately ascertained by Sykes' hydrometer, a sample of such spirit may be distilled . . ..'[15] British liqueurs, tinctures, or medicinal spirits, and only these liquids, awaiting drawback were distilled to ascertain their strength from 1880. Both the Commissioners of Inland Revenue, and of the Customs were give powers to deal with samples of these products, as they both obviously did from their respective sample returns, although relatively few were dealt with in the Inland Revenue Laboratory after the first year.[16]

The Act of 1881 came into operation on 5 April 1881, and sampling started on the same day. By the end of that year over 56,000 samples had been examined, (nearly 25000 in London). This nearly doubled the number of alcoholic samples being dealt with at the Laboratory. The extra work was noted as imposing 'much extra labour on the testing officers, who so zealously applied themselves to the work that serious cases of delay or inconvenience did not arise'. It was also noted that the new system saved the consumer money in that, in the case of liqueurs, the new duty payable on the actual proof strength was less than the flat rate of 14 shillings (70p) per gallon hitherto payable.[17]

The method adopted by the laboratories was to treat the sample as a wine, distilling it, making up the distilled spirit with water to the same volume as before, and then measuring the proof strength with the hydrometer. If this had been applied to every cask of spirits imported this would have meant that hundreds of thousands of samples would have had to be taken. To obviate this, instead of reporting the actual proof strength of the sample, the difference between the actual and the apparent strength (as found by direct testing of the spirit) was found. This difference, known as the obscuration, was the one reported. It was usually fairly constant for a given type of spirit even when the alcohol content was slightly different, so that the obscuration reported by the Laboratory could be applied by the Customs officer as a correction to the apparent hydrometer strength of casks of similar spirits, without the need to test them in the Laboratory.[18]

The very large increase in work occasioned by the Act of 1881 seems to have been the cause of a fundamental re-arrangement of the laboratory service of the Customs Department. This had been complicated in 1875 by a provision in the *Sale of Food and Drugs Act* of 1875 which placed on the Commissioners of Customs the duty of examining all tea imported into the United Kingdom. The tea was to be examined as to its fitness for human food by 'persons to be appointed' by the Commissioners, 'with all convenient speed'.[19] This duty had been laid on the Customs at the recommendation of the Select Committee considering the *Adulteration of Food Act* of 1872. The witness on behalf of the Customs, Frederick Goulbourn, the Chairman of the Customs, and Robert Ogilvie, Surveyor General, had said that there would be no great difficulty in their taking on this task, or of obtaining the staff. Mr Goulbourn said that there was considerable scientific expertise in the Customs, due to their operation of the wine testing laboratories and that they could handle all the necessary scientific work themselves, without referring any to 'Somerset House' (the Inland Revenue Laboratory), nominally on the grounds of saving time, but, reading between the lines, at least partly on the grounds of departmental pride. Mr Ogilvie suggested that staff could perhaps be trained in the Inland Revenue Laboratory: Dr Bell, Principal of the Laboratory had said in his turn that his laboratory would have no difficulty in carrying out the work with minimal cost. Customs estimated the cost at possibly an extra £5000 per year.[20]

The original high estimate of the cost of setting up and running the new laboratory was considerably reduced to £2000 after the firm decision to give the work to the Customs, but this proposed expenditure still horrified the Treasury. Customs were asked to consider ways of reducing this, including, even at this stage, allowing the work to be done at Somerset House. The Customs Board did manage to conciliate the Treasury without conceding the work to the Inland Revenue, (they boasted in 1885 that the cost was in fact £450 per year), and finally appointed three Inspectors and two Assistant Inspectors to inspect the tea imported into the Port of London. In other ports where little extra work was involved, the staff had this task added to their ordinary duties. The Inspectors were given an allowance of £50 *per annum* while actually performing their tea duties, the Assistant Inspectors were paid it only when they deputised for an Inspector. A Tea Analyst was also appointed, and he too received the extra £50 allowance. James Keene, Hydrometer Officer at the Custom House, was the first appointed. At first the new 'Laboratory' (properly so called as will be seen), and the 'Wine testing station', both in the Custom House and probably on the top floor, seem to have been regarded as separate establishments, and work done in the Laboratory was reported separately from that 'tested by distillation at the Custom House'. This distinction between wine testing and work in the Laboratory was still rather artificially being kept alive in 1891, when the Board directed Customs Collectors that the 'offices' in which wine testing was carried out should officially always be described as 'Testing Offices' and never 'Laboratories', since 'the only Laboratory properly so called in the Department being the Office in the Custom House in London in which analyses of tea, under the *Sale of Food and Drugs Act* 1875, and other similar work as directed by the Board, are carried on'.[21]

Tea imports continued to be examined by the Laboratory for HM Customs until the second world war when, in 1941, the duties were taken over by the newly formed Ministry of Food. They were returned to Customs in 1951 and the training of tea examiners in London continued at least until the mid-1950s. The test at the end of the training period lasted an hour with the written answers to a list of questions and a verbal examination on various specimens. The Government Chemist took over responsibility for staffing the chemical outstations then remaining in 1932. The tea was examined using special equipment. A standard weight of tea was measured on portable brass scales (kept in lightweight wooden cases designed for the purpose) and was infused with boiling water in a standard porcelain cup. This cup had a serrated edge to part of the circumference and was fitted with a porcelain lid. After a specified time the cup and lid were carefully inverted over a small porcelain bowl, allowing the infusion to drain into the bowl and the leaves to remain on the inside of the lid.[22] Both the liquor and the exhausted leaf were examined, the former for flavour and freedom from mould, the latter for excessive stalk and for foreign leaves. The original dry leaf was also examined for unacceptable amounts of sand or tea dust. A comprehensive collection of green, black and "near" teas, and of teas containing deleterious substances, was available for reference purposes, all of which it was necessary to recognise. In the 1880s, when gross adulteration was still being practised by the foreign manufacturers (at least), some condemned samples were described in the Customs *Report* as 'faced green teas, 500 [packets] of leaves other than tea, made up to imitate a tea known in the market as 'Imperial', and the remainder of decaying congous and fannings'.[23]

37   Tea testing cup and bowl

The slightly ambiguous wording by the Board Minute in 1891 concealed the fact that by then, and indeed for about ten years previously, the 'testing stations' in the Customs House in London had been amalgamated. The distinction had never been very real. Keene presumably still had his duties in respect of wine testing, and at least by 1878, and possibly earlier the Tea Laboratory had also, under an Act of 1876, tested imported substances in which ethanol was either present, or had been used in its manufacture. This work was very close to that done in the testing station. However, for whatever reason, as indicated above, the work resulting from the 1881 Act seems to have precipitated in London a formal union between the 'wine testing station' and the 'Laboratory', which took place in 1882. Late that year the Board decided that the staff of tea analysts could be reduced, and that since Keene was now fully occupied with his duties as Assistant Inspector General of Gaugers, to which he had been promoted in 1877, the Assistant Analyst, a Mr Excell, should be appointed Analyst. The Assistant was not to receive a £50 allowance in future, although Mr W Cobden Samuel (said to have been trained as an analyst and later, by 1889, Principal of the Customs Laboratory), was to receive the allowance if and when he deputised for Excell.[24]

The Customs Laboratory continued to go from strength to strength. Keene retired in 1886, by now Inspector of Gaugers but still with analytical responsibilities, and the Board commented in 1887 that 'By the retirement of Mr Keene the Crown has lost a public servant of exceptional ability, on whose advice we were specially dependent in all matters connected with Hydrometry, Gauging, and the scientific analysis of various compounds liable to duty'. Keene had invented a new hydrometer for use with wines, described in his book *A Handbook of Hydrometry* and a condenser which was still in use in 1896. Numbers of samples continued to increase, for example by over 1500 wines and spirits in 1886, due to a change in the levels of proof strength at which duty was levied, and in 1887, due to a change in sampling rates (caused by the 1886 change). Samples rose by nearly 40000 this time. By 1893, the year before the union with the Inland Revenue Laboratory, the number of wine and spirits (etc) samples examined was over 153000 (having reached as high as 177000 in previous years) of which less than 1000 were tea samples, slightly ironic in view of the fact that in a sense the 'Tea Laboratory' had absorbed the 'Wine testing station'. The variety of samples was large (but not of course as large as that of Dr Bell's Laboratory), and as well as wine, spirits, liqueurs and beer, included substances in which spirit was used in the manufacture such as transparent soap, chloroform, and acetic, butyric and sulphuric ethers.[25]

Dr Bell himself seems to have felt that there was really no need for two laboratories. From his evidence in 1874 to the Committee considering the 1872 Adulteration of Food and Drugs Act he obviously believed that his laboratory could easily cope with tea analysis with minimal difficulties. By his evidence to the Royal Commission on Civil Establishments, in 1888, he made it quite clear that his estimate of the abilities and expertise of the staff of the Customs Laboratory was very low. Bell agreed with the proposition that 'the Laboratory, so-called, of the Customs in London is not really a Laboratory in your sense of the word'. He thought that their work was all routine, and that only one or two of its analysts were trained (unlike his own staff). He was quite aware that most difficult scientific questions had been sent to his laboratory by the Customs for many years and seems to have felt that the Customs and their Laboratory were rather unscientific . None of these views were calculated to endear him to his opposite numbers in the Customs Laboratory, or indeed entirely fair.

The Committee on Civil Establishments was enquiring into the possibility of amalgamating the Inland Revenue and Customs departments as a whole, a move strongly supported by Bell's Revenue colleagues, by most of the trade witnesses, and perhaps most telling of all by Mr Gladstone, until 1885 Prime Minister, but drawing on his experience as Chancellor of the Exchequer. Gladstone had a very low opinion of the Customs Department in general, and the Laboratory in particular.[26] The Customs witnesses (who, oddly enough did not include the Principal of their Laboratory, which perhaps shows the lowly place he occupied in the estimate of the Board of Customs), all firmly rebutted the amalgamation proposal and the Commission finally rejected the idea, much to the satisfaction of the Customs Board.[27]

The antagonism shown by Dr Bell was reciprocated by the Board of Customs and the Laboratory. With an increasing work load, and obviously increasing experience and confidence, came a change in attitude towards the Inland Revenue

Laboratory, in a sense their mentor. As seen generous praise was given to the Inland Revenue Laboratory for its help in 1860, and this was repeated indeed in 1872 by the Customs Chairman, but leaving a feeling of being not quite so generous, although a considerable amount of work was being done. Phillips estimated in February 1869 that Customs work absorbed two Assistants full time and cost in addition to their salaries an extra £100 a year in chemicals and apparatus. No doubt was felt by another Customs witness in 1872 that the Customs department could analyse tea more quickly and with greater convenience than other departments.[28] The existence of bad feeling between the two Boards is revealed by several episodes between about 1860 and 1876. In one instance the Customs Board appear to have unilaterally altered an agreement with the Inland Revenue as to an allowance to be made on some imported wines. The Inland Revenue learnt of this decision with 'some surprise'. In another instance the Customs Board made an attempt to take over the lime and lemon juice work from Somerset House. Bell protested vigorously at this.[29] Such events cannot have improved relations and a feeling developed that the Customs laboratory could handle any scientific work without difficulty, and a further feeling of resentment that the Inland Revenue was apparently doing work they could quite well do themselves, and could indeed take over. This is shown well in an internal report of 1889 for the Customs by an internal committee concerning the drawback allowed for the export of 'offal' snuff (ie ground up waste tobacco), on which Mr Samuel, Principal of their Laboratory served. One of the recommendations made by this committee was that Customs should take over the analysis of the snuff from the Inland Revenue Laboratory, not only the routine samples, but also any disputed decisions, that is act as their own referees.[30]

In a memorandum presumably drawn up for this committee Samuel deals in a hostile manner with the part played by the Inland Revenue Laboratory in the discovery of frauds being perpetrated over offal snuff. In another memorandum apparently written 1890/91 Samuel rebutted Bell's evidence to the Civil Establishments Commission concerning the Customs laboratory. He remarks that the 'Customs Department' could have rebutted this evidence before the Commission's report was issued had they known the nature of it. This remark is rather curious when it is considered that several

Customs officers gave evidence, including the Board Chairman.[31]

The Inland Revenue and Customs laboratories continued to exist separately for some years after these events but the rejection of an amalgamation in 1890 did not the end the matter. Much of the evidence given to the Royal Commission was really implying that the Excise functions of the Inland Revenue department belonged with Customs, if not the rest of the Inland Revenue's functions, and that the two laboratories would undoubtedly be more efficient together. Both of these amalgamations eventually came about, as will be seen in later chapters, the second in only a few years, although the Customs Laboratory continued its independent path for a short while longer. It never occupied so prominent a position in government science as the Inland Revenue Laboratory, but served a very real purpose during its independent existence.

Part IV
The Government
Laboratory
1894–1911

CHAPTER 7

Formation of the Government Laboratory

THE *FERTILISERS AND FEEDING STUFFS ACT* of 1893 had given the Board of Agriculture power to appoint a Chief Agricultural Analyst (or a Chief Analyst) to analyse disputed samples under the Act. This Act was introduced to improve the quality of fertilisers and feeding stuffs sold and to try to prevent the adulteration which was endemic in the industry. A referee was needed to analyse a third sample if the local Agricultural Analyst and the retailer or manufacturer disagreed. Much concern was shown by the Chemical Manufacturers Association that the Chief Analyst should not be in any way connected with the trade, and should be 'above the possibility of suspicion'.[1] The Board wished to appoint the head of the Somerset House Laboratory for this purpose subject to his being a man of recognised standing in the chemical world. The Treasury were approached, and since Dr Bell was due to retire in the near future they appointed a Committee to enquire into the laboratory arrangements of both the Customs and Inland Revenue departments.

This Committee reported towards the end of 1893, and concluded that the chemical work of the Customs Department could not properly be transferred to the Inland Revenue Department (or indeed vice-versa), but that for the satisfactory performance of the chemical work of the government, and for uniformity in management and in the system of working, "a scientific man of high standing as consulting chemist or temporary Inspector-General of Revenue Laboratories" might be appointed to act as a referee in disputed questions. They recommended further that wines and spirits should continue to be tested on importation in the Custom House Laboratory, and that a competent chemist should be available there to advise the Board of Customs on the technical questions which constantly arose.[2] The Treasury, in a letter to the Board of Inland Revenue in March 1894 doubted whether such a referee would be satisfactory. They were also anxious that the Revenue Laboratory should be the Government Laboratory.

Other departments also needed to make decisions for which a knowledge of chemistry was required, and the Committee recognised "a natural law which governs all departmental development" which impelled heads of departments to prefer a number of small, inefficient technical establishments to one central and thoroughly equipped laboratory to which they could refer. This separative tendency had always been found difficult to resist and the Treasury was doubtful whether other departments would be content to submit their own chemical questions to chiefs of laboratories in either of the revenue departments with only the right of reference to a chemist of acknowledged eminence. The Treasury felt that a central laboratory was best conducive to efficiency and economy; so it was proposed to place the two laboratories under an "administrative chief", a single "chemist of eminence acknowledged by the scientific world", chosen not by the two departments but by the Treasury.

The Principal Chemist of the Government Laboratory would have an office in both branches of the laboratory and would be responsible to both Boards in their respective areas, either separately or jointly, according to the questions in hand. He would have a Deputy Principal at each branch, responsible to him alone; and the Principal Chemist would also receive references from other Government departments, to which he would report directly, keeping a register of the questions thus raised and making an annual return based on this to the Treasury. His salary would be met from the Treasury vote and he would not be allowed to take other appointments.[3] The "Government Laboratory", properly so-called, thus came into existence.

The term has Government Laboratory has two distinct but related meanings. It was originally used to describe the combined laboratory organization made up in 1894 from the Customs Laboratory at the London Custom House in Lower Thames Street (together with its outport chemical stations) and the Inland Revenue Laboratory at Somerset House in the Strand (together with its provincial testing stations), under a single, joint principal. It appeared on the brass name-plate which, in 1897, was affixed beside the front door of the new laboratory premises in Clement's Inn Passage and so also became the address of the headquarters premises of the organization located there. The Government Laboratory was thus the name of a laboratory organization during the period 1894–1911 and the name of the headquarters premises 1897–1964. The term is still used, loosely, at the present time. The title Government Chemist has been applied to the director of the Laboratory from 1911 to the present time, (see below). The Laboratory appeared as an independent entity in the Imperial Calendar (the list of Civil Service staff) for the first time in 1906; previously it had been listed with the Inland Revenue with names confined to those of the Inland Revenue Staff. The union between the Customs and Inland Revenue laboratories was always treated as one between equals: the Custom House laboratory was never regarded as an outstation. It was always designated the Custom House Branch, with the other at first called the Inland Revenue Branch.

The first Principal Chemist of the new Government Laboratory to be appointed was Professor Thomas Edward Thorpe, (said to have been known as "Tom" in his student days but Sir Edward after he was knighted in 1909). This appointment seems to have caused a great deal of intrigue and ill-feeling, and we are told that Thorpe was not the first person approached.[4] Be this as it may, Thorpe was sufficiently distinguished as a chemist to meet the chief criterion of the Treasury Committee report. At the time of his appointment, he was Professor of Chemistry in the Royal College of Science at South Kensington, where of course the Inland Revenue chemists studied. Thorpe was born in 1845, in Manchester, where he was educated, studying subsequently in Germany then the Mecca of all chemists, at Heidelberg and Bonn under the great chemist Bunsen. From Germany Thorpe returned to Owens College, Manchester, to work on photo-chemistry with Professor Roscoe, of whom he wrote a biography. After a period at the Andersonian Institution in Glasgow he was appointed to the chair of Chemistry at the Yorkshire College of Science at Leeds in 1874. Here he wrote his *Manual of Inorganic Chemistry*, first published in 1873. He also published in the 1870s his volumes of *Qualitative* and *Quantitative Chemical Analysis* (the latter with Pattison Muir), all of which went into at least one other edition. Just before his appointment to the Government Laboratory, in 1891, the first edition of his *Dictionary of Applied Chemistry* was

published.[5] This, in various editions, was for many years, (and in the form of its fourth edition still is to some extent), a major source of reference. His interests were very wide, he took part, before his appointment as Principal Chemist, in several magnetic surveys, and in four solar eclipse expeditions. The sea voyages connected with these may have led to, or stimulated, his love for the sea. He was certainly an expert and devoted yachtsman, and wrote several books on his voyages. He was elected a Fellow of the Royal Society in 1876.[6]

Thorpe was appointed from 5 March 1894 and took up his duties from that date. His Deputy at Somerset House was Richard Bannister (who had joined the Laboratory in 1862). Bannister was given an increased salary to recognise the increase in work in the Laboratory, and presumably to reflect also his increased responsibilities as well as his loss in 1891 of an allowance paid to him by the Indian Government. Bannister is said to have successfully opposed in evidence Dr Thorpe, (perhaps when Thorpe was, briefly, a Public Analyst), in a case in Londonderry in 1870 of alleged adulteration of tobacco with gum arabic, although this does not seem to have affected his subsequent relationship with Thorpe.[7] Bannister was undoubtedly a first class analyst, both he and Mr Lewin also of the Inland Revenue Laboratory are referred to specifically in the *Report* for the year ending 31 March 1884 in connection with a contested case involving the analysis of a soured sample of milk.[8]

The salaries of all of the professional staff at the two laboratories were raised in 1894, with the First and Second Class Analysts now being put into the charge of two Superintending Analysts, a new grade. The salaries were paid by the Board of Inland Revenue for the Somerset House staff, and by the Board of Customs for the Custom House staff.[9] This arrangement continued until 1911, with the formation of the Department of the Government Chemist, when a separate vote was granted to the Laboratory as a separate entity. It is perhaps surprising that the arrangement had lasted so long. The staff costs of the original Government Laboratory were divided between three votes, those of the Boards of Inland Revenue and Customs, as seen, with the Principal's salary being paid by the Treasury. This was obviously an untidy arrangement, and additionally, when in 1899 the Treasury

38   Sir Thomas Thorpe, Principal of the Laboratory, 1894–1909

agreed that the staff of the two branches could be regarded as interchangeable, some uneasiness became apparent in the attitude of the two Boards. This measure was seen by the Treasury as conducive to the development of a sense of laboratory unity and to the career development and professional standing of the staff of each branch, but the Boards did not like the prospect of an officer serving one board, but being paid by another.

In a letter dated 29 December 1903 the Treasury, responding to a representation by one of the Boards on this matter, agreed that there appeared to be some difficulty in considering the Customs Branch apart from the Inland Revenue Branch. It commented that little or no information was available as to the respective work of the two establishments, nor were the total costs and staff for either establishment separately shown. The Treasury accordingly requested estimates for both. Despite this payment of the staff appears to have continued to be divided between the two main departments until the formation of the Department of the Government Chemist in 1911. The last Inland Revenue staff list to carry the names of the Laboratory staff was dated 1 April 1908 and was the last staff list before the formation a year later of the new Board of Customs and Excise.[10]

8

A New Home

THE INLAND REVENUE BRANCH OF DR Thorpe's new Laboratory was still housed in Somerset House where it had been since 1859. When it arrived there first it occupied two rooms, it expanded to four in 1873, eight in 1884, and eventually to 20 in 1894. Even this seems to have been inadequate, and conditions in these laboratories do not seem to have always been satisfactory as is likely to be the case when rooms are adapted to a purpose for which they were not built. At about this time conditions in the Somerset House laboratories were described as 'wretched ventilation and miserable accommodation of many of the rooms' and went on to say that work seemed to proceed in an atmosphere often thick with choking vapours and reeking with nauseous gases and with the thermometer registering 80 to 90 degrees Fahrenheit. Detrimental to the health of the staff indeed as the source suggested. Perhaps at least partly because of this instead of trying to take over yet more rooms in Somerset House, or moving to another existing building and adapting it, as had been the situation since the foundation of the Inland Revenue laboratory, it was decided to design and have built a completely new suite of laboratories.[1]

The Treasury agreed to this, and the site chosen was in Clements Inn, one of the old Inns of Chancery. It was located immediately to the north of St Clement Dane Church in the Strand, in an area which until the close of the nineteenth century was characterised by narrow and dilapidated streets, just to the west of the City boundary. Within twenty years around the turn of the nineteenth century the whole area was demolished and reconstructed, and is bounded today by Kingsway, the Aldwych, and the Royal Courts of Justice. The building itself was on the eastern side of Clement's Inn Passage, which ran from the point at which Clare Market joined Houghton Street to the entrance to Clement's Inn.

This latter ran south to the Strand. It was officially known as 13 Clement's Inn Passage, although the number never appeared on the building.

The building itself was largely designed by Thorpe himself, and as he said in 1909 in his retirement speech, 'had been planned as all labotatories should be, from the inside outwards'. A contemporary noted that the 'exterior appearance' had been 'sacrificed to the elaboration of the interior' because the Treasury vote of £25,000 for the new building proved inadequate. Just previously to his appointment Thorpe had been involved in planning new laboratories at South Kensington. The building took some two years to complete and was not ready when expected, perhaps needless to say. It was occupied in early October 1897, the Inland Revenue issuing instructions that samples were to be sent to the new address on and after 4 October. Thorpe said with some satisfaction in his 1898 *Report* that 'the new laboratory has not only conduced to the comfort and healthier condition of the staff but also to the expediency, efficiency, and economy with which the work can be conducted'.[2] It was officially opened with considerable ceremony on 1 October 1897. On the same day a Conversazione was held on the premises. The events on this latter occasion were nothing if not eclectic. Proceedings were opened by the Whitehall Orchestra (conductor Mr Arthur Grudge) playing a selection of music in the Main Laboratory on the first floor, commencing with Wagner's march *"Under the Double Eagle"* and including the waltz *Amoretten tanze* by Gung'1. Other musical selections by staff and friends incuded harp and flute solos and a vocal quartet performance of *Annie Laurie*. Scientific exhibits, not all of them related to chemistry, were set out in many of the laboratories. Contributions came not only from the Laboratory but also from customer government departments, instrument manufacturers and suppliers of chemical reagents. The Chemical Society exhibited portraits in carbon of past presidents of the society and the Society of Chemical Industry portraits of Hofmann and Kekule. Professor Thorpe showed calculating machines and photographs of the Dutch waterways, Yorkshire, North Devon and of the total solar eclipse of 16 April 1893 taken in Senegal, whilst Mrs Thorpe showed samples of lace from the Azores. Other exhibits included the fluoroscope, the phonograph, an acetylene generator and lamp and apparatus "for

39   'Government Laboratory building in Clement's Inn Passage, occupied 1897–1964

The Government Laboratory

CONVERSAZIONE

ON THE OCCASION OF THE

OPENING OF THE NEW BUILDINGS

CLEMENT'S INN PASSAGE, STRAND

OCTOBER 1st, 1897

REFRESHMENTS

Are served in the CROWN CONTRACTS LABORATORIES (Rooms 14, 15 and 16),

ON THE GROUND FLOOR.

THE WHITEHALL ORCHESTRA will play the following Selection of Music in the MAIN LABORATORY—

(*Conductor*—MR. ARTHUR CRUDGE.)

1.	MARCH	"Unter dem doppel-Adler"	*Wagner.*
2.	SALONSTÜCK	"Das erste Hertz Klopfen"	*Eilenberg.*
3.	MOMENT MUSICAL		*Schubert.*
4.	SELECTION	"The Geisha"	*Jones.*
5.	WALTZ	"Amoretten Tanze"	*Gung'l.*
6.	MINUET IN A		*Matras.*
7.	TÄNZE	"Polnisch"	*Sarakowski.*
8.	SELECTION	"Romeo and Juliet"	*Gounod.*

During the evening the following Musical Selections will also be given in the MAIN LABORATORY by Members of the Laboratory Staff and Friends—

1. SONG . . . . "The Last Watch" . . . . *Pinsuti.*
   MR. E. G. B. BULLOCK.

2. HARP SOLO . . . . "Romance" . . . . *Thomas.*
   MISS AMALIE BROUSSON.

3. SONG . . . . "An Andalusian Maid" . . . . *Philp.*
   MISS SARAH DUGDALE.

4. FLUTE SOLO . . . . "Espagnola" . . . . *Maunder.*
   MR. SPENCER WEST.

5. VOCAL QUARTET . . . . . "Annie Laurie" . . . . *Cantor.*
   MESSRS. LATHAM, KING, KINGDON, AND DICKINSON.

At 9.30 p.m., in the MAIN LABORATORY, Mr. J. CONNAH, B.Sc., will exhibit and describe a series of Limelight Views, illustrating "Dr. Nansen's Voyage in the *Fram*," reproduced from the Original Photographs, and lent by Messrs. Newton & Co.

3

Ground Floor.

REFERENCE SAMPLE LABORATORY
(ROOM 19).

1. *Exhibited by Messrs. Negretti and Zambra.*
   Meteorological Instruments.

2. *Exhibited by the General Post Office Telegraph Department. (By permission of the Postmaster-General.)*
   Working Circuits.

3. *Exhibited by the Council of King's College, London.*
   Early Telegraphic, Musical, Optical, and Other Instruments from the Wheatstone Collection.

4. *Exhibited by the Swan Electric Engraving Company.*
   Photogravures and Swantypes.

RESEARCH LABORATORY (ROOM 23).

5. *Exhibited by Messrs. Baird and Tatlock.*
   Chemical and Physical Apparatus.
   Specimen of Bismuth—Alloy of Potassium and Sodium.
   Methods of Glass Blowing.

6. *Exhibited by the Government Laboratory.*
   Apparatus in Use in the Government Laboratory.

7. *Exhibited by Mr. L. Oertling.*
   Chemical and other Balances.

6

16. *Exhibited by Mr. Horace Seymour, Deputy Master of the Mint.*
    Medal struck in Gold, Silver, and Bronze, to Commemorate the Sixtieth Year of the Reign of Her Majesty the Queen.

17. *Exhibited by Messrs. Johnson and Mathey.*
    Rare Metals and their Salts.

SUPERINTENDING ANALYST'S ROOM, II.
(ROOM 31).

18. *Exhibited by Mrs. T. E. Thorpe.*
    Examples of Lace and Needlework, made by the Women of Fayal, Azores.
    Carbon Print : Summit of Pico, Azores.

19. *Exhibited by Professor Thorpe.*
    Photographs, printed in Platinotype, taken during a Cruise of the Steam Launch *Chilwa* in Dutch Waterways.
    Views in Yorkshire, North Devon, and at Champery, Switzerland.

20. *Exhibited by Messrs. Hopkin and Williams.*
    Specimens of Chemical Products.

TOBACCO LABORATORY, No. 2 (ROOM 30).

Apparatus for the Incineration of Tobacco and for determination of moisture.

showing the Rontgen Rays". Refreshments were served in the Crown Contracts Laboratories and at 9.30pm in the main laboratory, Mr Connah (a Second Class Analyst) showed some limelight views illustrating Dr Nansen's voyage in the *Fram*.[3]

The Laboratory itself was a large building with three floors, of some 7900 square feet, faced with red bricks, with string courses, and corners and window frames of Portland stone. Inside it was faced with glazed bricks, white above and brownish red below. This was partly to reduce fire risk. The entrance to the building, at the corner of Clement's Inn Passage and Grange Court, was up three stone steps and displayed to the right of the door a brass plate (which is now displayed in the Library in the LGC building at Teddington) with the two words "Government Laboratory". The plate was polished regularly and continued to identify the building for more than 60 years, finally being replaced in 1959. Inside, the Principal's room was located just along the corridor from the entrance and near the foot of the stone staircase which followed the inside corner of the building up to the main laboratory on the first floor. There was a small reception area for visitors, and a very small room on the left of the main entrance provided accommodation for an attendant or door-man. Initially the doorman on duty wore during the day (and later on special occasions only) a handsome uniform red frockcoat with black and gold trimmings and a black hat, also trimmed with gold. Also inside the entrance on the left was a small fitted wooden writing desk on which was laid the attendance signing-on book, withdrawn promptly at a critical time so that late arrivals had to go to the office of the chief clerk to register their arrival below the line which was drawn at the same time. On the other side was a similar book for ancillary staff. All staff below the rank of Superintending Analyst had to sign on.[4] Next to the left hand desk was a glazed, doorless wooden booth also with desk, above which was the Laboratory's single, candlestick telephone. The door attendant (or a messenger) could summon the principal or other senior staff by an electric bell and they had in their rooms a similar means of communicating in the other direction. In the basement boilerhouse the boilerman had communication with the booth by means of a speaking tube; communication was initiated by blowing down the tube, activating a whistle at the far end. Other rooms in the

41 Brass plate from Clement's Inn building

40 Conversazione programme, 1897

basement on each side of a central corridor included the porters and charwomen's rooms, the stores, and reagent and calibration rooms. In the reagent room standard solutions were calibrated for use in volumetric analysis.

On the ground floor in addition to those mentioned were also situated the Deputy Principal's room, laboratories (one of which, the Reference Laboratory, was still known as that, or sometimes the "Butter Room" from the original work carried out there until the Laboratory moved from the building in 1964), and the galleried library. The first floor consisted mainly of the "Main Laboratory" a lofty room some 50 feet high 43 feet wide and 49 feet long, rising to roof level, with a dormer lantern window. Superintending Analysts' rooms were also on this floor. The second floor was

42   Announcement of the
Laboratory change of address, 1897

GENERAL ORDER.

No. 14
1897.

Inland Revenue, Somerset House,
London, W.C.
27th September 1897.

REMOVAL OF THE LABORATORY.

Ordered that the attention of the Service be directed to the following matters :—

NEW ADDRESS.

On and after October 4th 1897 all samples and communications intended for the Laboratory should be sent to

The PRINCIPAL,
Government Laboratory,
Clement's Inn Passage,
London, W.C.

Labels and envelopes so addressed can now be obtained from the Controller of Stamps and Stores.

SEALING OF SAMPLES.

The arrangements at the new Laboratory will not permit of unsealed samples being enclosed in sealed parcels or boxes, but in future each sample must be separately and securely sealed with an Official Seal bearing a distinctive number. Such Seals, if not already supplied, should be obtained from the Controller of Stamps and Stores.

Any number of samples may be enclosed in one box as heretofore, provided they are individually sealed in accordance with these directions.

ADVICE OF SAMPLES.

The Board's instructions relative to advising the Principal of the Laboratory of the despatch of samples must be strictly observed, as great inconvenience and delay are caused at the Laboratory by the arrival of non-advised samples.

FORM 77.

After the date mentioned in the first section of this Order, Supervisors must not return Form 77 when marked " practically agrees " to the Laboratory, but after noting the result of analysis should forward the form direct to the Chief Inspector.

By the Board,
W. B. HEBERDEN.

14620—4000—9/97   Wt 11866   D & S   16

divided into two parts by the 'Main Laboratory', and included the Museum, Typewriting, and Photographic rooms.[5] Some of the rooms continued to be used in the same way for the next 67 years until the building closed.[6]

In design, the laboratories (the main laboratory in particular) clearly owed something to those in Dr Thorpe's old department, the Yorkshire College of Science. Here there is still a Thorpe Laboratory in what is now the Chemistry Department of the University of Leeds. Many improvements were incorporated in the new Clement's Inn laboratories though and they became in turn a model for many of those built later in other chemistry departments.

The lighting was by electricity from the 100 volt supply of the Strand Corporation and heating was by steam. A very efficient ventilation system for the fume cupboard extract was built into the structure of the building, worked by a fan operated by "a silent one-horse engine". The water supply was from the New River Company's high pressure main and there was a carbonic anhydride refrigerating machine for making ice. This was also used for cooling the water supply to the main laboratory in the summer months, principally for alcohol distillations. The refrigeration system was used to cool the 'sample room', to help preserve samples awaiting analysis. Bench tops were 1½″ mahogany 37″ above floor height with whiteware sinks and fitted with glass reagent shelves on a gun-metal framework. The fume cupboards followed the design used at the Yorkshire College. One fume cupboard together with examples of the benches, steam heated copper sand tray (later used as a hot water plate), with special steam heated drying oven, condenser and large earthenware receiver of distilled water were salvaged from the building before its eventual demolition. These, together with benches and wall tiling forms a faithful impression of one of the smaller laboratory rooms in a permanent exhibit now in the Science Museum, (see plate 45). The steam used to heat the drying ovens was condensed to form the hot water for the water baths and the hot water supply.

The new Laboratory, as the new principal national analytical laboratory, seems to have aroused a great deal of interest in the scientific community. Dr Thorpe, in his report for 1897–98 noted that "the arrangements and fittings of the laboratory have met with the approbations of the many eminent men of science, both of this country and abroad,

43   Steam oven, connected steam bath and distilled water container, Clement's Inn building

who have inspected them".[7] A new factor which entered the management of the laboratory in the move to Clement's Inn Passage was the need for support staff to run the building, a matter handled previously by the Inland Revenue (and which at the Custom House continued to be handled by the Board of Customs). In May 1897, before the move from Somerset House, Dr Thorpe asked for authority for an engineer (or "engine-man", as he was termed) to take charge of the heating, refrigeration, ventilation and lighting arrangements in the new building. The engineer worked from 7.00 in the summer and received 10d (4p) per hour extra if he worked more than 54 hours per week. To help him a Stoker was also employed. Thorpe also asked for a bookkeeper to attend at the door and the waiting room, to attend calls on the telephone and to keep a register of callers, and a storekeeper and night watchman for the new premises. The night watchman rested in the waiting room between his tours of inspection of the building. He recorded his tours with a portable

44 Fume cupboard in Clement's Inn building

recording clock into which he inserted keys chained to small metal boxes fixed to the wall in the corridors. It was said that one nightwatchman lost his job because he developed a weakness for dropping his clock down the stone staircases. The watchman was not really allowed to take off Bank Holidays nor extra days at Christmas and Easter, but was apparently allowed to absent himself provided no one questioned the matter. He was issued with instructions as to what to do in the case of fire.[8]

To keep the 37 rooms, wide corridors and about 120 stone stairs clean needed three charwomen who worked for an hour or so before and after official hours. The Board of Inland Revenue agreed to provide the engine-man and his substitute, when needed. His duties also included carrying coals removing cinders, maintaining the toilet facilities in a clean condition, wiping the outside brick work, and sweeping outside and inside and "making himself generally useful". The porters' duties included lighting the gas burners under water-baths on arrival at 9 o'clock, providing ice in the

45 Typical Clement's Inn laboratory as reconstructed in the Science Museum, London, from benches etc. taken from the building before demolition

laboratories, dusting bottles of chemicals and shelves, washing and drying butter and tobacco sample bottles, attending to boxes and packages received from railway and other carriers and distributing samples. Detailed schedules were drawn up for each individual, commencing at 8.15 and ending at 5.30 with half an hour (1.00–1.30) for "dinner" each weekday and a shorter schedule for Saturday. There were also Clerical staff, the 'Typewriters', Boy Messengers and Boy Copyists and up to four clerks. A considerable amount of time was spent in discussing the type and calibre of support staff needed. The Engineer needed to be a skilled mechanic for example but the Stoker (or coal porter) would need only to be strong and active, (an ex-Royal Marine was employed). The salaries were also much discussed, the Engineer received, in acknowledgement of his skill, 40 shillings (£2) per week, the Stoker 20 shillings, and the Door Keepers 22 shillings (£1.10) with 5 shillings (25p) allowance for Sunday duty. The Store Keeper received £2.10s (£2.50), the Charwomen 10/6 (53p), with the senior woman receiving a shilling more and the Boy Messenger 6 shillings rising to 15 shillings (30p rising to 75p).

In common with the rest of the Civil Service, all staff worked on Saturday morning. The hours for general staff were increased from six to seven per day in 1901, to 9.30 to 4.30 from April to September and 10.00 to 5.00 in the winter months. By February 1911 hours were 9.30–4.30 Monday to Friday, and 9.30 to 1.00 on Saturday at Clement's Inn, and 9.00–4.00 and 9.00 to 12.30 respectively at the Custom House Laboratory. The hours were strictly controlled. In 1911 a note was circulated to staff, drawing their attention to the fact that in the morning grace was given to 'meet occasional and unavoidable delays' but that this was not to be taken advantage of systematically. The proper time for arrival was 9.30 and not 9.40. The actual time of leaving the Laboratory also had to be recorded and initialled.[9]

The work of the Laboratory

THE LABORATORY WHICH DR THORPE HAD inherited was a busy concern. The number of samples reported in 1894 was 76513, and had increased to 176935 by 1909 just before he retired.[1] The Inland Revenue Laboratory had a variety of customers by 1894. As well as the routine revenue samples which were the staple of the Laboratory and samples received under the *Sale of Foods and Drugs Act* of 1875 and the *Fertilisers And Feeding Stuffs Act* of 1893, the Laboratory examined samples for the India Office, Trinity House, the Post Office, and the Board of Trade, as well as the Irish Office of Works. In addition they had been examining food and other victualling supplies for the Admiralty since 1891, and had more recently taken on the examination of certain items from the Army Clothing Department, and of drugs, chemicals and pharmaceutical preparations for the Army Medical Department. In August 1896 the General Inspector of Rations for the Army asked if the Laboratory could also help in the examination of food supplies. Dr Thorpe referred this question to the Treasury, (through the Board of Inland Revenue) saying that more and more work was being done in the Government Laboratory from departments other than the two branches of the Revenue Service, and that he wished to encourage this. He felt however that requests for the services of the Laboratory should be channelled through the Treasury. The Treasury agreed to this proposal, but once such agreement was obtained, samples should be sent directly to the Laboratory. They also requested that Thorpe's Annual Reports should in future be directed to the Treasury, not separately to the Inland Revenue Board.[2] A Report on the Customs Laboratory continued to appear separately, (the first combined Report appeared in 1897). This system had the dual effect of ensuring that the Laboratory's financial vote recognized the areas of work in which it was engaged and of maintaining some control over any tendency to proliferate small laboratory services in other departments. It also allowed the Laboratory to develop its expertise in analysis in several areas in a co-ordinated fashion and to make available the benefit of experience which it gained in working for one department to other departments who subsequently sought similar advice or services. Work on drugs and chemicals for the Crown Agents for the Colonies, for example, followed in 1910 and in 1897 the Director of Victualling for the Admiralty asked if rum could safely be packed in tins. This was not encouraged on account of the possible corrosive properties of the 'compound ethers' (esters) which rum contained.

The variety of samples examined was large, as may be gathered from the above figures. A large part of the work was of course still the analysis of routine Revenue samples. Interest in the wellbeing of the tobacco trade continued for example. This was manifested in the 1898 Report, in a calculation that the amount of tobacco consumed per head in Great Britain and Ireland had increased from thirteen and three quarter ounces in 1841 to just over one and three quarter pounds in 1898, the annual increase for 1897–1898 being nearly one ounce. This was reported with great satisfaction, with the comment that 'the consumption has increased in greater ratio than the population during the past ten years by more than five and a half ounces per head[3]. Not a great deal of adulterated tobacco was now being detected, Thorpe commenting in 1897 that 'there is reason to believe that tobacco was never more free from adulteration than at the present time, a measure of the success of the work of the Laboratory.[4] By 1899 the tobacco work had been completely transferred from the old Somerset House Laboratories to Clement's Inn Passage and by 1906 still accounted for nearly half of the total samples examined at the latter laboratory, most of them being received for moisture determination, to estimate the duty payable.

Much work was still being done on beer and other alcoholic beverages too. Very large numbers of wort samples were being received in the period under review, (wort is the malt before fermentation) for checking the declaration of the strength (the specific gravity), declared by the brewer, this figure determining the duty payable. Finished beer samples were also examined for adulteration, (dilution with water), still a surprisingly prevalent practice and still largely in

46 Refrigeration unit in the Clement's Inn building for cooling water throughout the building

London as it had been in the past.[5] Increasing quantities of exotic herb beers were also examined, 1133 of them by 1907. Herb beers were not liable to revenue control unless they contained more than 2% of proof spirit, and were frequently sold as temperance drinks. Samples included ginger beer, horehound beer, sarsaparilla beer, nettle beer, barm beer, hop ale and burdock and dandelion stout.[6] Over 10% of these (in 1907) were found to exceed the maximum spirit strength as had also been the case in the nineteenth century. Not perhaps entirely what the Temperance movement had in mind. At this date some 5% of the samples of beer and brewing materials examined were found to contain arsenic in excess of the limits laid down by the Royal Commission on

Arsenical Poisoning, (see below). The alcohol work continued much as before, except for minor changes, such as the new departure caused by the sanctioning of the use of pure duty free alcohol (or specially denatured spirit) in medical schools and scientific departments of universities by the *Finance Act* of 1902. Applications for this privilege were referred to the Laboratory for report. Methods were issued, showing the Laboratory procedure for determining the quantity of alcohol in medical tinctures, flavours, essences and perfumes for example. Another minor departure from previous practice was in the Board of Inland Revenue issuing details in 1899 of the tests used in the Laboratory for the evaluation of the naphthas used in methylation. These included bromine decolorisation, a methyl orange alkalinity test, an acetone reaction and the estimation of esters, needed for correcting the methyl alcohol content. The latter was determined on the wood spirit after reaction with iodine and red phosphorus and the measurement of the methyl iodide formed. This is in marked contrast from chromatographic methods which could be used today.[7]

The increasing amounts of new work caused by various Acts and Regulations did not in fact add very large numbers of samples to the annual returns, (excluding food samples, about 2000 in 1909, out of nearly 124000) but they certainly added to the variety; and undoubtedly added to the amount of work – unusual samples take up disproportionate amount of time. A considerable amount of time was spent on finding the active (poisonous) compound in the Yew tree, a problem referred to the Laboratory by the Board of Agriculture after a number of deaths of cattle caused by eating various parts of the tree. This work resulted in the identification of taxine, and was published in a paper in the *Journal* of the Chemical Society[8]. Other work was more routine, the determination of the gold content of badges for the Army Clothing Department for example, which was found to be two thirds of that called for, and with less than a quarter of the amount of silver[9]. Some of the work the Laboratory was called upon to do foreshadowed later developments. For example in the *Report* for 1900, a sample from the Post Office was examined which had apparently spontaneously burst into flames. It was found that it was finely divided zinc, a substance which is liable to rapid oxidation in the presence of air. The recommendation of the Laboratory was that such substances should be enclosed in air tight cases, not as here, merely in brown paper[10]. The question of the transport of dangerous goods is now a subject occupying much international time and effort.

As indicated above a considerable amount of effort was now being put into the analysis of foods. Each annual report contained a table describing the samples submitted under the *Food and Drugs Act* of 1875, which had caused a great deal of ill feeling with the Public Analysts, as has been described earlier, and subsequently under the 1899 Act. Those samples usually showed adulteration, with water in the case of milk for example, or frequently, in the case of butter, with boric acid. Some of the butter samples were taken under a survey of imported butters begun in 1896. This survey demonstrates an early example of international co-operation, with Dutch and Norwegian chemists visiting the Laboratory to be shown the analytical methods used, and samples of 'abnormal' butters being exchanged for examination in Holland and Great Britain. Methods were later exchanged with chemists in Denmark and Germany[11]. As a result of this exercise the number of adulterated samples fell rapidly. Partly as a result of this exercise Dutch butter was sampled only on an occasional basis, since it was produced under strict government controls and marked with official labels. This did not prevent adulterated butter being imported with false 'control' labels though. Considerable problems were caused by various butter substitutes with exotic names such as Cottolene (made from Cotton seed oil), Stearine, Lardine and Cotosuet. Preventing this turned on the definition and labelling of 'margarine' as a butter substitute. It had been originally defined in the *Margarine Act* of 1887, but it was finally necessary to pass a new Act to solve the problem, the *Butter and Margarine Act* of 1907.[12]

Food commodities were also examined for the War Office, to test their compliance with specification. These consisted fairly regularly of the whole range of goods being bought to feed the army, including calves foot jelly, presumably for the hospitals. Samples were received similarly from the Admiralty. This exercise was reported by Thorpe as undoubtedly contributing to saving money, as he said in his Report for 1901, 'There can be no doubt that the systematic examination of such articles tends to raise the average quality, and at the same time to lower the cost'.[13]

47   Crown Contracts laboratory, Clement's Inn, 1902

Very few samples were received under the *Fertilisers and Feeding Stuffs Act*, the first one not until 1894 and only four by 1897 and in neither case were they mentioned in the summary sample table of the *Report*. The samples that were received were usually adulteration cases. Part of the work under the Act was the approval or otherwise of chemists whom the local authorities wished to appoint as Agricultural Analysts for their district. Many early applications were turned down due to lack of experience or qualifications or both. The usual qualifications accepted were Fellowship of either the Chemical Society or of the Institute of Chemistry. Some of those appointed turned out to be highly unsuitable despite this care. In one case which went on for nearly seven years the Agricultural Analyst concerned issued testimonials on behalf of the manufacturers and promoted the use of a 'Patent Silicate Manure' which consisted mostly of ground flint. He later branched out into a similar substance known as

Normanby Patent Slag Manure, and into offering to provide 'an excellent trade certificate as to the value, purity, harmlessness etc. of your . . . [left blank] for the hair' for a fee of one guinea (£1.05), for which he also strongly recommended the preparation. There appeared to be strong legal reasons as to why he could not be dismissed, although everyone except apparently the local authority concerned felt that he should be stopped. He finally resigned seven years after the furore began.[14]

At this time, in April 1899, the Laboratory took over the routine analysis of London drinking water samples for the Local Government Board. This work had previously been carried out for the Board by Sir Edward Frankland, an expert on water analysis, and on his death in 1899 a special laboratory was fitted out and Sir Edward's assistant installed as 'Water Analyst'. Cholmondeley commented sourly that 'There was... no necessity to have a specialist for that particular branch of analysis, since his duties are well within the capababilities of the existing Inland Revenue staff.[15] This was undoubtedly true given the involvement of the Laboratory with water analysis for many years. The 'Water Analyst', one W T Burgess, who never became a pensionable Civil Servant, appeared on the staff list from 1900 onwards until he apparently retired some time in 1907–1908, when the work was absorbed into the main stream of the Laboratory work. He was paid at approximately the same rate as a Second Class Analyst.

Perhaps as a result of work for the Admiralty and the War Office the Laboratory was asked, as possible analysts, to examine in February 1901 a very long list of food supplies to be used by the National Antarctic Expedition. Dr Thorpe replied that he believed analysis was necessary in only a few cases, and that simple physical examination by the Admiralty Food Inspector would suffice in the others. He offered to act as food analyst to the Expedition, citing the considerable experience in examining food stuffs for the army and navy, provided the Treasury agreed. This agreement was speedily obtained by the Admiralty, and Captain Scott, commander of the Expedition, himself wrote to Thorpe asking if he would analyse samples of Pemmican (concentrated dried meat). This was not one of the items Thorpe had advised analysing, but Scott explained that Pemmican was particularly important, since it would be used by all of his travelling parties. It was

National Antarctic Expedition,
University Building,
Burlington Gardens, W.

19th. April 1901.

Dear Sir,

I should have acknowledged the receipt of your letter of April 16th. before. The analysis of the pemmican is certainly disappointing. I have informed the N.A.O. Food Co. of their failure, which, however, I think arises from ignorance rather than carelessness as they are certainly anxious to produce the right article.

With reference to the other samples, I think there is no reason for our not accepting the Copenhagen or American tenders on account of the proportion of fat. I am, therefore, grateful to you for proceeding with the examination of the four samples.

Since writing to you last, I have put Dr. Koettlitz, who will be doctor to the Expedition, in possession of the facts, as far as I know them, and have directed him to devote his attention to the matter. I suggested that he should call on you at the Government Laboratory. I regret that my time is so occupied that I cannot do so myself.

Thanking you for your letter,

Yours very truly,

Commander, R.N.

Dr. T.E. Thorpe,
Government Laboratory,
Clement's Inn Passage,
Strand, W.C.

48   Letter to Thorpe from Captain Robert Scott, 1901 (PRO DSIR 26/256)

supposed to be highly nutritious, and to contain 60% ox lard, with virtually no moisture. It was in fact found to contain 38% of starch. 20% fat and excessive moisture, (nearly 8%), as well as pieces of broken metal. Later samples from different manufacturers were found to be rather more satisfactory, and Scott wrote expressing his thanks. Further samples of other concentrated foods were analysed in 1902.[16]

The work of the Customs Laboratory continued much as it had before the amalgamation, with a certain amount of harmonisation of method.[17] Until the *Report* for 1899 a wholly separate report was made on the Customs work, in 1899 the first *Report* for the Government Laboratory, the work for each Branch was reported on under the same cover, but separate headings. This continued until 1911, the only change being the combination of customs and excise following the formation of the new Board of Customs and Excise in 1909. The actual Customs work (i.e. work in connection with samples of foreign produce) was done partly in Clement's Inn Passage, partly in the Custom House Laboratory, and partly at the testing stations at the docks and out stations. Custom House work was defined in 1899 (and subsequently) as consisting 'mainly in the determination of the dutiable value of samples of imported wines, spirits and beers; in addition a great variety of other products are examined for alcohol or for the evidence that alcohol has been used in their preparation.'[18] By the terms of the arrangement setting up the Government Laboratory the Custom House laboratory was in charge of a 'Deputy Principal'. For the first seven years this was William Cobden Samuel, who had been in charge there from 1890. He retired in January 1901 and was replaced by Mr James Connah, a First Class Analyst from the Inland Revenue branch. He was promoted to Superintendant Analyst in February 1904 and remained in charge at the Custom House until 1919 when he was made Deputy to the then Principal.[19]

The workload varied, as did the work of the Clements Inn Laboratory, but was perhaps more susceptible than that Laboratory to changes in the law. In 1901 for example the Budget Act of that year imposed duties on sugar, molasses, glucose and saccharin, (saccharin had been introduced commercially as a sweetener in 1900), including manufactured articles containing sugar, and the work load of the Custom branch doubled in consequence. To avoid delays in reporting the results, two new laboratories were fitted up, and eight additional temporary assistants appointed to deal with this work. The rate of examination of other samples was reduced where possible, for example where variations in composition had previously been found to be small. At the same time the polariscope had been legally sanctioned for testing raw and refined sugars, and several Customs officers were instructed

49   James Connah, Acting
Government Chemist, 1920–1921

in sugar examination at a number of ports (including London in Lower Thames Street, separate from the Custom House, Liverpool and Glasgow). A selection of samples tested by these officers were re-tested at the Custom House, and found to have been accurately analysed. The instruction of Customs officers for this work continued in subsequent years. The molasses samples caused some difficulty, and considerable work had to be done on these to determine suitable rates of duty.[20] The smuggling of saccharin, the duty on which was £1 per pound, was not apparently unusual. In the same 1902 *Report* a sudden decline in the number of tea and tea dust samples sent for examination was reported. This was attributed in part to the action of the Russian government in buying large quantities of tea dust for compression into 'brick' tea for their troops.[21] Tea, 90% of which was imported from India, had been an important part of the Customs Laboratory work since 1875 (as described above) and continued to be examined at the ports by Tea Inspectors appointed under the *Sale of Food and Drugs Act* of 1875. Samples of doubtful purity were sent to the Custom House Laboratory for further testing. Those rejected for home comsumption were allowed to be exported or, after rendering inedible with asafoetida and lime, used duty free for the manufacture of caffeine. [22]

10 Committee work

THE INLAND REVENUE LABORATORY HAD long been increasingly recognised as a centre of analytical expertise and knowledge. As has been seen it had been appointed official referee under the Sale of Food and Drugs Act of 1875 and as early as 1872 Dr Bell, then Principal of the Laboratory, had given evidence to the Select Committee on the *Adulteration of Food Act*[1]. Work on committees, and work by the laboratory for committees became increasingly common after the appointment of Thorpe however. This was both in recognition of his own eminence, and because the appointment of select committees and Royal Commissions to investigate adulteration of commodities, or the toxic effects of others, was increasingly resorted to in the late nineteenth and early twentieth century. These committees reflected an increasing governmental concern with the effect of chemicals on the environment or the population. At one point indeed, in 1901, no less than four reports were issued in which Thorpe and the Laboratory had played a part[2].

Two matters in which Thorpe himself was particularly interested were lead in pottery, and phosphorus in matches. Both lead and phosphorus form highly toxic compounds and the terrible jaw necrosis (phossy jaw) suffered by many people in the match industry was well known. Thorpe and a co-worker (A.E.H. Tutton) had done a considerable amount of work on the oxides of phosphorus at South Kensington in the 1880s and had then isolated white phosphorous oxide for the first time. It was later shown that this compound was the cause of the jaw necrosis when it was formed in match making. Its formation could be avoided by the use of red phosphorus instead of yellow (also known as white).[3] Despite this discovery in the early 1890s, it was considered necessary for the Home Office to appoint a committee in 1899 to investigate the use of phosphorus in the match industry.

Thorpe together with two others was appointed to this committee and undertook a series of visits to British and foreign match manufacturers. In addition work was undertaken at the Laboratory on the nature of the 'fume' from phosphorus and its effect on teeth. The committee recommended that the use of yellow phosphorus in the manufacture of matches be subject to very strict conditions.[4] As a result of these recommendations work was subsequently carried out in the Laboratory into the relative character and ignition temperatures of matches made with yellow phosphorus and those made without phosphorus, or with less toxic phosphorus compounds. 'Strike anywhere' matches were the chief problem: such matches were commonly in use and very popular before 'safety matches' became common (these use non toxic red phosphorus) and difficult to make without yellow phosphorus. The use of phosphorus sesquisulphide eventually solved this problem. Yellow phosphorus was subsequently forbidden, and a paper was later published by Thorpe on the detection of yellow phosphorus in matches[5].

50 Main Laboratory, Clement's Inn, 1902

The other subject in which Thorpe had long been interested was the question of soluble lead in pottery glaze, and in the years 1899–1902 he and his staff spent a great deal of time on this matter. It seems to have been largely at Thorpe's instigation that a Home Office committee, consisting of himself and Professor Thomas Oliver (a medical man), was appointed. Regulations as to the employment and medical examination of workers in the pottery industry existed, but cases of lead poisoning were still occurring in the industry and in the public at large. This was due to the use of soluble compounds of lead in glazes. Thorpe and his colleague visited a great many manufacturers of pottery, both in the United Kingdom and on the Continent, and a comprehensive series of investigations were carried out in the Government Laboratory. The results of this work was issued in the form of a series of Command Papers, the conclusions being that lead free pottery was perfectly practicable in most cases. In the few instances where it was not practicable to stop the practice then the use of insoluble lead silicates in the glaze would prevent lead poisoning. This conclusion was bitterly resisted by some members of the pottery industry but without effect, since the Home Secretary adopted the recommendations and took steps, which were ultimately successful, to implement them. In the first of the Reports Thorpe, rather unusually for that time, acknowledged the help of Charles Simmonds, a Second Class Analyst, and H V Templeman and A More, Assistants, members of his staff at the Laboratory.[6]

Other matters which much exercised the Laboratory at this time were enquiries into the questions of preservatives and colouring matters in food[7], (something which is very much a contemporary problem also), and that of arsenic in beer, fortunately not now a problem. The latter enquiry was caused by an alarming outbreak of arsenic poisoning in the Manchester area in November 1900. This outbreak was traced by the Laboratory to the beer being consumed in the area and further traced to the glucose used in its manufacture. This glucose was found to have been supplied by a Liverpool firm of brewing sugar manufacturers and beer made from their product, containing arsenic, was eventually found over much of the northern and midland counties. The arsenic was traced to the sulphuric acid used in the preparation of the glucose from cane sugar. A Royal Commission was appointed to investigate arsenical poisoning, and after deliberation and the taking of much evidence, recommended that no brewing materials should contain more than one hundredth of a grain of arsenic per pound. It had also been found that malt itself often contained small amounts of arsenic, from the coke used to dry it. Thorpe was a member of the Commission and gave evidence to it, as did E.G. Hooper, one of his Superintending Analysts. As a result of the recommendations of the Commission the Board of Inland Revenue appointed a committee with Thorpe as Chairman and Thomas Cheater, one of his Assistant Analysts, as secretary. This committee was to develop tests for small amounts of arsenic. This was eventually done by staff of the Laboratory.[8] In this report Thorpe again acknowledged the 'skill and care' with which Cheater, and Stubbs also an Assistant Analyst at the Laboratory, had 'worked out the details of the several analytical processes described'. As soon as the Report was published chemical instrument makers hastened to make arrangements to come and see the method working, so that they 'may be be in a position to supply' the apparatus.

Some of the work carried out by Thorpe during his tenure of office as Principal Chemist was as a result of his involvement with the wider community of science. He played a considerable part for example in the compilation of the first table of atomic weights of the elements. The concept of atomic weight had been originated by John Dalton in 1803 and was the subject of much debate during the nineteenth century. Both the American Chemical Society in 1892 and the *Deutsche Chemische Gesellschaft* in 1898 published comprehensive lists of atomic weights, but there were no internationally agreed lists. The Germans had recommended that an international committee should be formed to consider the subject and a committee of 57 was appointed in 1900. This conducted its business by correspondence, and to speed up the work a smaller committee of three consisting of F W Clarke (USA), K Seubert (Germany) and T E Thorpe was appointed. This group published the first international atomic weight table in 1903 and reported annually (except for 1918) until 1921.[9] It in fact continued as part of the International Union of Pure and Applied Chemistry (founded in 1919), with Dr Thorpe as honorary president of the committee until his death in 1924.

Due to his particular skills in semi-micro analysis

Thorpe was asked in 1906 by the Royal Society to contribute in a practical way to help in this work by carrying out an exact determination of the atomic weight of radium. This he did, in collaboration with Arthur Francis, a member of his staff and later Deputy Government Chemist. The determination was made on 64 milligrams of radium chloride extracted from 413 grams of pitchblende concentrate. The efficiency of recovery was about 60% and involved some 9,400 recrystallisations, mainly to separate barium and radium. The results were given in a lecture to the Royal Society in 1907 in which he acknowledged the help of Mr Francis who had done the recrystallisations, of Madame Curie who gave advice on the chemical separation, of Professor Rutherford who made some of the activity measurements, of the President of the Royal Society who had checked for absence of barium and of Mr Oertling for the loan of an accurate balance. A number of samples of radium in milligram amounts were sent by post between Manchester and London, the only precaution taken in those days being to register the parcels. There is an amusing letter from Lord Rutherford in which he apologised for an accident in which he spilled one of the specimens he had standardised. The spillage occurred on a clean table and he was able to sweep it up without loss. Subsequently Thorpe and Francis used a similar recrystallisation technique to redetermine the atomic weight of strontium. All of this work was carried out at Clement's Inn and correspondence between Francis and Thorpe relating to it continued after the latter's retirement in 1909. The results of the radium and strontium determinations were published in 1908 and 1910 respectively.[10]

Another important set of tables also appeared at this time. They were a revision of those published earlier for the determination of the original gravity of beers. These had originally been developed by Graham, Hofman and Redwood (see above p.53) in 1852, but had been found to give inaccurate results when used as prescribed in the *Inland Revenue Act* of 1880. In 1909 the Board of Customs and Excise (who as related took over the responsibility for the Excise from the Board of Inland Revenue on 1 April 1909), decided to commission a new set of original gravity tables, and nominated Thorpe, in collaboration with a nominee of the Institute of Brewing, to construct such a set. The Institute nominated Dr Horace Brown, a Public Analyst[11]. In parallel

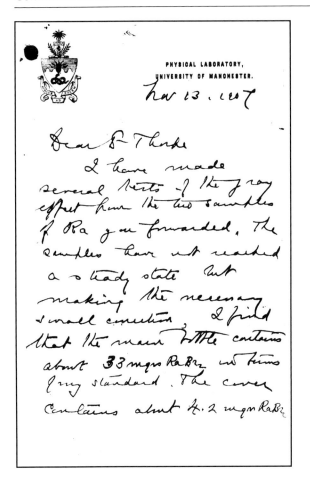

51   letter to Thorpe from Professor Rutherford, 1907 (PRO DSIR 26/256)

with this programme, which was partly carried out at the Government Laboratory, work was taking place on a revision of Sikes' Tables for the determination of spirit strength and for producing trustworthy tables correlating specific gravity of aqueous solutions of ethyl alcohol with their alcoholic content. This work was carried out at the Customs House Laboratory under James Connah (then Superintending Analyst there), and John Holmes at Clements Inn Laboratory, under powers granted to the Boards of Customs and Inland Revenue by the *Finance Act* of 1907. The work was necessary since the only legal way of determining spirit strength was by means of Sikes' hydrometer (see above, p.000), was not entirely accurate under extreme conditions. The use of the specific gravity bottle was to be preferred in many cases as Thorpe himself had pointed out in 1894.

52   Dr A G Francis, Acting
Government Chemist, 1944–1945

'Thorpe's tables', enabling the determination of the strength of spirits in terms of 'proof spirit' were sanctioned for use by the *Finance Act* of 1915, and by the *Strength of Spirit Ascertainment Regulations* of 1916, and remained in use until metrication in 1979. The original gravity tables for beer were legalised by the *Finance Act* of 1914. Thorpe also developed a table of weights per gallon, enabling the content of a cask to be determined from the hydrometer indication and the weight of the spirit. This was necessary since the table originally produced by Bate in 1835 and enshrined in various Acts of Parliament was not entirely accurate. It had already been checked by the Laboratory in 1863 at the request of Adam Young, then Secretary to the Board of Inland Revenue, and some inaccuracies found. They were reported by George Kay, Phillips' deputy, as being not serious.[12]

A few years before this important programme of work, Thorpe served on another committee which had far reaching effects. This was a Treasury committee set up in 1898, under the chairmanship of Lord Rayleigh, 'to consider and report upon the desirability of establishing a National Physical Laboratory'. This committee decided in a very short period that such a Laboratory should be set up, under a Governing Body which was independent of any Government Department control[13]. This precedent of such independence may have played a part in the reorganisation of the Government Laboratory a few years later[14].

Change

Repairing fences

THE RELATIONS OF THE LABORATORY with the Public Analysts had not been good since the 'chemical officers' of the old Inland Revenue Laboratory had been appointed referee analysts under the 1875 *Sale of Food and Drugs Act*, as described above.[1] The position of a referee, who must always disagree with one party or the other to a dispute, and sometimes with both, was not always appreciated by some of the Public Analysts. Relations seem to have become worse in the last few years of the nineteenth century, frequent disagreements with the results of the Public Analysts are noticed in the Reports of the 1890s, either directly in that analytical results were disputed, or in the inability of the Laboratory to draw the same conclusions as to adulteration[2]. In some ways this dispute came to a head in the proceedings of the Food Products Adulteration Committee of 1892–1894, when Mr Hehner, Public Analyst for Derbyshire, launched a powerful attack on the Government Laboratory, its methods and its standards. This was returned with interest by Richard Bannister, Deputy Principal of the Laboratory, directly contradicting the claim of the Public Analysts that the Laboratory refused to work in any way with them. Bannister was as harshly critical (with as much justification, and possibly more) of the Public Analysts as they were of the Laboratory[3].

One of the problems still causing friction was the analysis of sour or decomposed milk. This was of great importance since the Laboratory always had to make allowance in its results for the changes on decomposition when deciding if adulteration had taken place, while the samples of the same milk analysed by the Public Analysts were usually fresh. In 1892 the Laboratory had devised a scientific method to replace the original empirical method of determining the decomposition products and relating this back to the original composition. This was first revealed in 1894[4].

The Final Report of the Adulteration Committee noted that relations between the Laboratory and the Public Analysts were not all that could be desired, and also noted the complaint of the latter that they could not obtain information from the Laboratory on matters of analysis, a complaint which the Committee upheld. This complaint had some basis. They recommended that a new Court of Reference be set up to deal with the whole question of food standards and the wholesomeness of foods. This Court (on which Laboratory and Public Analysts would be represented), would they thought obviate the complaints, although they noted that the Laboratory had performed its duties as referee with credit and had no reason to believe that the standards or limits adopted by them could not be justified by the law. They also recommended that Public Analysts' qualifications should be more strictly enquired into[5].

The Special Court of Reference was never appointed, because of practical difficulties in determining its powers, but after this time relations between the two rival parties slowly improved. The Laboratory staff began to publish papers on the result of their researches at about this time, and Thorpe himself published a paper in 1905 in which he described the methods adopted to obtain figures for fresh milk from the analysis of decomposed milk[6]. Dr Thorpe was invited to speak at the Annual Meeting of the Society of Public Analysts in 1896, as Dr Bell had been before him, thus showing that relations between the two parties were not as strained as they sometimes appeared to have been. Despite some rather harsh words from the President of the Society he was received generally with expressions of esteem, and in his speech Thorpe stated that he wished to work 'cordially and harmoniously' with them[7]. This 'harmonious co-operation' certainly began speedily to take place, and in 1900, soon after the passing of the new *Sale of Food and Drugs Act* (1899) the Council of the Society of Public Analysts appointed a committee of senior members to confer with the Principal Chemist on prosecutions arising from section 8 of the new Act. Butter was often mixed with margarine to improve its appearance and flavour and the new Act limited the amount of butter which could be mixed with margarine to 10 per cent. A detailed specification for the design of the apparatus to be used was agreed, together with a table relating results to percentage butter-fat[8]. From this point onwards the relationship between the two bodies has always been relatively harmonious.

As has already been discussed, on a part time basis Laboratory staff had long received formal training in theoretical chemistry outside their working environment, as well as receiving practical training in chemical techniques in the Laboratory. Dr Thorpe lost no time after his appointment in changing this system, transferring the practical instruction to the Royal College of Science (now Imperial College) at South Kensington. This change was apparently resented by the first batch of students to be affected, and Thorpe visited South Kensington to explain the reason for it. According to Cholmondeley it was done so that they could attain better 'manipulative skills', (a branch of chemistry which he says South Kensington was notoriously weak in), but according to Bannister in 1896 it was due to lack of space in Somerset House for the practical training. At nearly the same time as this change (in 1895), the course of study which had been devised by Professor Tilden (see above), was increased from one year to two. The College charged no fee for this arrangement since it was run by the Science and Art Department.[9]

The course itself was a fairly intensive two years. Lectures on inorganic chemistry were given in the first term, with laboratory work and demonstrations, together with mathematics (according to requirements) twice a week and with the study of physical chemistry in the second year in view. Students were examined in February in inorganic chemistry and practical and theoretical analytical chemistry and went on in the second term to take lectures and practical classes in organic chemistry (together with mathematics as above). Both subjects were examined in June, students obtaining a mark of 80% in organic chemistry were excused further attendance in this subject. In the first term of the second year there were lectures and practical work in physical chemistry, with examinations in the following February. In the second term there were more lectures on organic chemistry. For those excused these lectures, reading at the direction of Professor Tilden and reporting to him monthly was prescribed, together with practical work. The final examination was at the end of June, on 'qualitative and quantitative analysis, physics and chemical experiments or determinations done during the session'. Professor Tilden reported to Thorpe each month on the students, including a note of their attendance. After 1896 the Science and Art Department made Tilden report to them and they passed the report to the Board of Inland Revenue to give to Thorpe.[10]

The first six students to have followed this new course were noted in the *Report* for 1897 as achieving a 'satisfactory' final examination result, but no further mention was made of any such students.[11] Cholmondeley says that in fact the results were extremely good. He further criticises the new system by saying that its result was that the assistants who went on these courses now took less interest in the Laboratory, since for at least eight months of the year they had no connection with the Laboratory. Additionally the College

53 Apparatus for determining the original gravity of beer, 1913, with 'Thorpe' revenue condenser

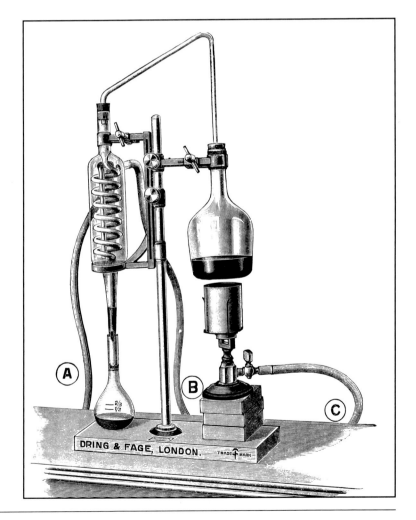

regarded them as 'casual students' and did not 'exert themselves' to the same extent as with more regular students. Since
this system was an early example of a now familiar process,
with students returning to the Laboratory in the summer to
work there (a process which works well), it seems possible
that Cholmondeleys remarks were at least partly prompted
by his dislike of Thorpe's apparently rather autocratic ways.
The students received their full salaries while at College, plus
a grant of £10 for books and apparatus. They had received
this grant from 1873, £3 of it being allowed during the first
year. The books bought had to be reported to the Laboratory. They enjoyed the normal College holidays at Easter and
Christmas, together with 14 days during the summer vacation. After successfully completing their course the students
were required to remain as Temporary Assistants at the
Laboratory (with normal pay and leave) for not less than
three years in default of which the cost of their training was
refundable. Assistants not subsequently appointed to the
permanent staff were drafted to outdoor departments of the
Inland Revenue or Customs, including the provincial 'chemical stations'. Training at the Royal College of Science
continued until 1911.[12]

Those eligible for the Government Laboratory studentships were members of the Inland Revenue and Customs
departments who had entered the Civil Service in the usual
way by competitive examination, and had served for six
months. They could then compete for admission to the
Laboratory, by taking an examination in elementary inorganic chemistry, elementary organic chemistry (with special
reference to the chemistry of brewing, distilling and other
industries subject to Revenue control), elementary physics,
algebra up to quadratic equations, and Euclid, Books I and
II[13]. Papers are in existence covering the period 1881–1908,
and they remained much the same throughout, although the
time allowed changed from five hours initially to seven hours
in later years. Some years at least there seems to have been
considerable competition for places, since a proof of the
paper for 1897 is endorsed in manuscript '100 copies to be
sent under seal to R Bannister, Room 110 Inland Revenue
Laboratory, Somerset House'. Students were required to sign
a copy of the conditions given above, to certify that they
agreed to abide by them if successful in the examination[14].
Prospective students could take the examination at their own

Station, supervised by a suitable officer of the Revenue, and the paper was then returned to the Government Laboratory, accompanied by a certificate that the conditions of the examination had been complied with, (see Appendix 6 for examples of the examination papers).

Customs Assistants were governed by conditions similar to those which applied to Assistants of Excise. The first set of regulations for their employment were issued in 1897. Others were issued by the Board of Customs in May 1904 and related to a two year course at South Kensington together with six months training at the Clement's Inn branch. A bond (£100) was called for, realisable in the event of leaving the service before the expiry of five years from the commencement of their Laboratory Studentship. This differed from the Inland Revenue students who had to refund the whole of the cost of their training, (presumably greater than £100), in the event of their leaving. Until 1898 established staff had been asked to provide a bond of £54 (calculated as the cost of their training) against the possibility of their leaving the service soon after finishing their training. This was dispensed with in 1898 by permission of the Treasury.[15]

The alteration in the study arrangements was not the only staff change made by Thorpe in his period as Principal Chemist. Towards the end of the century he began with the assent of the Treasury to recruit to the Laboratory staff scientists who had already had some university training, instead of relying on retraining revenue officers. This change greatly annoyed some of the assistants of Excise formerly at Somerset House. Cholmondeley much regretted the change, which he condemned as 'a wretched innovation'. This increase was explained in the Report for 1902 as due to the large increase in samples caused by the imposition of duties on 'sugar and cognate substances' in the 1901 Budget. Apparently not more than three were to be recruited, at least at first, at salaries of not more than £120 *per annum*.[16]

Similarly a number of non-established qualified chemical assistants were recruited to the Customs House Laboratory in 1904, pending the training of regular temporary assistants recruited from the Customs service and sent to the Royal College of Science for two years. This was due to an increase in the number of permanent analysts[17]. The existence of these 'Assistants not on the Establishment' (i.e. non established Civil Servants) was noticed officially on the 'conditions' signed by candidates for the Laboratory examination by 1905.[18] Applicants for these places had to have had a good chemical training in an accredited school of chemistry and be proficient in organic and inorganic chemical analysis. Their application had to be accompanied by certificates, or testimonials from their teachers.

As frequently occurs in such matters the recruitment of these non-established Civil Servants was preceded by rumours that it was to occur and a deputation of Assistants went to Thorpe to enquire as to the situation. He apparently denied that it was to happen, just before it did, and afterwards, according to Cholmondeley, 'virtually' threatened to abolish the Saturday half holiday if complaints were made to the Board of Inland Revenue. Despite the threat complaints apparently were made.[19]. The Board seem to have been sympathetic to the Assistants, but in practice could do nothing about the original grievance since Thorpe had Treasury permission to act as he did. This loss of control over their Laboratory, inevitable from 1894, must have caused the Inland Revenue to acquiesce more readily in radical changes not long in the future, (see p.168 below). Numbers of Supervisors of Inland Revenue continued to receive one month's training in the Laboratory throughout this period, 34 passing through in the year ended 31 March 1900, for example, and Customs assistants similarly to qualify them for work in the outstations[20].

The Assistants of Excise who had been trained in the Somerset House Laboratory and sent subsequently to a chemical station in the provinces were frequently worried about their continuing career in the Inland Revenue service. They had little or no prospect of getting onto the permanent laboratory staff, their pay was relatively poor and their future prospects unclear. From time to time an assistant would apply for transfer back to his excise station (presumably the post from which he had come into the laboratory service) and when several of them did this in 1895, Dr Thorpe sought and obtained authority for an allowance of £20 per annum to be associated with ten of the posts, presumably the main provincial chemical stations and to be given to those of special merit. From 1901 all Second Class Officers attached to the Laboratory as temporary Assistants were given an allowance of £15 a year unless they were in receipt of the other allowances.[21].

12 Reorganisation

IN THE NEW YEAR HONOURS OF 1909, DR Thorpe was knighted, having been made a CB in 1900, and he retired as Principal Chemist at the end of September in the same year. On his retirement he was presented with a silver tea and coffee service and a silver cigarette box, subscribed for by former colleagues and by associates in other government departments.[1] As Sir Edward Thorpe he returned to South Kensington, to resume his post as Professor of General Chemistry in the Imperial College, newly formed from the Royal College of Science.

Thorpe formally resigned as Professor in 1912 but continued to work actively in literary and scientific fields for the rest of his life. He had received many honours in the course of his career, he was elected to the Royal Society in 1876, and served on their Council, and as Foreign Secretary to the Society. He was President of the Chemical Society (of which he was a member for 54 years) and of the Society of Chemical Industry, and received numerous awards and honorary degrees. He died at his home in Salcombe on 23 February 1925[2].

Sir Edward Thorpe was succeeded as Principal by Dr James Dobbie, who took up his appointment on 2 December 1909. Dobbie was born in Scotland in 1852 and had studied under Kolbe in Leipzig. He had previously held the chairs of chemistry at Glasgow University and at the University College of North Wales at Bangor. Immediately prior to his new appointment he was Director of the Royal Scottish Museum in Edinburgh[3].

The work of the Government Laboratory continued under Dobbie much as it had under Thorpe, most of the work being part of a continuing process. In January 1910 for example the *White Phosphorus Matches Prohibition Act, 1908*, came into force. This Act had been passed as a consequence of the work done by the Home Office committee (described in the previous chapter), and of a Convention agreed by principal European countries, to protect workers in the match industry from the terrible disease known as 'phossy jaw'. White phosphorus was already banned in the United Kingdom and the Act speedily had the effect of stamping out its use in imported matches. Simple tests had been devised in the Laboratory, and instructions on its use were issued to Customs officers to enable them to detect white phosphorus[4]. One new area of work did appear at about this time, due to the passing of the *Old Age Pensions Act* 1908. Claimants for pensions had to produce evidence in the shape of birth or marriage certificates. This was not possible in some instances since civil registration of births did not start until 1837 and not all births were registered before

54   Sir James Dobbie, Principal of the Laboratory and Government Chemist, 1909–1920

1875 so that family Bibles and Prayer Books were sometimes produced to prove the claimants were of the age they claimed. In some cases the Local Government Board (administering the Act) believed that some entries had been altered to secure a pension. These documents were submitted to the Laboratory where it was shown that the ink used was of recent date[5]. Similar examination of an Account of the Master of the Revels (Sir Edmund Tilney), for 1604–1605, showed that an early reference to Othello in that document was in fact contemporary with the rest of the manuscript, a fact which had been disputed[6]. This type of handwriting and ink examination work became a regular part of the Laboratory work, as indeed it still continues today.

An entirely new departure early in Dobbie's tenure of office was participation as part of the UK delegation in June 1910 by Dobbie and James Connah, the head of the Customs House Laboratory, in the International Conference on Methods of Analysis of Alimentary Products. The Treasury had authorised Dobbie and Connah to attend, on the basis of 'travelling expenses plus 30 shillings (£1.50) per night'[7]. Other members of the delegation included representatives of the Local Government Board (including Dr. Monier-Williams), the Society of Public Analysts and the General Medical Council. Many other countries were represented, including most of Europe, but not Austria, Germany, Russia or the United States. The object of the conference was two-fold, firstly to agree a common chemical language in which to express results of analysis by whatever method they were arrived at, and secondly to agree on common analytical methods. The first objective embraced such measurements as temperature, density, rotary dispersive power, saponification value and iodine number, and the units in which these were expressed. The final communiqué of the conference announced that all delegates were agreed on the desirability of unifying the expression of results and supported the setting up of Laboratories where unified analytical methods could be prepared. Elaborate rules for the expression of results were drawn up[8]. However, despite the announced agreement Dr Dobbie's view was that it was not desirable for chemists to be restricted to one mode of expressing their results but that whatever method was used should be definitely and clearly stated. He was also against standardised methods, as being incompatible with his statutory referee functions; he was generally supported in this by the Public Analysts, who claimed that the only two official methods then in use (the Excise method for the determination of the original gravity of beer and the official method for determining the proportion of butterfat in margarine) gave incorrect results. After the conference, the Foreign Office was advised not to sign the convention which was proposed and the government decided not to participate in the establishment of an International Bureau of Analytical Chemistry, stating in 1914 that it regarded fixed standards and particularly obligatory methods of analysis as impracticable, as would be the recognition of an international committee which issued authoritative statements as to the value of analytical methods.

A continuing problem inherited by Sir James was the staffing and organisation of the Laboratory. As has been seen, the recruitment of new junior staff was rather complicated, and could be by several routes. Additionally Revenue and Customs officers trained at the Laboratory and sent to outstations were not entirely happy about their minimal career prospects as noted briefly in earlier chapters. In 1904 some of the temporary assistants at Customs outpost testing stations drew up a draft petition to draw attention to their career grievances and to the anomaly that two separate testing facilities (customs and inland revenue) still existed in some ports. They were supported in some cases by the Collector at the port, who at Liverpool the following year suggested that an allowance be given to the chemical officer there. The suggestion was given serious consideration but it was concluded that, partly because the scope for combination of stations and economies therefrom was limited (the duties at some of the smaller ports also included general duties), the time was not yet appropriate for the combination of provincial laboratories at Belfast, Bristol, Cork, Dublin, Glasgow, Liverpool, Newhaven and Southampton. The facilities at Cardiff, Folkestone, Greenock, Hull, Leith, London Docks and the West India Docks were for customs purposes only and those at Burton-on-Trent, Edinburgh and Leeds related only to Inland Revenue[9].

Whilst the immediate outcome of the 1904 petition (which apparently never got beyond a draft, although it was treated as a basic document) was limited, it appeared to sustain a continuing interest in the terms of appointment of junior laboratory staff and of the role of the outstations. The special regulations issued by the Civil Service Commission for the competitive examination for Inland Revenue Assistants of Excise in January 1907 included an optional paper on elementary chemistry. In due course, on 6 April 1909, the Treasury appointed a committee to consider the recruitment and remuneration of laboratory staff, the members being F.S. Parry (chairman), Dr. T.E. Thorpe, J.P. Crawly and M.G. Ramsay. The committee reported a month later that it would be best to start with a model scheme and to see how best the existing personnel could be absorbed into this, rather than to base any new structure directly on the existing arrangement. It therefore proposed that consideration should be given to the ideal organization for the work and how far this should be centralised or not. The committee met frequently thereafter, at the Treasury and at the laboratories, taking oral evidence from the Deputy Principal Mr. Davis and from several of the Superintendents and First Class Analysts together with written evidence from officers in the provinces. Mr. Davis said that at busy periods the permanent staff were often at work from 9.30 in the morning until 8 o'clock in the evening; Mr. Woodward, who was responsible for the tobacco section at Clement's Inn Passage, said that the South Kensington training courses were unnecessary and tended to give men 'a swollen head'. Mr. Helm considered that it was easier to train Revenue officers in chemistry than chemists to do revenue duties. The decision to stop sending students to South Kensington was probably made easier by the fact that the Imperial College of Science and Technology, which the Royal College of Science had now become, intended to charge fees, although the difference in cost would only have been £54 per year for each Assistant.[10]

Much of the detail of this enquiry now seems lost, but note of it seems to have been taken by the Treasury where it was apparently felt that with the growth in demand by government departments of the services of the Government Laboratory, it was not a satisfactory position to have the latter staffed entirely by Revenue Officers, and to be under the control of the Board of Customs and Excise. To place all departments on the same footing therefore with regard to using the services of the Laboratory, and (as in 1894) to promote centralisation of government chemical work, the Treasury in a letter to Dr Dobbie dated 15 December 1910 announced that funding of the Laboratory would be by a separate Government Chemist Vote from 1 April 1911 and called for financial estimates as a basis for this. The letter went on to express the hope that Dobbie would continue to maintain harmonious relations between himself and the Board of Customs and Excise and that he would bear in mind the continuing importance of the Revenue work. It stated that he would in future have independence under the broad aegis of the Treasury, the implications being that he could within limits pursue his work in the way he thought best. Such independence reflected and surpassed that granted to the new National Physical Laboratory in 1899.

There were many details to be sorted out, from the numbers and recruitment of staff to permanent and temporary posts and the question of overtime, retirement age (settled at 62), and the names of officers who could sign and countersign on the new Government Chemist account. One important question arising was that the *Sale of Food and Drugs Acts* of 1875 – 1907 specified reference of disputed samples to the 'Chemical Officers' of the Commissioners of Inland Revenue. When the new Department of Customs and Excise had been set up in 1909 it was ordered that the samples should be sent to the Commissioners of Customs and Excise instead of the Commissioners of Inland Revenue, while retaining the references in the Act. In the new Department the staff were no longer connected with either the Inland Revenue or the Customs and Excise though. This was solved by the Commissioners of Customs and Excise appointing the officers of the Laboratory their Chemical Officers for the purposes of the Act. Dobbie replied in very great detail to the Treasury with his proposals for sorting out these questions. One of his letters deals with the question of the salaries of the 'Subordinate Staff' and their salaries which it had been agreed would be on the new Vote. This Vote was £19088 for 1911–1912. The subordinate staff included two boy clerks (to receive £42 each), three Police pensioners (£231 total), five Porters (£375 total), one Boy Messenger (£39), and four Charwomen (£120 total). All of this was finally agreed on 10 April 1911. A few days before on 1 April the Department of the Government Chemist had formally come into being with Dr Dobbie as the first Government Chemist.[11]

Part V
The Department of the Government Chemist 1911–1959

13 The New Department

IN HIS FIRST REPORT AS GOVERNMENT chemist Dobbie gave a brief outline of the history of his Laboratory since 1842, and discussed the changes in adminstrative arrangements and method of staffing of this new Department (then and for some time one of the smallest in the Civil Service).[1] Laboratory posts were no longer filled exclusively from the Customs and Excise Service, as they had been from 1842. Revenue officers continued to be employed on the purely revenue work of the Laboratory, but other chemists were in future to be recruited from the open market for non-revenue work. At this time the main chemical staff consisted of the Government Chemist and his Deputy, four Superintending Analysts, eight First Class Analysts, and thirteen Second Class Analysts. In addition to these there were a varying number of Temporary Assistants, (62 in 1911), composed partly of those on loan from Customs and Excise from whose point of view they were 'unattached officers', (Temporary Revenue Assistants as they are called in the staff list), and partly of those appointed by the Government Chemist, (Temporary Chemical Assistants). There were then 29 outstations in various parts of the United Kingdom for testing work for revenue purposes, 18 from Customs and Excise and 11 from the Inland Revenue. These all became part of the Customs and Excise Department. Elaborate arrangements were made for the staffing and the training of the staff of these. The staff were, as before, to be given specialist training in the laboratory in the examination of tea, tobacco and alcoholic liquors. Separate stations for the two departments of Customs and Excise still existed in Belfast, Bristol, Liverpool, Newhaven, Southampton, Glasgow, Cork and Belfast and the process was commenced of amalgamating these.

In addition to the staff engaged on analytical work there were a number, (a slowly increasing number), of support staff. The Laboratory had enjoyed this type of support from the earliest days, beginning in the 1860s with a single Bookkeeper. A Keeper of Chemicals was employed in later years at Somerset House, with a deputy later, as related, although there was only one Keeper again in 1901. In 1911 there were also fifteen clerks headed by a minor staff clerk, (with two Second Division Clerks, ten Assistant Clerks and two boy clerks), six Porters, four Charwomen, three Door Attendants and Night Watchmen, and a Boy Messenger. Dr Thorpe in his last *Report* in 1909 proudly drew attention to the increase in staff in his years as Principal Chemist, from 64 to 91, but he appears to have been counting only scientific staff, there were a total of 119 in 1911, of whom 89 (including the Government Chemist and his Deputy) were scientific staff [2].

Most of the staff listed were employed at the Clement's Inn Laboratory. The Custom House branch (or Thames Street, as it was occasionally called) was much the smaller of the two. In 1909 (for which year we have a list prepared for the purposes of the enquiry setting up the Department of the Government Chemist), the staff at the Customs House were in charge of James Connah, a Superintending Analyst. He had been in charge there since 1901. Connah's deputy in 1909 was a First Class Analyst, (Thomas Cheater), and additionally there were three Second Class Analysts, eleven Temporary Assistants, and three clerks (plus a 'Boy Clerk'). Five Customs Watchers were employed as Laboratory Porters.[3] There was thus a total of 25 staff, as opposed to 85 appearing on the same list for the 'Clement's Inn Branch'. On this basis the staff costs for the Customs House laboratory were less than a third of the Clement's Inn Laboratory. Total staff costs had risen in the years since the foundation of the joint Laboratory in 1894, although in general salaries had not. Dr Dobbie was paid £1200 *per annum* rising to £1,500 after five years, his Deputy received £700, rising to £800 in four years. The basic salary of the Superintending Analysts (of whom there were three) was £600 rising to £650, of First Class Analysts £400 rising to £550, and Second Class Analysts £160 rising to £350. The Assistants received a maximum of £150 *per annum*. These salaries were exactly the same as their predecessors had received in 1894. In the more lowly grades the Assistant Clerks received a maximum of £150 *per annum*, the Boy Clerks 15 shillings (75 pence) per week, rising to the princely sum of 19 shillings (95 pence) after four years. The hours worked by the staff were exactly the same as in 1894 (see chapter 8, A New Home). Assistants received 21 days 'Vacation', 'at periods in accordance with the general leave table of the laboratory'.[4] More senior staff received 48 days leave.

As to the actual conditions of work in the Laboratory, we are lucky in possessing accounts by two staff whose names appear on the 1911 staff list, Richard Church, an Assistant Clerk, and Albert Rayment, a Temporary Revenue Assistant, both employed at the Customs branch. Rayment was one of the Revenue Assistants who had taken the chemistry option in his entry examination, and applied to be seconded to the Government Laboratory staff as soon as the opportunity arose. He was originally appointed to the Clement's Inn branch but never worked there, being sent immediately to the Customs House branch in his opinion the first such recruit to be sent there. He describes the various mid-summer smells of nearby Billingsgate fish market, the horse traffic and the Laboratory itself. He also described the blue uniformed Customs watcher who had charge of the laboratory attendance book, which was returned to the Superintending Analyst's room shortly after he had looked carefully for late comers down the long flights of stairs from the top of the building.[5]

Rayment gives a most interesting account of the duties of a Revenue Assistant, both in the 'Beer and Wine' Laboratory, where he worked initially, and then in the 'General Import' section. He gives a graphic description of the calibration of two specific gravity bottles with which he was issued in the latter room, a routine which many junior members of staff of the Laboratory have undergone until very recently. In the General Import Laboratory samples were tested to see if they contained dutiable sugar or spirit, and if so, how much.[6]. He goes on to describe something of general life in the Laboratory, of the disposal of samples for example. These had legally to be destroyed when finished with, and after the results had been duly recorded. If edible they were inevitably 'destroyed' by being shared out amongst the staff. This was a custom which endured for many years. When he was seconded to the Government Laboratory, Rayment had hoped to be one of the students seconded to Imperial College. He was told that this had been stopped recently, since many

55   Specific Gravity bottle (1000 grain) and counterpoise

of those who had obtained a degree in this way found that they could then obtain posts at a considerable increase in salary outside the Civil Service. The 'authorities' had therefore decided to train staff entirely in post[7]. Rayment was eventually, after less than his expected two years in the Laboratory, sent to take charge of the Customs test office at Greenoch, the Laboratory then being in the process of staffing these test offices with its own 'seconded men'. He spent the rest of his career in Scotland.

In the course of his description of his work in the 'Import Section', Rayment describes Church, the second of the two staff members whose descriptions of life at the Customs House exist. Richard Chuch was an Assistant Clerk in the Custom House, having joined in April 1911. He left in 1920 and is known now as a poet and novelist. His description of the administrative organisation does not entirely coincide with that given by Rayment, but it is not incompatible. Rayment describes Church, whom he knew and talked to. They do not appear to have been friends, Rayment says that Church had 'rather long' black locks and a pale face, and as a would be poet found the day's work too mundane for his taste[8]. Church's account of the Laboratory occurs in two novels, *The Porch* and *The Stronghold* and Rayment's description is a mixture of John Quickshott, the hero of the book, generally thought to have been a self portrait of Church and Mouncer, Quickshott's fellow clerk, and a would be poet.[9]

Church's book and its sequel traces Quickshot's career from the day he joined until he left to train to become a doctor, through the 'Sample room', the 'Sugar' room, and the 'Tea room'. He describes the staff of each room, and his fairly unflattering portraits were accepted on publication as depictions of actual members of staff at that time.[10] It seems very likely that the descriptions owe much to poetic licence but they were certainly readily recognised when the book was published, beginning with the 'chief', a 'stout elderly man in a morning coat', 'cold blue eyes behind gold-rims' and an 'aggressive moustache'. This was James Connah (see plate 49), the Superintending Analyst. Also described are 'Mr Lyon' the 'deputy chief', a small man with black hair and beard, and 'close set furtive eyes', dressed in a morning coat which was 'greenish with age', and an obsession with punctuality. This was an unsympathetic and apparently not wholly accurate picture of Thomas Cheater, at that time a First Class Analyst. The clerk in the Sample Room, called Christopher Bembridge (actually Frank Osborne), was more sympathetically described.[11]

It is not now possible to identify with certainty any other members of staff.[12] It seems probable though that 'Francis', in charge of the 'Tea Laboratory' was the George Sheppard whom Rayment says was in charge in the Beer and Wine Laboratory in 1911. These two departments were in fact in the same room, and when James Halpin was transferred from here (where he worked in 1909) on his promotion in 1911 to the Import department, where Rayment met him, Sheppard (also here in 1909) would have become the senior person in the room. 'Francis' is described by Church as a 'bald dark eyed person with chronically flushed cheeks'[13]. The laboratories as described by Church are strikingly similar to those of Rayment. Both men made a point of mentioning the heat in the summer caused by the mass of Bunsen burners and gas furnaces being used, a feature of laboratories of course before air conditioning and electric heating mantles. One interesting piece of equipment in use described by Church is the Arithmometer, an early hand operated mechanical calculator[14].

The analytical work carried out in the laboratories under Dr Dobbie's control remained much the same at this time as it had for many years. Analysis of samples of beer, cider, spirits, wines, tobacco, sugar, coffee, chicory, cocoa,

56 Electric heaters for wine and spirit distillation, Clement's Inn 1917

chocolate, the calibration and testing of hydrometers, saccharometers, thermometers and graduated vessels, and referee analyst functions under the *Sale of Food and Drugs Acts* of 1875 and 1899 continued to be carried out. In the *Report* for 1911 this work was analysed as usual, and in addition rather more information than usual was given on the principles adopted in the work, and the law which lay behind it. For example technical details on the charging of duty on beer, how the specific gravity was assessed, and the practice at breweries was given[15]. Beer was still being tested for illicit dilution and beer and brewing materials were still being examined for the presence of small amounts of arsenic, (see above, p.150). The tobacco work, which at this time was not showing any signs of changing in volume was generally carried out to determine the amount of moisture contained and the form of the tobacco, upon which the rate of duty depended. Samples also included such items as the examination of 'joggery', which was apparently a mixture of tobacco and opium with sugar and molasses, used, as was said primly, 'by Asiatics,' and samples of explosive cigarettes[16]. The rate of duty levied on the latter is not given.

A considerable number of samples were being analysed for other government departments, an increasing number of whom were referring their analytical and chemical problems to the Laboratory. This was of course one of the objects of the recent re-organisation. For this reason the Treasury made some attempt to persuade all government departments to submit their problems in analytical chemistry to the Laboratory. This was on the grounds of saving money by such centralisation and on the grounds too of greater efficiency in that a central laboratory had greater expertise. There was strong resistance from at least the Local Government Board and from the Stationery Office on the grounds that their analytical problems were such that the Government Laboratory could not cope with them. This proposition was strongly disputed by Dr Dobbie. After this exchange the Treasury virtually conceded that the idea of one central laboratory would not be adopted by other departments and did nothing further to promote the principle. The work which was received from other departments frequently involved special investigations, or was purely consultative for which no sample examination was required. A considerable increase in this work took place from about 1913 and it is still an important feature of the work of the Laboratory today.[17]

Examples of new areas of work being submitted to the Laboratory are seen in the determination of the salinity in samples of sea water undertaken for the Board of Agriculture in connection with the international exploration of the sea. It was decided at about the same time that the chemical work for the Geological Survey should be carried out by the Government Laboratory, instead of by the Survey itself. Prior to this time any chemical work for the Geological Survey was done by a District Geologist with an interest in chemistry at the Museum of Practical Geology in Jermyn Street, on a site now occupied by Simpsons. A competitive examination for the new post of Analyst held in 1914 resulted in the appointment of F R Ennos, a Second Class Analyst who joined the Laboratory at Clement's Inn on 15 June that year. He was conducted to Jermyn Street by bus by Dr Dobbie. He later described how Dr Dobbie was dressed in his (apparently) usual costume of a tail coat and top hat. Both the salinity and geological survey work continued for many years, until quite recently[18].

Two of the more exotic investigations undertaken were on behalf of the Public Record Office and of the Office of Works. In the former the Deputy Keeper of the Records asked for guidance on the better preservation of wax seals under his charge. These varied in dates from the thirteenth to the sixteenth centuries, and were found in most cases to be bees wax. In two instances, of the fifteenth century, they were probably wax of East Indian, not European origin[19]. In the case of the Office of Works advice was sought as to the measures to be taken for the preservation of the roof of Westminster Hall, where the oak beams were suffering from rot and death watch beetle. Extensive (and ingenious) experiments were underaken, both in the Government Laboratory, and by Dr Westergaard, Professor of Mycology at the Heriot–Watt College, Edinburgh, but were inconclusive, no sufficiently powerful insecticide being available at that time which could both be guaranteed to kill all of the grubs and be easily and safely applied. The solution finally suggested by the Laboratory was fumigation with sulphur dioxide[20].

The work on Westminster Hall showed up the lack of reliable, easy to apply substances to use as pesticides, fungicides, etc., which were not also too dangerously toxic to human beings. At this time the Laboratory was regularly examining some of the compounds used, such as samples of copper sulphate (frequently sold as Blue Vitriol or Blue Stone), which was used as a dressing for wheat, and 'liver of sulphur', (fused alkali or alkali carbonates and sulphur), which was used as a fungicide by fruit growers. One of the first samples of pesticide residue analysis was carried out in 1914, on a sample of apples which had a slight residue on the skin. This was found to be copper, which was thought to be due to the fruit having been sprayed with Bordeaux mixture, and not an arsenical spray as had been feared[21]. Other work with a health aspect which was carried out at this time was a considerable amount on lead in paint and whether or not there was a health hazard associated with it. A Departmental Committee on the use of lead compounds in painting public buildings was set up in 1911 by the Home Office. It resulted in several White Papers in the next ten years, and led to other investigations, including work on paint scrapings from Buckingham Palace in 1921. The Laboratory did a considerable amount of work for the Committee, both practical and in the preparation of reports. An early form of spectrograph was used in work for this Committee.[22]

14

The First World War and its aftermath

THE ANNUAL *REPORT* FOR 1914, ISSUED ON 31 March, was the last to appear before the 'Great War'. As might be expected the war brought significant changes to the Laboratory, both in work and in its staff. Initially established civil servants were not expected to volunteer for the armed forces, and indeed were discouraged from doing so, but as the war progressed, and the need for manpower increased, so the junior ranks of Temporary Revenue and Chemical Assistants became depleted and on 18 July 1916 the first woman scientist, a Temporary Revenue Assistant, Miss E.M. Chatt, was appointed. This was as part of a general move to place women in the lower ranks of the civil service to replace men being drafted to help with the war effort. The reaction of such an action on the older members of the civil service may be imagined. Until then the service had been almost entirely a male preserve, particularly in the professional posts. Women had been qualifying in scientific subjects since the 1870s, at least at Cambridge University although neither here nor at Oxford University could they be given a degree until much later. London University granted degrees to women (in Chemistry as in other subjects), from 1878, and Manchester University from 1880. The effect of the arrival of Miss Chatt on the junior staff was perhaps more favourable, and Richard Church gives (in *The Porch*) a graphic sketch of this arrival. He describes her as being watched in silence by the male staff as she was first brought into the Laboratory, an intimidating experience for anyone.[1] By the end of the war, in 1920, the Laboratory was employing 40 women out of a total of 78 in the junior analytical grades. Twenty three of these were in the new grade of Temporary Laboratory Assistant (the total in the grade) and the rest Temporary Chemical Assistants or replacing Temporary Revenue Assistants[2]. Some idea of the effect of this change can be seen in a photograph of the Tobacco department, taken on 23 April 1917 (plate 57). A total staff of 28 are shown, including 21 women. In 1909 the staff of this department was nine, all men. Some if not all of the women shown were probably 'Temporary Women Clerks', since these began to appear in the Tobacco section in August 1915 and a list of staff needed in the section in 1918 included 18 such clerks (together with one Superintending Analyst, one Analyst First Class, one Clerk Second Division and four Customs and Excise Assistants from a total of 25. Someone who worked there at the time commented later that the arrival of these women occurred at the same time as the introduction of 'morning coffee breaks, afternoon tea breaks and smoking', presumably not cause and effect.[3].

At the beginning of the war Dobbie anticipated that changes were likely in the pattern of the work, and on his own authority reorganised the Laboratory in order to be able to cope with it. This reorganisation worked admirably[4]. He was knighted in 1915 in recognition of his work. The work of the Laboratory in general continued much as before, with samples being examined for the same wide range of Government Departments, but numbers of traditional samples

57   Staff of the Tobacco Department, 1917. Mr J Woodward, the officer in charge is in the centre

tended to fluctuate, usually downwards because of the bad trading conditions. A large increase was noted in the number of samples of tobacco submitted for the assessment of drawback payable on export, intended for use by members of the armed forces in various parts of the world. The numbers of these samples continued to increase throughout the war, with a sudden drop in 1919 after the Armistice[5]. This increase more than counterbalanced the drop in the numbers of Customs samples received. A general increase in work was experienced through additional demands by the Admiralty and War Department, as might have been expected. Throughout the war many thousands of samples were examined for the armed services (and by 1918 this included the Air Force), mostly foodstuffs of every description, and contract stores and drugs. During 1917 for example over 19,000 samples were received from contractors' deliveries which were in course of transit to the Expeditionary Forces.[6] Those were examined by staff at Clement's Inn, or at branch laboratories established at ports. Staff also worked at army supply bases.

Samples were also received from new Departments set up during the war. One of these was the War Trade Department, established early in the war to control the export of goods affected by various restriction measures. A licence was needed, granted by the Department, to export goods falling under a prohibition and in doubtful cases samples were sent to the Laboratory to check if they were subject to restriction.[7] Other new departments for which work was carried out were the Ministry of Munitions, the Coal Controller, and the Central Control Board of Liquor Traffic. The samples from the first of these were fairly innocuous (the work on munitions proper was mainly carried out at the Royal Arsenal, Woolwich), and included special glue and petrol substitutes, as well as a so-called substitute for platinum, which turned out to be an alloy of nickel, chromium and tungsten.[8] Mineral waters were also examined for liability for duty during the period 1916–1946, those for medicinal purposes (as opposed to table waters) being exempt. Samples were received from sources throughout the United Kingdom, including the famous spa resorts.

Work for the Foreign Office, The Treasury Solicitor and the Admiralty Marshal, as well as for the War Trade Department on questions relating to contraband trading with the enemy were first mentioned in the *Report* for 1917, although some work in these fields had been done earlier. In the years 1918–1920 further samples were examined for the Treasury Solicitor and the Admiralty Marshal. These represented materials seized as prize (or Prize as the *Reports* have it), which had to be identified and graded for sale. They were unexciting and were mostly oils, soap, resins, paints and pigments, but also included thorium nitrate used for gas mantles[9]. In addition to these samples, another indication of unusual wartime conditions is seen in 1915 when the Chief Commissioner of Police asked the Laboratory to investigate the products of combustion discharged into the atmosphere from the exhaust of a motor omnibus which was burning coke fuel. Another sign of the unusual was the request in 1918 to test for the effect to the contents, if any, of cardboard containers proposed for the storage of condensed milk. This was necessary because of the shortage of steel. Some German rations, one of tinned black pudding and one of 'sausage substitute' were tested in 1917. There is no information on why the testing was done, or indeed how they came into British hands.[10]

During the war, Dr. Dobbie served on various Committees which were set up to investigate problems caused by war conditions. These included the Research Committee of the War Cabinet, and the Nitrogen Products Committee, necessary to find the best way of making artificial fertilisers and other nitrogen compounds by fixing nitrogen from the air. This was needed because of the vastly increased need for nitrogen products for use in munitions as well as in agriculture. Much of the industrial production of fixed nitrogen was in German hands, and the supply of natural fertilisers from overseas was not certain in wartime conditions. Connected with these committees was the Fertilisers Committee. One of the tasks of the latter was the finding of a method for the extraction of potassium from feldspar, again necessary because most of the potash salts used in the United Kingdom before the war came from Germany. Extensive work was done for these committees in the Laboratory, and on related problems such as the value of ash from various waste materials as potash fertiliser.[11].

After the Armistice in 1918 conditions in the Laboratory returned to normal. The number of samples decreased immediately, due largely to fewer samples of foodstuffs and other materials for the War Office and the Admiralty being examined (although for some reason the number of samples for the Air Board increased massively in 1919). Many fewer samples of tobacco also were submitted for drawbacks on export to the expeditionary forces: the armed forces were reduced in size as fast as possible after the end of hostilities. Reassuringly though even in 1919 there was an increase of 42,123 samples of wine submitted after import restrictions were lifted, and the total number of samples increased again in 1920 as trade revived. Some changes were caused by the war, for example in 1919 the tea samples, which had been examined on importation since 1875 to check that they were genuine tea, were transferred to the new Ministry of Food. However, samples from consignments which were not passed as fit for consumption continued to be sent to the Laboratory. Reject tea had always been allowed delivery duty free for caffeine manufacture, with the tea first being rendered unfit for consumption by denaturing with asafoetida and lime under the supervision of Customs and Excise. Samples continued to be examined in the Laboratory to ensure that this had been done effectively[12].

Soon after the war Sir James Dobbie was able to resume his own research, which had been in abeyance during hostilities. Dobbie was particularly interested in the absorption spectra of molecules, chiefly for what could thus be determined about their structure, but also as a means of determining trace amounts of one compound in another. He regarded the determination of the constitution of a substance by means of the absorption spectra as very useful, but only as ancillary to normal chemical processes. His original interests had been in the alkaloids, and after arriving at the Laboratory he had produced papers on some of these with colleagues, including Dr. John Fox. After the war he continued to collaborate with Fox, but now applying the technique to gaseous forms of elements. The results of this important work were published in the *Proceedings* of the Royal Society[13].

Dobbie continued to serve on committees for the government after the war, serving for example on the Royal Commission for Awards to Inventors. He had passed normal retirement age in 1917, staying on at the request of the Treasury. His health had suffered as a result of his work during the war though and he was forced to retire on 20 July 1920. On his last day of service (writing from home) he wrote

a charming letter to James Connah his deputy, thanking him for all his help and support. He was remembered as a most courteous gentleman, who usually dressed in a tail coat and a top hat. Sir James Dobbie died on 17 June 1924 at the age of 71.[14]

Sir James was succeeded as Acting Government Chemist by James Connah, previously Superintending Analyst at Customs House, who had been Deputy Government Chemist since July 1919. The permanent replacement for Dobbie was Sir Robert Robertson, who was appointed on 7 March 1921. Robertson was a distinguished scientist, he had been elected a Fellow of the Royal Society in 1917 and appointed a Knight Commander of the Order of the Bath a year later. Born in Scotland on 17 April 1869, Sir Robert had worked for a short period early in his career with the Glasgow City Analyst Robert Thomson. He transferred to the Royal Gunpowder Factory at Waltham Abbey in Essex in 1892, where he learned factory operations as well as undertaking original investigations. By 1900 he was in charge of the main laboratory and subsequently had a distinguished career in explosives and as director of the research department of the Royal Arsenal in Woolwich during the first World War. In the Government Laboratory, where the adminstrative burden was carried by others, he was free to devote himself to work which was congenial to him, that is taking part in the activities of many committees and also to carry out research, his own and that of his Department.[15]

58   Sir Robert Robertson,
Government Chemist, 1921–1936

CHAPTER 15

Increasing Versatility 1922–1939

THE LABORATORY SOON SETTLED DOWN into much the same routine as before the war and after the inevitable disturbance caused by the appointment of a new Government Chemist. The annual reports continued to come out regularly, as they had done since 1894, and would continue to do without a break until 1939, giving an account of the work of the Laboratory, listing each department sending samples. In the 1922 Report for the first time the Customs and Excise department was put in its correct alphabetical place, instead of first where it had been historically. From 1921 the large foolscap format was reduced to octavo. This change coincided with the inclusion towards the end of each report of a few pages of more personal detail relating to the professional activities of senior members of staff. Lists are given of some of the committees, to whose work the Laboratory contributed and the publications which resulted, as well as details of the scientific papers published by individual members of the staff, or given at the internal colloquia. These colloquia are a feature of the Laboratory's activities from 1923 onwards, (see next chapter).

Sir Robert Robertson had been appointed Government Chemist at the beginning of a period when legislation was being introduced which sought to protect various industries, to control the quality of consumer goods and to protect the public health. These and other factors lead to an increase in work of all kinds, ranging from routine analysis to research. The annual *Reports* showed this in a steady and apparently inexorable increase in the volume of samples, culminating in nearly 431000 in 1939 (from nearly 195000 in 1921, both excluding samples analysed at the Chemical stations). The point about increasing numbers of samples was made strongly in June 1921 in a letter to the Treasury requesting more staff (see p xxx below).[1]

The samples included those types which had been

received for many years. Regular reference was still made in the *Report* for example to the examination of food supplies for the Admiralty, fertilisers and feeding stuffs (under the *Fertilisers and Feeding Stuffs Act*) and sheep dips for the Ministry of Agriculture and Fisheries, foodstuffs for the Crown Agent for the Colonies, beer and brewing materials, cocoa etc, wines and spirits, sugar and other sweetening agents, tea and tobacco for the Commissioners of Customs and Excise, lime and lemon juice for mercantile marine use for the Board of Trade and water and household supplies for the Office of Works. Large quantities of alcoholic samples were tested to check that the ethanol content was as declared, and that alcohol for scientific and industrial purposes, which was denatured to make it non drinkable, contained the correct amount of denaturant. Some of the 'traditional' samples received, for example arsenic in brewing materials, had been examined since the poisoning scandal of 1900. By about 1930 an electrolytic method had been devised for measuring very small amounts of arsenic, down to a millionth of a gram (a thousandth of a milligram). The arsenic was reduced to arsine in a cell and this was reduced again to arsenic in an electric furnace after sweeping out the cell with a stream of hydrogen. The arsenic was allowed to form a mirror and compared with standards. Other new work in old fields was in the Customs area, where not only traditional samples such as spirit varnishes and transparent soap had to be analysed for the spirit content but also newer forms of lacquer where nitrocellulose was the base. This was dissolved in mixed solvents of great variety and complexity consisting of alcohols, esters and hydrocarbons. From these mixtures the dutiable solvent or solids had to be separated and determined using such techniques as fractional distillation, extractions and chemical condensations. Speed, accuracy and considerable analytical skill was usually needed. Sugar goods had to be analysed in a similar fashion to determine the mixture of sugar, molasses and glucose, all dutiable at different rates. A great deal of money was frequently involved in these samples dealt with for the revenue. The referee function of the Government Chemist under the various *Food and Drug Acts* was also still a regular feature of the Laboratory's work.[2]

Not only were these 'traditional' samples received in increasing numbers however, but so were new types in the

59   Cartoon of *c*.1930, showing Sir
Robert Robertson. Sir Robert liked
this so much that he used it as a
Christmas card

traditional areas, reflecting changes in the law and also
changes in society. For example in the *Report* for 1921
appears the first reference to 'motor spirit'as a dutiable
article. A duty of 3d (1.2p) per gallon had originally been
levied in 1909 and samples had presumably been received
before 1921, although it does not appear in the list of
Customs and Excise samples received, being subsumed under
'miscellaneous'.[3] The duty was increased to 4d (1.7p) in the
Finance Act of 1928. Sir Robert served on a Committee in
1926 (a Committee which included the future Sir Henry
Tizard as a member) to revise the definition of motor oil. The
duty was in fact imposed on all hydrocarbon oils, but rebated
for any oils other than light oils when used for home
consumption. Duty was thus in effect payable only on motor

spirit, white spirit, turpentine and other light hydrocarbon oils. These involved composite products such as enamels, paints, varnishes and lacquers and large numbers of samples (12000 in the first year, 1929). They were subjected to a carefully standardised distillation process to determine the 'light oils', (as defined in the Act) and thus the duty payable or the drawback refundable.[4] Mineral jelly and bitumen paid different duty on importation if they were liquid or solid. The differentiating test was the penetration by a standard cone and needle under specified conditions using a special penetrometer with a novel electrical release mechanism, all constructed in the Laboratory. This work was carried out in the 'main' laboratory at Clement's Inn for many years. Other new Customs samples in this period were 'lager beer', not only brewed in Britain, but also exported, and samples of home grown tobacco leaf taken from cultivation areas in Ireland, the south of England and East Anglia, the latter from Methwold in Norfolk, where the Ministry of Agriculture was assisting in culture trials.[5]

One very important new area of work which began just after the war, and which increased in importance subsequently was the analysis of samples received in conection with the *Dangerous Drugs Acts* of 1920, 1925 and 1932. These Acts were all passed in response to regulations agreed at various Conventions, the first of them at The Hague in 1912. The first of these Acts made it for the first time an offence to import or export without a licence various drugs of addiction such as raw opium, morphine, cocaine, ecgonine and heroin, (or indeed any medicinal preparation containing more than 0.2% of morphine and 0.1% of the others). It forbade completely the import or export of prepared opium, ie opium prepared for smoking purposes. Suspected cases were seized by the officers of Customs and sent to the Laboratory for analysis. The first such samples were received in 1921. Subsequent regulations and the 1925 Act tightened up the situation further, adding to the list of prohibited drugs *inter alia* 'coca leaves' and 'Indian hemp and resins obtained from Indian hemp', and preparations containing any proportion at all of morphine and heroin and their salts. The first sample of hemp under the new Act was not received until 1928–29. The 1932 Act extended considerably by the list of drugs to which the 1920 Act applied.[6]

Methods for the determination of some of these drugs

were developed in the Laboratory during this period, as the samples coming in demanded it.[7] A steady trickle of samples were received in the 1920s and 1930s, varying between 22 and 105 of which (in those happy days), never more than 24 (and usually far less), contravened the Acts. A very small amount of drugs could be determined. In one instance a trace of drug on a hypodermic syringe was identified by observing the formation of characteristic crystals with a reagent under a microscope. Until the Police Laboratory at Hendon was established police officers submitted the samples suspected of containing drugs direct to the Laboratory. The samples sometimes also included blood samples for the determination of alcohol content, (by the reduction of acid dichromate solution), or for the detection of carbon monoxide. This was determined using the Hartridge Reversion Spectroscope. Occasionally blood stains were shown to be present by demonstrating the presence of the absorption spectra of haemoglobin.[8]

In these years after the war many new Acts and Regulations were brought into being to safeguard British industry, or the consumer, or both. These included the *Dyestuffs (Import Regulations) Act* of 1920, and the *Safeguarding of Industries Act* of 1921. The first of these was used to check for synthetic dyes or intermediaries, products of which were forbidden except under licence and did not produce many samples. The *Merchandise Marks Act* of 1911 was also used at this time to ensure that goods were what they claimed to be. That this action was much needed is seen from a sample of 'Biscuit meal' which was found to consist of sour smelling mouldy lumps of biscuit which had been damaged by fire and water and mixed with cement, plaster, burnt wood and paper, coarse string and glass. It sounds as if it could be condemned under other headings too, but it was in any event reported to be totally unfit for use as a feeding stuff.[9]

From the Merchandise Marks Act (re-enacted in 1926) there developed in the 1930s various 'National Mark' schemes laying down specifications for national produce. Such items as National Mark Honeys were involved, (here for example pollen analysis was used to confirm that Heather honey was correctly named), but the Laboratory even examined chemically and bacteriologically water from water-cress beds for a proposed scheme to market water-cress under such a scheme. Other work in the food divisions in these inter

WAS IT SOUR?

60   Laboratory cartoon of *c.*1928 showing most of the senior staff of the Laboratory including Sir Robert Robertson and Dr Fox (picture 5)

war years was done under other new Acts. Samples were examined for liability to duties under the *Import Duties Act* and the *Ottawa Agreement Act* for example and by 1934 numbered over 3200. Under these Acts arose problems of the correct descriptions affecting liability of imports to duty. Under the Ottawa Agreement the Laboratory considered such questions as did the term 'cod liver oil' apply only to the liver oil of the cod fish or also to oils with similar qualities from the livers of other fish of the *gadidae* species: haddock, saith whiting etc. It was agreed that it did. The same laboratory also carried out work in connection with the *Wheat Act* of 1932 and various regulations under the Act, checking that flour complied with the Act in ash or soluble solids according to its grade. These inter-war years saw a considerable increase in foodstuffs standards and for example samples of many kinds of foodstuffs were received for checking under various Regulations as to whether permitted or non permitted preservatives had been used or used in excessive quantities. In the 1930s around 1300 samples of dairy products were analysed annually to check that they conformed to the legal standards of composition and purity. Cheeses were analysed to prove that skimmed milk had not been used in their manufacture. Dr J R Nicholls, Superintending Chemist (later Deputy Government Chemist) used analytical data from over 2000 samples of ten English cheeses to show that their composition fitted into a general formula which included a factor showing whether the cheese had been made from whole or skimmed milk.[10]

Not many samples were received under the *Merchandise Mark Act* or the Scheme, but the *Safeguarding of Industries Act*, designed to protect industries from foreign competition (Empire goods were exempted from duty) was a different matter. The Board of Trade issued lists defining the articles covered by the Act, the first list of chemicals gave some 5600 synthetic organic or fine chemicals liable to duty – all of which could be sent to the Laboratory for analysis. The average number of samples received under the Act in the years 1921–1939 was about 10000. When the samples were first received in 1921/22 (some 3000 in the last six months of the year) it was noted that it was possible by readjustment of the duties of the staff to deal with the samples expeditiously.[11] These samples were examined to see if the chemicals or products came into the class liable to duty, the Key Industry Duty as it was called. The department in which these samples were analysed was known (after the second world war at least to the mystification of junior staff), as KID. A considerable amount of thought and discussion went on in 1921 into the practical aspects of this Act. Regulations as to the sampling of the chemicals involved had to be laid down also for dealing with composite articles. A somewhat similar field of new work arose from the *Import Duties Act* of 1932, which with other similar Acts imposed a 10% *ad valorem* duty on all imported goods, with few exceptions. The range of samples received under these Acts was immediately numbered in the thousands, and involved difficult analyses and investigations. The analyses were complicated by different duties being involved in what were sometimes quite complicated mixtures of organic chemicals, which were difficult to separate by the methods available. The duties involved were not simple either. Specific duty, such as that on ethyl alcohol, took precedence over key industry duty and both took precedence over simple *ad valorem* duty. An example of the complexity involved would be given by an imported solvent which contained ethyl acetate and ethyl alcohol, both subject to specific duty and butyl acetate and butyl alcohol, both liable to key industry duty. The ethyl compunds had first to be separated using methods developed in the Laboratory and then the butyl compounds. The ethyl compounds would then be separated using very efficient fractionating columns. Much effort was expended in determining whether or not a sample contained any of the many chemicals included in the lists in the Acts and if so in what proportion. The samples covered an enormous range, compound dyes, proprietary medicines, cosmetics, flavouring essences, pigments, lacquers and essential oils among others.[12]

One of the features of much of the new work was the very large numbers of samples involved. For example silk and artificial silk were added to the list of dutiable articles in 1925, the Government Chemist being consulted beforehand as to possible chemical links between the different types of artificial silk, in an effort to simplify the definitions of the substances to be included. Sir Robert assured the authorities that there was no such common link. Over 12000 samples were received in that first year, increasing to nearly double that number in the following year, at which level it remained. Only artificial silk was dutiable (on export or import) and complex mixtures had to be analysed, sometines without damaging the article. The analysis was carried out chiefly with the microscope although in cases involving mixtures of dutiable and non dutiable fibres complex chemical methods had to be devised.[13]

The increasing volume of work meant that early action had to be taken to reinforce the permanent staff of analysts, following the minimum recruitment which had taken place during the war years. In June 1921 Sir Robert asked permission to recruit second class analysts. This was granted and the Civil Service Commission issued regulations for the competitive selection of Chemists Class II in the Government Laboratory in August 1921, (they were issued in a revised form in January 1925), and several candidates were called for interview later in the year. The six successful candidates all remained at the Laboratory for the remainder of their career, most of them becoming Superintending Analysts.[14]

By 1922 the permanent staff of the Laboratory consisted of the Government Chemist, his Deputy, five Superintending Chemists (before 1921 Superintending Analysts), nine Chemists Class I and 23 Chemists Class II, (a total of 39), together with 36 Temporary Assistant Chemists. These Temporary Assistant Chemists were chosen by a Board which included the Government Chemist and the President of the Institute of Chemistry. For a post as Chemist Class II the temporary Chemists applied to the Government Chemist who forwarded suitable names to the Civil Service Commissioners. The Commissioners inteviewed selected candidates to 'assess their personality', took into account the record of their work as Temporary staff and examined them in English. Vacancies were filled from a list of these candidates, arranged in order of merit. At the age of 30 if a Temporary Assistant Chemist had not been offered a permanent post he had to leave. The Government Chemist's view was that if a Chemist had not been successful in obtaining a permenent post by the time he was 30 he could not be any good.

There were also 100 Customs and Excise officers seconded for service to the Laboratory.[15] Discussions were held with the Treasury in 1923–24 on the replacement of these Customs officers with a grade of Laboratory Assistant. These seconded officers (known as 'Unattached Officers', unattached to a Customs and Excise station that is), carried out the routine examination of Customs and Excise samples. It was proposed to reduce their number from 103 to 30, for duties at the chemical outstations and for training, and to recruit a number of Temporary Laboratory Assistants from ex-servicemen at 29 shillings per week (£1.45), rising by 2 shillings (£0.10) per week to 35 shillings (£1.75). This was agreed, and a number of temporary clerks on loan from the Customs service, working in fact as assistants in the Laboratory, including all of the women clerks, were returned to their own department, and a new recruitment scheme put into operation in the following year. The matter continued to be discussed on the National Whitley Council (the Civil Service negiotating body) for several years, the Customs staff maintaining that for them service in the Laboratory constituted valuable training. The unattached officers were finally phased out of Laboratory work in 1932, as a result of further reorganisation, which took place through much of the 1920s and 1930s. Sir Robert also argued strongly to the Treasury in 1925 that he needed more staff to investigate new methods. This was necessary because of the increased complexity of the work, the increased volume of which meant that he could not spare any of his current staff for the investigations. He was given permission to take on one Chemist Class I, a Chemist Class II and one Clerical Officer.[16]

This reorganisation was as a result of the report of the Committee on the Staff of Government Scientific Establishments (the Carpenter Committee), which met 1929–1930. Oddly enough although the Government Chemist was on the Committee, the new grades introduced for professional staff were not those recommended by the Report although those for ancillary scientific staff (the Assistant grades) were adopted. The Carpenter Committee had been appointed partially as the result of pressure by the staff side of the National Whitley Council for an enquiry into the conditions of service of scientific staff in the government service. One problem which it addressed was that the Government service was not attracting the 'cream of scientific workers'. As implemented in the Laboratory, under the Superintending Chemists were now Senior Chemists, Higher Grade Chemists and Chemists and Assistants, grades I, II, and III. Older junior grades co-existed with Assistants for some time.[17] During this time the number of non-technical staff was growing all the time, since as the work increased, the scientific staff increased and so did the need for support. As seen above, in 1896 there were ten support staff, by 1939 there were 32 clerks, four typists, and 24 subsidiary staff (porters, messengers, and stores staff), as against 218 scientific staff.[18] The salaries recommended in the Carpenter Report were generally followed in the Laboratory, including the award of special salary increments to selected chemists of high ability.

These salaries had risen since before the war, although not by a great deal, nor indeed for all staff. The Government Chemist for example received the same salary in 1930 as had his predecessor in 1894 (£1500 *per annum*). The difference was that this amount was now worth only a little over half its value in 1894. The salary of the Deputy had risen though, from a little over half that received by the Government

Chemist to £1000 and Superintending Chemists now received £850. A First Class Analyst, or his new equivalent received a maximum (all figures given are the maximum of the scale) of £650. In general salaries rose dramatically during the war and then dropped afterwards. Civil servants had received less than the national average salary before the war, and still did so. Scientific staff received less than the corresponding administrative grades. For example in 1914 the average salary in the administrative civil service was £550, that of his scientific equivalent, in the form of a First Class Analyst in the Laboratory, £480. With the outside world the Laboratory salaries compared well. The 1931 Institute of Chemistry income survey showed that nearly 50% of their senior grade members (Fellows of the Institute, Superintendent level) received between £500 and £1000 a year, and 62% of Associates received between £250 and £500.[19]

The apparently inexorable increase in staff meant that pressure on the Clement's Inn Passage premises became even greater. They had been designed with some expansion of the work in mind, but were occupied to full capacity very soon after building. Thus in 1916 the first of many areas of 'overflow' accommodation was established in the vicinity. This was a single storey temporary building in the grounds of the Royal Courts of Justice (adjoining the Laboratory building), immediately to the south of the Bankruptcy Court, and from its structure invariably referred to by the staff as the 'tin hut'. It was in fact a World War I army hut. This building was referred to by the Principal Architect to Royal Palaces as the 'iron building' in correspondence with Sir James Dobbie in November 1915. It had first been used by the Public Trustee.[20] It was fitted with chemical benches and work relating to brewing control was transferred to it. This work continued to be carried out there until the building was destroyed in 1940 by enemy action. Other overflow accommodation actually in the Bankruptcy Court building in Carey Street was used from about 1928. A suite of rooms on the top floor of this building were used, these premises being adjacent to the Laboratory, on the other side of Grange Street which ran along its south side. Access was from a gateway in the iron railings not normally available to the public. A small luncheon club, to which Laboratory staff were admitted, operated in the Bankruptcy buildings surviving until well after the war. The work for which the rooms were taken was the examination of silk and silk containing materials for liability for duty, (see above). A subsequent increase in the work and development of methods used in the examination of tobacco made further new accommodation essential, and this was provided in 1934 by HM office of Works building a new storey on to the top of the headquarters building. The laboratories in this new storey contained the latest equipment, including electric ovens. These ovens contained fans, designed by the staff, to ensure uniformity of temperature.[21]

The principal site used at this time as overflow accommodation was at 5/6 Clement's Inn (usually known as '5 and 6') parts of this building being occupied in the 1930s.[22] Clements Inn was a narrow lane with a gate at each end running due south from the junction of Clement's Inn Passage and Grange Court (and opposite the main entrance to the Laboratory) to the Strand. It was closed to the public on one day each year to preserve in law its private character, that is to say, Clement's Inn was not a public right of way. The area underwent considerable change following the move of the Laboratory to Cornwall House.[23]

The actual building used for the 'overflow' was red-brick former legal chambers. The units were addressed in pairs, numbers 1 and 2 at the Strand end, numbers 3 and 4 in

61 Tobacco section in the new storey of Clement's Inn, *c.*1935

the middle, and 5 and 6 immediately adjoining Clement's Inn. It was thus geographically very convenient although far from suitable as laboratory premises. Laboratory benches were installed in many of the rooms, but some of them were still heated by open fires. Access to the upper floors was by means of stairs and also a lift from the entrance hall. The lift was operated by hydraulic power, activated by the operator pulling a rope which ran the full height of the building and through holes in the roof and floor of the lift itself.

Non-revenue outstation laboratories staffed by the Laboratory had also been taking some of the work for many years. These were at the Geological Survey Museum in South Kensington (firstly in Jermyn Street, Piccadilly) for example and at the Supply Reserve Depot at Deptford in south-east London for the War Office. The latter was destroyed in 1940 and the work transferred to Glasgow. To these were added in about 1928 laboratory premises at Park Royal for the examination of stores for the Office of Works, under Mr Gilbert, the Inspector of Stores. This too was destroyed by enemy action during the war and staff accommodated in

62   Laboratory cartoon of *c.*1935 showing Edwin Nurse, just appointed Inspecting Officer of the outstations

THE NEW CAPTAIN COMES ABOARD

Dean Bradley House in Horseferry Road, Westminster until new premises were found at Barry Road in Harlesden, north London.[24]

Since 1917 chemical stations (for Revenue work) had been only 12 in number with briefly the addition of Belfast (closed in 1916) from 1921 to April 1922. The station at London Dock was run from the Custom House Laboratory and that in Liverpool was in the charge of three of the Laboratory's chemists. By 1929 work at Liverpool was extended to types of samples not dealt with at the other outstations. This included coffee, matches, silk and artificial silk and various food samples. In 1932 the senior chemist at Liverpool also supervised the Manchester station. This closed in March 1934. The remainder of the outstations, at Bristol, Burton, Glasgow, Greenock, Hull, Leith, Manchester, South-ampton and Dublin (which appeared in the Annual Report for the last time in 1923) were operated by Customs officers previously trained in the Laboratory. The residues of selected samples tested in the outstations were retested at the head-quarters laboratory as a check and were normally found to have been dealt with in a highly efficient manner. The number of stations steadily dwindled until by 1939 there were only six, Burton, Greenock, Leith and Manchester having been closed. The Scottish work was centralised in Glasgow. In the year 1933/34 the use of specially trained Customs and Excise officers in the Custom House and chemical outstations, which had been a feature of the Laboratory since its Inland Revenue days in the middle of the nineteenth century, finally came to an end. The 1934 Report was the first to say that the chemical stations were now staffed by officers from the central Laboratory. The return of Customs and Excise officers to general duties had com-menced in the previous year, and no such officers appeared in the staff lists after 1934. The phasing out process had been under discussion since 1920.[25]

CHAPTER 16 Research

THE NEW FIELDS OF WORK AND WIDE ranges of new samples now being received demonstrate the increased involvement in science by Government at this time. These changes were twofold, fuelled by an ever increasing involvement of the state in everyday life, be it improving the public health, controlling the quality of consumer goods, raising revenue or in protecting various industries. The Laboratory was itself increasingly involved in what can be described as the government science network. It no longer stood alone as the only major analytical establishment. The part that the Government Chemist played in in the foundation of the National Physical Laboratory in 1898 has been described and other laboratories had been founded since that time. The Department of Scientific and Industrial Research had been set up by the Government in 1916 and had itself founded the Chemical Research Laboratory in 1925. Its role was to carry out pure research in such areas as the corrosion of metals, water pollution and coal tar. Some of these were areas in which the Laboratory was active too.

Members of staff of the Laboratory helped to establish other laboratories in the inter war period. In 1935 two senior members of staff were seconded to the Metropolitan Police to help set up the first forensic science laboratory at Hendon and others went to the Water Pollution Research Board, the Medical Research Council, the Ministry of Agriculture and Fisheries and abroad to assist the Egyptian Government.[1] Another feature which developed at this time was the training of analysts for Colonial posts, with subsequent liaison visits to London.[2]

The Department of Scientific and Industrial Research also began to take up staff time, chiefly in support of committees.[3] This activity demonstrates the increasing contribution of Laboratory staff to concerns outside their own Department. The involvement of senior staff with Royal

Commissions began very early in the life of the Laboratory and work at this level continued, but Departmental committees, and indeed international groups, now proliferated. This proliferation continues particularly on an international scale. The number of committees was noted in the Annual *Reports* from 1922 to 1939 (after the Second World war this was impractical), and rose inexorably from seven in the former year to 50 in 1939. This does not include 'numbers' served on by the Government Chemist, and not clearly specified. Nor 'various committees dealing with Departmental questions and regulations' of the Board of Customs and Excise. Also part of the trend is the noting for the first time that as well as samples, 'numerous papers were submitted for consideration and elucidation of technical matters'. The Laboratory was now increasingly acting as a source of advice to many departments of state as well as providing them with analytical services. Help was also given in drafting specifications and devising new methods of analysis.[4]

Some of the Committees involved dealt with far reaching and important issues. Sir Robert Robertson (and briefly before him, Sir James Dobbie), chaired an international technical committee to advise the Board of Customs on denaturants for alcohol for use as road fuel (so-called 'power alcohol'). Denaturants were necessary to avoid fraud by the use of the alcohol for other purposes. Eventually a mixture of

63 Main Laboratory, Clement's Inn, 1935

methanol, wood naphtha, pyridine bases and benzene with a red dye was recommended. A considerable amount of work went into finding a suitable formula. The situation was similar to that in 1853/1854 when George Phillips was trying to find a denaturant for alcohol to enable it to be freely available for industrial purposes and produced 'methylated spirit'. As then the main object under consideration by the Committee in 1923 was to find markers which could not be removed by distillation and which left alcohol non potable no matter how ingenious the distillation or other process to which it was subjected. The final product was yet another form of methylated spirit in which there was apparently at first a great lack of interest, a disappointing result after so much hard work.[5] Interestingly, at a meeting of the Committee in 1923 Sir Robert mentioned a movement in Germany to change the name of methyl alcohol to 'methanol', adding that the suggestion seemed to have something in its favour.[6] A different aspect of road fuel arose in 1923, when the Laboratory worked with the British Engineering Standards Association (now British Standards Institution), on standards for both motor spirit and aviation spirit. This committee was chaired by Dr John Fox, later Government Chemist, but then a Superintending Chemist. Much work was involved in developing and testing the specification, which had to ensure that, for example, there was a sufficient proportion of the constituents which volatilised at a fairly low temperature, necessary for ease of starting an engine, and that kerosene and other higher boiling oils had not been added to petroleum spirit. One of the two aviation fuel specifications developed specified a freezing point of not less than $-60°C$.[7] This Committee continued its work into the 1930s.

Also in 1923 the Colonial Office requested the help of the Laboratory in evaluating technical information which it had accumulated together with the government of Palestine. This information was relevant to the economic development of the chemical and mineral resources of the Dead Sea. This had become important due to the League of Nations Mandate over Palestine held by the United Kingdom. A committee was formed to consider the matter and the possibilities were recognised of recovering carnallite (a mixed chloride of magnesium and potassium), potassium chloride and bromide.[8]

Sir Robert Robertson chaired a committee on methods

for the estimation of fibre in feeding stuffs for the Ministry of Agriculture in 1926. This was in order to secure agreement on the method which was subsequently prescribed under the *Fertilisers and Feeding Stuffs Act* of that year. Members of the staff were seconded for work in connexion with the Ethyl Petrol Committee, the Atmospheric Pollution Committee and for river surveys (for pollution) carried out by the Ministry of Agriculture and Fisheries in 1928. By 1930 Sir Robert was able to refer to a very esoteric selection of technical committees on which staff served, including the Empire Marketing Board, the General Stores Technical Co-ordinating Committee, and the African Liquor Traffic Control Committee (for the Colonial Office).[9]

To the various official committees on which the Government Chemist served (or was represented) was added in 1933 the newly constituted Poisons Board; this board was then actively engaged in drawing up revised lists of poisons. In the same year the Society of Public Analysts (now the Analytical Division of the Royal Society of Chemistry) published its first report of a standard method for the determination of unsaponifiable matter in oils and fats, to which the Laboratory contributed. Standard methods were developed through the activities of many of these committees and a considerable amount of analytical work was needed. Sir Robert, in his *Report* for 1930 said, 'a considerable amount of work has been done in connection with the revision of existing methods and the investigation of new methods of detection and determination of substances.' The work was reflected in the papers and reports published by staff during the year.[10] Similar remarks were made in other *Reports* at this time.

Much of the work could and frequently did involve the appearance of staff as expert witnesses in court. This role had been developing since the nineteenth century. In 1935 such cases involved prohibited drugs, alcohol in blood, tobacco and saccharin smuggling, illicit spirits, beer adulteration, the illegal admixture of petrol and kerosene, salmon poaching and the watering of milk and the abstraction of cream, a mixture of traditional laboratory areas and new ones reflecting changes in the lifestyle of citizens and also changes in the law.[11]

As has been described much of the work involved with the new samples being received in the 1920s and 1930s and of course the committee work, required considerable research, as indicated by the above quotation. The neccessity for research was expressed in such remarks as 'to keep abreast of advances in manufacture and industry where these concern matters dealt with in the laboratory, and for the improvement of processes of analysis', written in 1931.[12] The consequence was an increased importance given to research by Sir Robert Robertson. On his appointment in 1921 he introduced at the end of the Annual *Report* a list of papers published in the scientific literature. There were three in 1921 rising to an average of 15 in the ten years before the second world war. Two years later he instituted periodic meetings at which staff read research papers, also listed in the *Reports*. These internal papers evidently originated in the instruction given by Sir Robert soon after he became Government Chemist in 1921, instituting two new series of records, of Government Laboratory Reports and Papers. These two series were much the same, the Papers being less formal than the Reports. Both lists in the *Reports* frequently contain papers by Sir Robert himself, and by Dr Fox, his successor. Interestingly enough, Fox, a research chemist, immediately abolished the periodic research meetings when he became Government Chemist in 1936 probably because he had considered them to be of little practical use. The meetings lasted up to an hour and were held after official hours. They were therefore not very popular with staff.

The range of topics covered in these internal meetings was considerable, in some cases the papers were general surveys rather than in depth research, but in others a considerable amount of work lay behind them. Sometimes the object was obviously to be useful practically to other Laboratory staff, as in 'Some Problems arising from our daily work', in 1930 or 'Adding up to 100%' in 1924. Others were of general interest, such as 'The Explosives exhibit at Wembley' by Sir Robert Robertson, an expert in explosives, in 1925, or sometimes they were papers on new methods of analysis, as in a series on the infra red by Sir Robert and Dr Fox (1926 onwards), or the 'Raman Effect' by Dr Fox in 1933.[13] Analytical work by staff also resulted in books of great importance, one by Fox and T H Bowles on paint analysis was for long a standard work, and *Alcoholometry* by F G H Tate is still of considerable value.[14] The development of new or improved methods of analysis, to meet the needs of the public service had been a feature of Laboratory activities

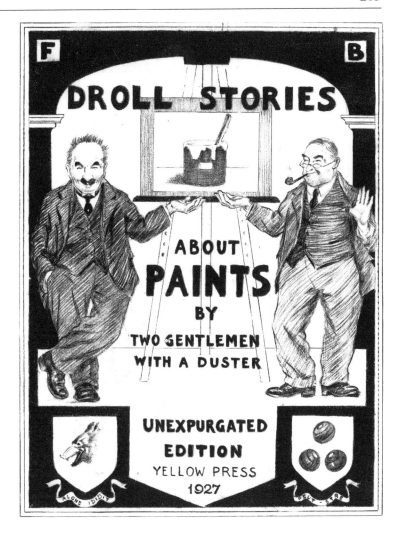

64 Laboratory cartoon of 1927 after the publication of Fox and Bowles book *The Analysis of Paints, Pigments and Varnishes*

for many years but the increase at this time was very considerable.

Sir Robert Robertson retired as Government Chemist on 16 April 1936 but continued with his work at the Royal Institution. Later during the second world war he returned to work on explosives at the Royal Arsenal in Woolwich after which he again returned to the Davy–Faraday Laboratory in the Royal Institution. A broad-shouldered man of small stature but strong personality and with a deep voice, he was a rigid but just disciplinarian and his autonomous powers of appointment and promotion of staff commanded considerable respect. He encouraged research by members of staff, giving them every facility to do so, provided they worked in their own time. He had a knack of getting his own way, but had a good sense of humour, and a hearty laugh. He died at the age of 80 on 28 April 1949.[15]

Sir Robert was succeeded by Dr J J (later Sir John) Fox, who had joined the staff 40 years earlier and had risen first to the rank of Superintending Analyst (later called Superintending Chemist) and in 1929, to Deputy Government Chemist. In his early days he assisted Sir Edward Thorpe in the work of the Departmental Committee set up by the Home Office to ascertain how far the danger of plumbism might be diminished or removed by substituting for white lead either less soluble compounds of lead or "leadless" glaze. During the first world war he was in charge of the Crown Contracts section, where his wide knowledge of chemistry enabled him to advise Departments on many matters relating to the war effort, such as captured materials and seized cargoes. When the *Safeguarding of Industries Act*, which placed a 33% duty on synthetic chemicals, came into operation in October 1921, he was called upon to organise the section of the Laboratory dealing with this work. It was largely owing to his encyclopaedic knowledge of organic chemistry and his sound judgement that the administration of the Act proceeded smoothly. His attention was also directed to such matters as the causes of decay of buildings, pollution of rivers by drainage of tarred roads and the cleaning and restoration of walls and paintings in buildings under the charge of the Office of Works, notably in the Chapter House at Westminister and in the Orangery at Hampton Court.

Fox continued to recognise the need and value of research and investigation in supporting the work of the Laboratory after he became Government Chemist. As seen above, he had a considerable interest in infra-red techniques. In his first full year as Government Chemist he published a paper in *Nature* on 'Some Infra-red bands in the 3 region', and other research papers continued to be published.

Experiments with infra-red spectroscopy had commenced some time before 1927 under the direction of Sir Robert Robertson. Dr Fox had discussed the purchase of a new instrument with Adam Hilger Ltd as early as 1922. They had quoted £125 for the supply of an infra-red spectrometer. This was fitted with a rock salt prism and a thermopile detector. The optics were completely open to the atmosphere, the prism being varnished with a solution of pyroxylin in amyl acetate which 'allows most of the infra-red rays in the region over which this spectrometer is calibrated (0.5–0.9 µm) to pass almost unabsorbed'. Work on analytical infra-red spectroscopy only really started in the mid 1930s though and a new spectrometer with a 6 inch rock salt prism was constructed. This prism was originally varnished and gave disappointingly poor resolution. When the varnish was removed a wealth of fine structure was seen. The removal of the varnish meant that the prism needed protection from moisture, so the optics were enclosed in a glass case which was subsequently replaced by perspex in the post-war Grubb Parsons spectrometers. Recording a spectrum was a very laborious procedure: each wavelength had to be selected by hand, the slit settings adjusted to compensate for the non-uniform emission of the Nernst filament and the output from the thermopile detector measured with a mirror galvano-meter. A complete 'finger print' spectrum would require several hundred readings. In another field, work on emission spectrography led in 1935 to the production of a method for determination of small amounts of magnesium in aluminium alloys. A spectrograph made by Messrs Hilger was used with a home-made microphotometer to measure the intensity of

65 Hilger infra-red spectrometer, 1922

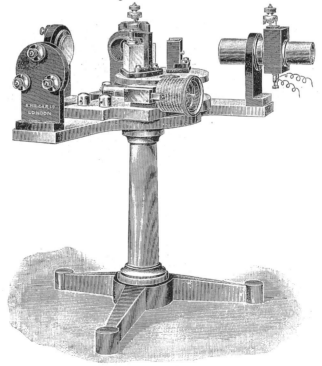

lines on the photographic plate, the problem of non-reproducibility of the emission being avoided by comparing the intensity with a line of aluminium as an internal standard. This work was later of value during the war.[16]

A new instrument, the Tate Saccharometer, designed and developed in the Laboratory for use in breweries and distilleries, was introduced in 1936 after exhaustive testing under practical working conditions. It was chromium plated and had no poises (counter-weights) to bother with. Additionally it had a longer stem than the Bate saccharometer (which it was intended to replace) so that it could be calibrated more accurately. It was in use until about 1976 when all metal hydrometers were phased out. Another project carried out in 1936 was an interesting exercise of pre-synthetic insecticide days. It was the search for a suitable acaricide to control bed-bug infestation. The substance selected as being readily available, harmless to humans, cheap, simple to apply and effective was heavy naphtha with a boiling range of 160–190°C.

Advisory work included the provision and revision of technical specifications for stores items for the Office of

66   Tate saccharometer

Works and food supplies for the War Office. Various research committees of the Research Associations came into prominence at about this time, including those concerned with cocoa and sugar confectionery, food manufacturing generally, laundering, paint and varnishes and the electrical and allied industries, on all of which the Government Chemist was represented.[17]

One entirely new area of work carried out during this period arose directly out of the first world war. This was the recovery of radium from gun sights, compass cards and other items which had been treated with radium salts to make them luminous. Samples of these were first received in 1919–20 and continued until 1924, although work continued until 1939. Not many samples were involved, never more than 40 or so, but the recovery posed considerable difficulty. The process involved first the removal of the 'luminous' ingredients (a mixture of zinc sulphide and radium bromide) from the large excess of paint. After a lengthy series of precipitations, fusions and extractions a mixture of chlorides containing about 1% radium was obtained and this was further concentrated by fractional crystallisation of the bromides. About 200 mg of radium bromide accumulated each year up to 1934 and purification continued up to 1939. The radium pool was made available on loan to other scientific institutions. Although luminous paint was widely used in the second world war there was no call for the recovery exercise to be repeated, partly because the atomic energy programme provided other sources of radioactive materials but also because the long-lived isotope radium-226 had been at least partly replaced by the much shorter lived radium-228. In 1960 it was decided to dispose of the remaining stocks of impure radium, although some purified material was retained for use in the Radiochemistry division for standardisation purposes.[18]

In an extension of this radiochemical work in 1930 the Directorate of the Meteorological Office approached the Government Chemist with a request to plate copper electrodes with polonium in instruments used to measure the electric potential gradient in the atmosphere. These were for use at weather stations at Kew, Eskdalemuir and Lerwick. The source of the polonium was a collection of old radon tubes made available by the Middlesex Hospital. By treatment of these with acid, a solution was obtained containing *inter alia* radium D and polonium-210. Immersion of the copper rods in this solution caused deposition of a layer of polonium and by the time they were due for retreatment after three months the activity of the solution had been regenerated by decay of the radium D and its conversion into polonium-210. This work is described in the 1955 *Report* and continued until 1975, when it was transferred to the Radiochemical Centre at Amersham.[19]

Life in the Laboratory was not all as serious as might have been thought from a discussion of its research activities however. The community spirit noted in the 1890s was still very apparent, shown by the staff dinners which were organised at intervals throughout the 1920s and 1930s. These were organised by the staff, once at least by the women clerks and typists. They were first revived after the first world war, in 1919, in the grandly named '1st Annual Dinner'. This one was presided over by Sir James Dobbie. This event was much the same as the previous similar dinners with a concert in which members of staff played and sang, but later events seem to have been solely dinners. These dinners were revived after the war in 1948 and subsequently in the 1960s and 1970s. They were then organised by the Laboratory Staff Club. A later series for retired and older current members of staff was founded by Dr Egan.[20]

The popular press too seem to have regarded the Laboratory as a community, albeit a community of slightly eccentric individuals carrying out mysterious but undeniably valuable operations. The *Strand Magazine* published a description of the laboratory and its work in 1902 and they did the same in 1939. The article gives a good description of the routine work as well as the more unusual samples such as the lump of 'ambergris' which turned out to be beef fat. The staff were pleasantly surprised by the accuracy of this article when it was published. This publication of popular articles on the Laboratory had begun as early as 1877 when an article by Charles Dickens appeared, slightly whimsical in tone but containing much solid fact. Just after the second world war another article, more popular in tone appeared, and others appear at intervals to the present day. The Annual *Report* too, in the 1920s and 1930s in particular, was frequently quoted in the popular press, who usually seized upon something unusual to describe. The Laboratory was and indeed is, perpetually being 'discovered'.[21]

17 War Again

AS IN SO MANY OTHER WALKS OF LIFE, THE second world war from 1939 to 1945 marked the end of an era for the Laboratory causing many changes, although much of its work continued at both Clement's Inn and the London Custom House throughout the period of hostilities. There was to begin with some dispersal of staff and work. Detailed arrangements were made for the evacuation of parts and then the whole headquarters staff to an undisclosed destination. A school in Bath was considered but sites in Bristol were finally decided on. In the event no general dispersal took place, although some senior staff were seconded to other departments and others were dispersed to branch laboratories which were set up in Aberystwyth in October 1940 and a smaller group (under Mr A T Parsons) in Colwyn Bay. These outstations were concerned with specific areas of work related to programmes for other government departments who had themselves moved out of the centre for the period of war. The Aberystwyth branch had moved with the Ministry of Food to the Zoological Department of the University College of Wales. It consisted of 12 technical staff under Mr C A Adams (later under Mr E H Nurse, (a Superintending Chemist and Senior Chemist respectively), supported by a small group of locally recruited staff. They did not return to London until January 1946.

Other members of staff associated with food supplies for the War Office were transferred to supply reserve depots in this country, or in uniform, to various stations overseas to continue this work nearer the battlefront. They were usually given the rank of Captain, and later Major, while on this work. One such person was Mr E I Johnson (later Deputy Government Chemist), who set up small laboratories first at Redon in Brittany, then in Beirut in the Lebanon and Sarafand in Syria. Others set up laboratories in Haifa (Palestine) and Alexandria in Egypt. Later in the war chemists from

outside the Civil Service were drafted to make up the numbers required. The system was continued after the war and overseas laboratories were established at supply reserve depots in Cyprus, Singapore and in 1955 at Munchen-Gladbach in the Federal Republic of Germany. The latter subsequently moved to Viersen near to the Netherlands border and the Singapore laboratory was transferred to Hong Kong in 1971. Army food supplies had been examined by the Laboratory since before the first world war and later staff had been attached to the supply reserve depot in Evelyn Street, Deptford (see above, Chapter 15). In 1940 the examination was transferred to first floor premises in accommodation nearer to Clement's Inn Passage in Kean Street, (off Drury Lane on the other side of Kingsway). Some 14 years later it was transferred to laboratories in the basement of 5/6 Clement's Inn Passage. At the same time the Laboratory's close association with armed services food supplies was marked by the Chief (WD) Analyst being provided with an office in Kingston Barracks, the then headquarters of all this activity.[1]

The Laboratories at Clement's Inn and the Custom House continued to work throughout the period of the war, with the work load remaining steady. The same applied to most of the outstations, although at Hull the work fell off to such an extent by March 1940 (due to fewer ships entering the port) that it was closed on 1 May and the work transferred to Liverpool for the duration of the war. The hours of attendance at the London laboratories were staggered, with a relatively early start (8.00 or 8.30) and, in common with other premises, windows were screened and lamp shades (the old-fashioned standard conical white glass variety) painted black on top. Although no extensive damage to the fabric of the building occurred it did not escape the effect of enemy bombs completely. One night a bomb exploded in the grounds of the adjoining law courts, blew in the front door and damaged windows of the Laboratory. The loose and broken windows were removed the following morning and the gap filled with woodwork and plastic sheeting pending more permanent repairs. 'Work as usual' then resumed without interruption. The Bankruptcy Court building was extensively damaged and use by the Laboratory staff of the bicycle parking area adjoining this in the outside courtyard in Grange Court was discontinued. Damage to the

67   Sir John Fox, Government Chemist, 1936–1944

Laboratory was also caused by a unexploded anti-aircraft shell which fell through a large water tank above one of the laboratories and caused a considerable mess, destroying some books and papers.

Able bodied male staff were rostered for 'fire watching' duties. Those involved remained on the laboratory premises overnight, on approximately one night each week, in order to alert and help the fire and other emergency services in the event of damage from incendiary or high explosive bombs (some of which did not explode). Simple sleeping arrangements were provided in the Storekeeper's day time office in the basement. Two camp beds were provided. Since fire-watchers were rostered in pairs, any sleep was taken 'in turns' but both were expected to be on the alert during periods of air raid warnings, which varied from a few minutes to the whole of the night. Many of the staff also did similar duties at their home location, so that they might be on 'fire watch duty' on two or three nights each week. A small allowance, a few shillings for each period of duty at the Laboratory, was paid on a four-weekly basis. During the day an alarm system was organised. When a bell rang for about 10 seconds it warned staff that enemy aeroplanes could be expected in about 20 minutes. 'Imminent danger' was signalled by seven short rings of the bell. Planes were then about five minutes away.[2]

Much of the wartime work of the Laboratory was a continuation of the pre-war work, and always depended on the variety and numbers of samples and enquiries received. War time shortages and regulations increased the work in various ways though. The newly established Ministry of Food issued a vast array of regulations regarding every aspect of food supplies. Ministry inspectors brought to the Laboratory samples for checking for identity and conformity with war time specifications. Evidence was frequently given in court as to whether or not a cooked article (for example) conformed with specifications which included size and shape. Economies in the use of food had been instituted and in London kitchen waste was collected for feeding to pigs. Samples of this were sent to the Laboratory for determination of the food value which was found to be closely related to the dry matter content.

Many commodities of all kinds were rationed and practically all were in short supply. This led to the sale of substitutes, samples of some of which were sent to the Laboratory because they were found to be fraudulent. For example a sample of 'poultry food' which had caused hens to froth at the mouth and die in agony was found to be common salt disguised by mixing with millers' offal sweepings and an article described as good for washing woollens turned out to be a water softening substance which would have caused the clothes to shrink seriously.

The Government Chemist himself joined the staff of the Director of Scientific Research in the Ministry of Supply, as Chief Adviser to the Controller of Chemical Research. He also retained his duties at the Laboratory and served on many new committees including the Committee on Enemy Oils and Fuels and the Infra Red panel investigating methods for analysing enemy fuels by infra red and ultra violet methods. Much of his new work involved ensuring that chemical supplies needed for the war effort were forthcoming, toluene for the manufacture of TNT for example. In fact a considerable amount of confidential/secret work had been done in the Laboratory before the war. A very large amount of effort had gone into the analysis of alloys for the Air Ministry in the 1930s for example, mostly on aluminium alloys. Considerable research was carried out into spectrographic methods as has already been mentioned. In 1942 (on 5 April, his 65th birthday) Fox was due to retire but was given informal permission to continue in office. This was to be continued one year at a time. In effect the Head of the Department (the Government Chemist) gave Dr Fox permission to continue.[3]

An early task arising from war conditions came about with the imposition of the new Purchase Tax on 21 October 1940, which distinguished between luxuries and necessities. This new tax called for a critical interpretation of formulations, for example of various proprietary articles (including medicines and other remedies) and required the analysis of these in certain cases. Examples of the many other analytical investigations undertaken included the examination of zinc alloy bomb tail-fins in 1941 to determine why some had fractured. In 1943 a study was made of the deposits in hydrogen compressors for the Department of Hydrogen Production of the Air Ministry. Hydrogen for barrage balloons was produced in large quantities by the action of caustic soda on ferrosilicon. The deposits were identified as iron phosphate due to the presence of phosphorus in the ferrosilicon giving rise to phosphine; this material when held under pressure in the presence of water was oxidised to acids of phosphorus which then attacked the iron. Deposits of another kind were the subject of a request from the Ministry of Works in 1944 to investigate the dust in the air supply to the War Cabinet underground fortress. Another project was the production of an oxygen generator for use in submarines. One based on the use of potassium chlorate had reached the production stage when the war ended.[4]

Other samples related to the effect of enemy action. Early in 1941 the Public Record Office, on behalf of a private finance house, asked the Laboratory to devise a means of reading several thousand blackened record cards following a fire at the firm's offices. A simple method was recommended; to dip the cards in alcohol and examine by oblique lighting while still wet. A letter of thanks from the firm has a pencilled footnote '[telephone] exchange out of order'. Fire precautions were also the subject of a request for advice from an Admiralty rum store regarding foam fire extinguishers. It was confirmed that the normal foam producers of the degraded protein type were ineffective against burning alcohol and offered no advantage over water. Occasional requests were also received from the Ministry of Supply for advice on the value of salvaged materials, one unusual one being a consignment of 'packaged heat', which was mainly powdered iron and a request from the Ministry of Food to help in the salvaging of a cargo of butter, which after a difficult voyage had ended up blackened with engine oil. The large number of casualties during the war led to vastly increased demand for artificial eyes and the Ministry of Pensions considered the mass production of plastic eyes containing a printed iris sandwiched between two pieces of perspex. It was found that the dyestuff Monastral Blue faded, due to the solvents in the cement used to bond the perspex. As both the plastic and the dyestuff were ICI products the firm was asked to produce alternative products. Another area of concern was the conservation of resources. The Ministry of Supply asked the Laboratory for a process for recovering silver from spent photographic 'hypo' solutions; the estimated cost of the recovered silver was reckoned to be less that 1½d (0.5p) per gram. The Armaments Research Department of the Ministry of Supply sought advice on the suitability of American asbestos for use in Gooch crucibles. A more substantial project was the production of flax in Britain, which led to the establishment of a number of processing factories and a prolonged investigation by a committee into the problems caused by dust in these. This work continued until 1945.

In 1944 the Ministry of Aircraft Production requested the Laboratory to undertake the examination of some special alloys for high temperature use for the Royal Aircraft Establishment in connection with the development of jet propulsion. This entailed the determination of several elements in about twelve samples per week. There was some reluctance to take on this work at first since it would require the full time services of four experienced analysts who were already engaged on high priority work. Once it had been taken on it could not be started as soon as had been planned since there was an unforeseen delay. On 14 June the Laboratory apologised to the Ministry for this, explaining that 'the enemy early this morning blew in our windows and did other, apparently minor, damage'. Another hint of important things was a request from DSIR to carry out the 'determination of traces of X in certain specimens'. The details were given verbally but it is evident that X was uranium, some of the draft letters on the file, including the result of a literature search for methods, mention uranium, but all formal communications confine themselves to 'X'. The report to DSIR was directed to the Directorate of Tube Alloys, the code name used for the atomic bomb project. The specimens included post mortem remains from Guy's Hospital; no uranium was found, the detection limit being 0.1 ppm. Work on uranium went on after the war for a considerable time. It finally merged into general radiochemical work.[5]

Some minor work arising from war conditions which is of interest in showing the efforts made to find substitutes for common items was the submission of samples of tinned iron and tinned brass cutlery. These were sent to see if they would be adequate substitutes for stainless steel. The advice given by the Laboratory was that brass would be alright if it was regularly cleaned and verdigris not allowed to build up, on the grounds that verdigris is poisonous. One sign of war shortages was that all draft letters were written on the back of old reports, and new reports were on very poor quality paper.[6]

The bomb which fell in the grounds of the Royal Court of Justice caused the work on brewing and alcohol control (conducted in the 'tin hut' in the grounds of the Royal Courts of Justice since about 1917) to be moved, first to what was termed the 'special room' on the top floor of 5/6 Clement's Inn and, in 1946, to premises in Foster Lane, near to St Paul's Cathedral in the City of London. Later it went to rooms on the top floor of Cornwall House in Stamford Street where a few rooms had been in use for several years (since about 1928) for the conduct of analyses which called for freedom from contamination by trace elements of atmospheric origin. During the war one room was in use for testing food for contamination with mustard gas. Other work also involved equally hazardous substances, such as developing a technique for dealing with unexploded bombs by dissolving a hole in them with strong acid so that the TNT could be blown out

68   Rock salt prism, used late 1930s and early 1940s in infra red work

with steam, or disposing of pure TNT accumulated during work on its manufacture from toluene by burning it in the Laboratory boiler furnace.[7]

As the war drew to its conclusion Sir John Fox, the Government Chemist, who had been knighted in 1943, died in office on 28 November 1944. He was in his 71st year. Dr A G Francis, his Deputy became acting Government Chemist for the next few months. Sir John had served in the Laboratory for his whole working life, joining as an Excise Assistant in 1896. He attended the courses at the Royal College of Science, then recently organised by Dr Thorpe (he obtained a scholarship 1897–99), and at East London (later Queen Mary) College, but for some reason only graduated BSc 'by research' in 1908. He obtained the DSc in 1910. His scientific interests changed over his career. He was initially interested in organic chemistry, later in spectroscopy and its relation to molecular structure, and in the UV absorption spectra of such substances as the alkaloids, sulphur and the halogens. He finally turned to infra red spectroscopy.

Fox was a man of enormous energy, with a boundless enthusiasm for chemistry. He undertook a great deal of research in his life, publishing his first paper with his Professor of East London College in 1901 and his last in 1940, 68 of them altogether, including one book and several obituaries. He had a photographic memory, with an encyclopaedic knowledge of chemistry, and frequently cleared up problems by remarks founded on his own wide knowledge. He was kind and generous to younger staff if slightly overbearing at times, encouraging them to undertake research. He also went out of his way to get the special 'Carpenter' awards for exceptional staff, pressing the Treasury for an increased number of awards. He was awarded an OBE in 1920, the CB in 1938, and elected FRS in 1943.[8]

18

Recovery

THE ENDING OF THE SECOND WORLD WAR coincided with the end of an era for the Laboratory. Sir John Fox had joined the staff just before it moved from Somerset House, and when the Deputy Principal was Richard Bannister who had served for many years with George Phillips, the first Principal. He was thus a link with the very beginning of the Laboratory.[1] Fox's successor, appointed on 1 September 1945 was Dr George Bennett, an academic. He had been first a lecturer, (from 1924), subseqently Firth Professor of Chemistry, at Sheffield University, and then Professor of Organic Chemistry at King's College, London. He was elected a Fellow of the Royal Society two years after his appointment as Government Chemist. Bennett was thus a complete break with the past, although it was not the first time an academic had been appointed head of the Laboratory. It was thought at the time that his transition from university life to government circles was somewhat strange (he made the change at the request of a senior Treasury official) and it was certainly unexpected. He met the administrative challenge well however, and contributed fully to the work of the Laboratory. He had a mastery of detail, an alert brain, and a wide knowledge of chemistry, (he was a specialist in crystallography), and took a great pride and interest in the running of the Laboratory, although a heart condition meant that he seldom went far from his office on the ground floor, not far from the front door of the building. His presence was thus felt less by staff than that of Sir John before him.[2]

Bennett's arrival coincided with an enquiry into the structure of the scientific Civil Service. There was also a subsequent and thorough review (by Mr J B Legg of the Treasury) on the objectives, activities and organisation of the Laboratory. The former enquiry, carried out by a committee under the chairmanship of Sir Alan Barlow, resulted in a

69   Dr George Bennett, Government Chemist, 1945–1959

White Paper, which recommended a new structure and salary conditions for the scientists in the Civil Service. The government of the time accepted the recommendations.

The Barlow Report recommended that the salaries of the most highly qualified members of the scientific Civil Service should be brought into a relationship with those of the Administrative class, and new classes were to be formed to make this easier. The old titles for senior scientific (graduate) staff: Superintending Chemists, Senior Chemists, Chemists Higher Grade and Chemist were abolished, and a new uniform structure of Scientific Officer (SO), Senior Scientific Officer (SSO), and Principal Scientific Officer (PSO) was introduced throughout the scientific Civil Service. They had already been used in some establishments before the war. Higher grades of Senior Principal SO, Deputy Chief SO and Chief SO were to be created to give a regular career structure. The junior grades of Assistants Grade III to I were abolished and in their place appeared Assistant Experimental Officers, Experimental Officers (EO) and Senior Experimental Officer. It was to be theoretically possible to gain promotion from EO class to SO class. Technical Officers and Chemist grades were abolished. There were also Assistants (Scientific), newly formed from Laboratory Assistants. Broadly speaking these latter, and the new Experimental Officer grades, were employed on development, or routine, work, and the 'Scientific' grades on research. Accelerated promotion was again to be possible (as it had been under the pre-war Carpenter Report) and the 'man of more than usual ability' was to have a reasonable expectation of reaching PSO grade in his early thirties. These changes were implemented from 1 April 1946. In fact despite these splendid new prospects a shortage of qualified chemists made it difficult to obtain staff, particularly in the Scientific Officer class.[3] This was compounded from 1949 when staff numbers were rigidly controlled by the Treasury, and indeed had been cut generally in the Civil Service.

The shortage of staff was not improved by the need to supply senior staff to other departments to act as advisers. Thus at the same time secondments were made to the Office of the Lord President of the Council as well as to the War Office, to the Board of Trade and to the Iraqi government. Other overseas connections also developed from this time; Dr J R Nicholls, the Deputy Government Chemist, was part of a

UK delegation to visit the United States in 1951 in connexion with laboratory methods for the examination of cocoa products and again in the same year for the 75th anniversary meetings of the American Chemical Society and the 16th Conference of the International Union of Pure and Applied Chemistry. He also attended a World Health Organisation meeting on drugs in Geneva and a meeting to discuss unified methods of analysis for fertilizers and feeding stuffs in Paris, whilst another senior member of the staff, Dr John Longwell, visited the United States in connexion with methods of water fluoridation. Senior staff were also much in demand again for research boards and their technical committees and were active in the affairs of the professional societies including the Chemical Society, the Society of Chemical Industry, the Society of Public Analysts (Dr Nicholls was elected President in 1951), and the Royal Institute of Chemistry. The list of organisations upon which the Laboratory was represented appended to the Report of the Government Chemist for the year ending 31 March 1952 was longer than any previous list, before or after the war and the total number of samples examined during the war again rose, to a post-war record level of more than half a million, (see also below).[4]

The review of the Laboratory by Mr J B Legg mentioned above resulted in a report in three parts in September, November and December of 1947. The first was concerned with the main aspects of objectives, activities and organisation and indicated that the place occupied by the Laboratory in the field of government scientific work did not appear to have been defined. On the basis of the position as it was, it was agreed in the following year that its objectives were:

— to provide any government department with services in the field of chemical analysis and investigation and with advice on chemical questions and

— to examine and analyse samples and to provide expert evidence in accordance with the statutory duties of the Government Chemist as a referee.

Although these objectives were broadly compatible with the work of the Laboratory as it had developed from its earliest days more than a century before, it is remarkable that formal 'terms of reference' had not been set out before, or if they had that they had been so little publicised that no-one could quote them.

The review then established that the current activities of

the Laboratory were chemical analyses and advice on chemical questions, the statutory obligations of the Government Chemist, developing new and improved methods and general research and investigation of *ad hoc* problems identified by other government departments, Royal Commissions, parliamentary and department committees and various quasi-government departments, membership of a wide range of national and international technical committees, delegations, working parties, etc, as well as analytical service and advice to other government departments. The advisory services were regarded as particularly important and were seen to extend to general services to the scientific community, through membership (including office bearing) of learned societies, the publication of books and original papers in the periodical scientific literature and participation in national and international scientific meetings. It was also recognised that in the past the Laboratory had on occasion engaged in research in areas not directly bearing on its practical needs, the results of which had in some cases proved subsequently to be applicable to its normal work. Staff shortages prevented much research work being undertaken at the time of the review. It was considered that research work undertaken by the Government Chemist should be closely related to the practical needs of the Laboratory, in reducing the cost of the work, improvement in accuracy (where this was necessary) or increasing the speed of the work. The initiation of a general research division was in practice being delayed owing to the difficulty then experienced in recruiting suitable Scientific Officer staff and in providing proper accommodation, so that it was several years before an effective research programme could be developed. The report also suggested that an effort should be made to separate (and where possible reduce) repetitive from non-repetitive work, and to centralise the specialised analytical techniques within the Laboratory.

In the report the work was rather simply classified as analytical and reporting, the former being either repetitive (or purely routine) or *ad hoc* (and more complex) and the latter either routine or non-routine. The non-routine work involved a critical appreciation of analytical results and the application of scientific knowledge and judgment in the interpretation of the Customs' tariff. The work of most of the divisions included both repetitive and specialised analysis. Another major conclusion of the report was that the Laboratory's research activities needed to be defined more closely in relation to its other work, with the strong implication that the research area should be revived and of an applied character. The final report went on to conclude that the work should be regrouped into repetitive and non-repetitive, with a Test Branch and an Advisory and Research Branch corresponding to these. It recommended that the intake of repetitive work should be reviewed regularly, together with the possibility of further centralisation of some of the specialised techniques used such as bacteriology. The need for revenue outstations at Bristol and Glasgow should be reviewed. Other recommendations related to the transfer of responsibility for

70 Polarograph, 1964

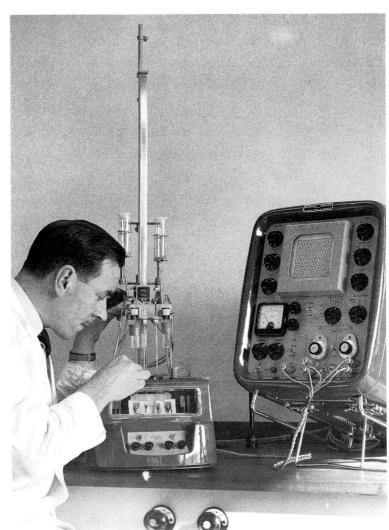

stores, messengers, porters and cleaners from the Physical Methods division to the Establishment and Finance section and that the latter should be in charge of a non-scientific officer responsible directly to the Deputy Government Chemist.

The second part of Mr Legg's report was concerned with the running of the Laboratory's stores and reagent rooms and the third part with the Establishment and Finance section. The ordering and issue procedures for apparatus were found to be generally satisfactory but some improvements to the stock-taking procedures were recommended as was transferring much of the clerical work to an Office Services section based (in accordance with recommendation of the second part of the report) on the Establishment and Finance section.

Mr Legg's proposals were widely discussed in the Laboratory, and raised the anger of some senior staff at least. The Principal Scientific Officers, with, it seems probable, the active encouragment of the Government Chemist, circulated a copy of the proposals for reorganisation to the outstations, who were particularly affected. The general opinion was that Mr Legg had not understood the nature of an analytical chemistry Laboratory and did not entirely understand what he was criticising. It was said that this was particularly true of his suggestions concerning repetitive and non-repetitive work. The very valid point was made that even repetitive work could be highly skilled and the analogy between similar clerical work was entirely false. Another criticism made was the fact that the major customer of the Laboratory (HM Customs and Excise) had not been consulted. This was odd in view of the strong recommendation to close the outstations at Glasgow and Bristol which did largely Customs work.[5]

Although various procedural changes were made as a result of the Legg report, the main recommendation was not implemented. This was the organisation of the Laboratory into two separate branches for Advisory and Research Work and for Testing, with a third branch under the Deputy Government Chemist (DGC) responsible for Technical Services and Office Services. There was some division of opinion amongst senior staff as to whether a research programme should be exclusively oriented to the needs of the work or might, in part at least, be totally academic in character. The importance of a revival of a research programme was universally agreed however. The view was also widely shared that opportunity for research should arise in both the advisory and testing areas and that these should be integrated as far as possible throughout the Laboratory. Very little was in fact done about reorganising the Laboratory and its organisation remained more or less as it had been when Mr Legg started his investigation, with the work being done by 'divisions' grouped into three branches, Revenue, Non-Revenue, and the Deputy's. This latter was roughly research but included the Library, Establishment and Finance, all reporting directly to the DGC. Although little came of the Legg Report, in the long term it did have considerable effect. The Linstead Committee, whose report (see below) was very significant for the Laboratory, made much use of 'Legg'.

Each 'branch' was in charge of a Senior Principal Scientific Officer, the eighteen 'divisions' being in charge of Principal Scientific Officers. It is of interest to see the work of each division at this crucial point in Laboratory affairs.[6]

Revenue Branch

1   *Customs Division*
Handled the examination of spirituous liquor and sugars, also the chemical work which arose in connection with the Customs duties on cocoa, composite articles containing cocoa and sugar and the examination of tea and matches from the public health aspect.

2   *Chemical Stations*
General revenue work on samples received from ports and Customs stations in the vicinity, for the convenience of Customs and to prevent delay in transmission of samples to headquarters. The Principal Scientific Officer in charge at Liverpool was the Inspecting Chemist co-ordinating the work of the various stations with headquarters.

3   *Excise Division*
The control of the use of alcohol in industry. Examination of denaturants, denatured spirit, preparations containing denatured or "rebated" spirit and spirits suspected to be of illicit origin. Questions relating to brewing and spirit distillation were also dealt with and chemical problems arising in connection with the duties on beer, spirits, glucose, wine, saccharin, hops, etc. Examination of lime juice under the *Merchant Shipping Act* 1894.

4   *Tobacco*
Examination on behalf of the Board of Customs and Excise of imported unmanufactured tobacco to determine the appropriate rates of duty, and of manufactured tobacco for export and drawback on export. Examination of retail tobacco for adulteration and conformity with the regulations and also of materials such as denaturants, insecticides and flavourings connected with manufacturing operations.

5   *Key Industry and Oil Division*
Advice to the Board of Customs and Excise on matters arising in connection with duties and drawback on key industry chemicals imported under the *Safeguarding of Industries Act* (1921). Hydrocarbon oils, and the duty on imported spirituous preparations other than beverages. Also examined samples in connection with the above duties.

6   *Ad Valorem and Silk*
Examination of imported goods liable under *Import Duties Act* (1935), silks and artificial and other natural and synthetic fibres and questions related to Purchase tax and Board of Trade Control Orders.

Deputy Government Chemist Branch

1   *Technological Development Division*
Special long-term investigations on behalf of the Admiralty, Ministry of Supply, DSIR, etc, for example fundamental studies of the mechanism of gas absorption, (as in air purification in submarines). Investigations on noxious gases for the Factory Inspectorate, Ministry of Labour and National Service.

2   *Special Chemical Investigation Division*
Investigations into special problems that were referred to the Department, for example the development of methods for the estimation of insecticide spray residues in agricultural products.

3   *War Office Adviser and Army Supplies*
Advice on the suitability for purchase, storage and distribution of all foodstuffs, water purifying, hygiene and anti-malarial chemicals used by the Army in home and overseas theatres, the examination of samples connected with these, the framing of specifications and related duties.

4   *Research Division*
When established was to undertake independent scientific research having a direct or potential bearing on the work of the Department.

5   *The Library*

6   *Establishment and Finance*

Non-Revenue Branch

1   *Physical Methods*
The application to scientific problems of X-ray diffraction, radiography, infra-red absorption, visible and ultra-violet absorption, emission spectrography, and fluorimetry. Responsible also for the Stores, calibration of new thermometers etc and workshop services. In the Workshop was carried out the design and construction of special apparatus of very varied types from complex blown glassware to mechanical devices and electronic instruments and the maintenance and repair of all apparatus used in the Laboratory. It also provided advice on the suitability of apparatus to be bought in relation to the work required of it and the economics of such purchases.

Calibration and Standardisation checked the calibration

71   Ultra violet spectrophotometer, 1966

of volumetric glassware, thermometers and weights, tested hydrometers, saccharometers, thermometers and graduated vessels as used by officers of Customs and Excise in assessing various duties. Sikes' hydrometers were also tested for the National Physical Laboratory.

Stores kept stocks of chemicals, glassware and other apparatus required in the Laboratory, tested deliveries of chemicals and made standard solutions.

2   Crown Contracts

Tested supplies used by other Government departments, involving analysis of an enormous variety of materials such as paper, inks, postage, stamps, oils, rubber, soap, wax, paint, bitument, creosote, disinfectants, building materials, coal, metals, etc. Gave advice on the safe carriage of dangerous chemicals on board ship and on aircraft, particularly as regards packing and stowage. Examined articles and materials in connexion with forgery and fraud and of articles subject to prohibited or restricted importation or export.

3   Water

Chemical and bacteriological examination of drinking water supplies from Government undertakings and all kinds of water for other purposes, including boiler sediments to determine treatment and to ensure absence of scaling of corrosion in boilers. Examined sewage and industrial effluents to assess suitability for discharge into streams or rivers. Determined salinities of sea and estuary waters in connection with a scheme of oceanic research. Bacteriological examination of disinfectants for approval under the *Diseases of Animals Act* or on tender for government use.

4   Geological Survey Division

Advised on all chemical matters relating to the work of the Geological Survey and Geological Museum. Carried out complete analysis of rocks and minerals for inclusion in published Regional Memoirs.

5   Coal and Oil Products Division

Analysed by-products from the coal-carbonising industries or allied products made from petroleum and examined some residual products in connection with the problem of oil pollution of the sea.

6   *Food and Agriculture Division*

Examined non-revenue imported samples in connexion with statutory obligations held jointly by the Department of the Government Chemist, the Ministries of Health and Food and the Board of Customs and Excise under various Acts of Parliament and Statutory Instruments. General analysis of foodstuffs and special examination of products for Ministries for example in connexion with a flour survey undertaken jointly with the Ministry of Food and samples connected with various dietary surveys. Acted as referee analyst of disputed analyses in connexion with the Drug Testing Scheme of the *National Health Service Act* 1946.

7   *Food Enforcement Division*

Quality control for medicaments and foods (human and animal) with special reference to spoilage due to microbial action, oxidative rancidity and to contamination.

8   *Drugs and Medicines Division*

Examined medicinal preparations for compliance with the provisions of the British Pharmacopoeia, the British Pharmaceutical Codex and the requirements of government departments. Collaborative investigations, preliminary to the publication of revised monographs for new editions of the British Pharmacopoeia and the British Pharmaceutical Codex.

All seventeen sub-divisions gave expert technical advice to other government departments. A major part of the post war work of the Laboratory was the continued and increasing emphasis on advisory work in the fields of health, environment and revenue.

The work of the Laboratory was still dispersed amongst several different buildings, in addition to the outstations (see below). Attempts were made in the post-war period to find new accommodation for all of the various divisions: the report for the year ending 31 March 1952 referred to a site which had earlier been set aside for a new Laboratory in the City of London, but which had been frustrated by financial and building restrictions, and there was little liklihood of a start on a building there for some years.[7] The need for a new building was stressed in most Annual *Reports* from the first post-war *Report* in 1950 until a decision to rehouse was finally taken in 1959. The *Report* of 1952 also mentioned that as well as in the headquarters building at Clement's Inn work was being carried out in the Custom House, Cornwall House (on the South Side of Waterloo Bridge), Dudley House (an ex-workhouse in Endell Street, Covent Garden) and 5/6 Clement's Inn Passage.

At '5 and 6' as it was usually known, water analysis was being carried out on the first floor by 1946 and the basement was used as storage accommodation for old laboratory equipment. Physical methods of analysis, an embryonic version of what later was to become the Research Division of the Laboratory, were housed on the top floor but by 1950 moved to the first floor, the top floor then being occuped by water analysis and later, when the latter moved to Dudley House, by the Technological Development section working on problems related to the atmosphere in submarines for the Admiralty. This later became the Toxic Gases section, concerned with problems of factory industrial hygiene. This work on the fourth floor was then joined by a new Flour Section, formed to carry on the survey of nutrients in wartime 'national flour' at the time when the former Cereals Research Station at St Albans in Hertfordshire ceased to come under the control of the Ministry of Food. Many thousands of samples were involved in this work. Welfare orange juice examination was transferred from the main laboratory to a small room on the fourth floor in February 1948. As already indicated, the armed services food supply work moved into the basement in about 1954 on the closure of the premises in Kean Street: a brass plate on the basement entrance indicated that the rooms had once been used by the Albert Club, of which no other record appears to have survived.

Dudley House was originally a London County Council workhouse, built in what was vulgarly known as the 'public lavatory' period of architecture of plain, dull brickwork with wooden floorboards covered (in some places at least) by brown linoleum. The large rooms on the top floors were used for bacteriological and chemical examination of water samples, received from many government owned sites throughout the country from Windsor Castle to remote archaeological areas possessing a small well. The physical methods section of the laboratory also moved to Endell Street in 1952, except for the photographic rooms which remained at 5 & 6 Clement's Inn. All of this work and that at 5 & 6 continued until, with the headquarters laboratory, it was transferred in

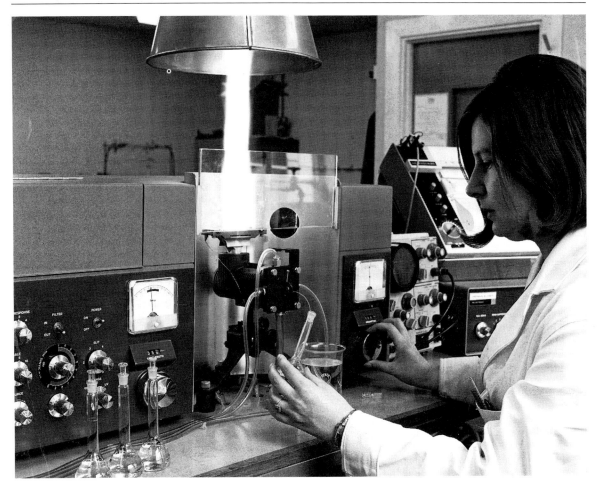

72 Atomic absorption
spectrophotometer, 1973

1964 to Cornwall House. The laboratory at the London
Custom House, which dated back to the latter half of the
19th century, was also transferred to Cornwall House at the
same time. The Custom House in Lower Thames Street was
also at one time the headquarters premises of the Board of
Customs and adjoined Billingsgate, the traditional fish market (until its removal to the Surrey Docks area in 1980) of the
City of London. The laboratory occupied part of the upper
floor of the building. The original conversion of offices and
other rooms in the 1870s was still evident in later years: most
of the rooms had open fire places (the larger rooms had two
fire places) with an adequate supply of house coal. Petroleum
ether extractions were carried out at the end of the room
furthermost from the fire.[8]

Dr Bennett's initial 'annual' report related to the year
ending 31 March 1950, and was the Laboratory's first since

1939. It covered in summary form the work of the years between 1939 and 1950. In the *Report* Dr Bennett noted that the annual number of samples examined during the war years had dropped sharply, particularly at the Custom House and at the revenue outstations, but began to increase again in 1944. There had been a similar effect after the first world war. No account of the samples was given, they were merely summarised in a table (see Appendix 8). The overall number of samples examined annually fell from 555,045 in 1939, to a low point of less than half of this in 1943, and recovered to 414,172 in 1950. The Revenue divisions, and the outstations continued to deal with large numbers of samples, dropping to 40,373 in 1943, but recovering to 347,059 by 1950, (higher than before the war).[9]

Although the war had brought many new and unfamiliar areas of work to the Laboratory, requiring in many cases special investigation and the development of new techniques, time was again made available in the 1950s to investigate some of the basic problems associated with the traditional work. Work on the changes in composition of tobacco on drying, an operation which went back to the early years of the original Inland Revenue Laboratory, led to the development and introduction of new and improved drying ovens in which a uniform temperature was maintained throughout the drying area. Also in connexion with tobacco examination, a new proportional balance was introduced, designed to indicate directly the percentage of paper in cigarette samples. National flour was surveyed (for vitamin B_1, nicotine acid and trace iron contents) as well as imported butters (as previously, for moisture content and butterfat characteristics as indicated by Reichert and Polenske values) and agricultural limes.[10]

The main areas of revenue work continued largely unchanged, involving alcohol control, sugar and cocoa products, tea (the control of purity under the 1875 *Sale of Food and Drugs Act* had been handed over to the Ministry of Food in 1941 but returned to the Laboratory ten years later), sugar, silk and artificial silk and samples taken in connexion with the *Safeguarding of Industries Act* 1921 and the *Import Duties Act* 1932.[11]

Chromatography, then a relatively new technique, came into its own in the analysis of starch conversion products, together with radiochemistry which was used in the examination of cigarette tobacco for K-40 content for the Radiological Protection Service. A very early use of chromatography, in the form of paper chromatography, was employed at the end of the war. Gas chromatography was used from a relatively early period, and a 'home made' gas chromatograph is described in a method for the determination of volatiles in coal products. The column was packed with brick dust on liquid paraffin, or more orthodox Apiezon grease. The detector was a hydrogen flame detector.[12] The laboratory also undertook work in the field of toxic gases and vapours in industrial atmospheres for the Department of Scientfic and Industrial Research. The use of lead salts for clarifying solutions of sugar products prior to polarimetry occasioned laboratory investigation work with the International Commission for Uniform Methods of Sugar Analysis. The Food, Drugs, Agriculture and Water branch continued to provide referee services under the *Food and Drugs Act* 1938 (and, later, the 1955 Act), with new work arising from the *Agriculture (Poisonous Substances) Act* 1952. Nutritional analyses made up an important part of the new food work, with the National Flour Survey, the examination of margarine for vitamin A (added in accordance with the government's acceptance of World Health Organisation advice) and of orange juice and dehydrated vegetables for vitamin C. A substantial contribution to a new edition of the Medical Research Council's Food Composition Tables (published in 1978) was supported by extensive compositional analyses and analyses for trace minerals and vitamins.

The examination of pharmaceuticals supplied under the *National Health Service Act* 1946 (these included capsules for vitamins A and D), together with referee functions under the test prescription scheme, were also responsible for a substantial analytical programme as well as the continuing need to examine materials suspected of being, or containing, substances prohibited under the *Dangerous Drugs Act* 1935. The testing of antibiotics also began, and increased considerably in this period, as well as analysis of therapeutic injections. One of the requirements of the British Pharmacopoeia in the analysis of these substances was for them to be tested on animals. Animal testing had never been carried out in the Laboratory and in 1957 it was seriously proposed that, when the Laboratory moved, such testing should be started.

Perhaps fortunately this proposal was rejected.[13] Work on the examination of questioned documents, mainly for handwriting or typing forgery of the alterations of date stamp impressions, was an important growth area, the main customers being the General Post Office, the Board of Customs & Excise and the Inland Revenue.

The Industrial and Mineral branch continued to examine a wide range of samples for the Ministry of Works and other government departments. The laboratory which had operated at Droylsden was closed in June 1951 and all of the work handled subsequently at the Barry Road Laboratory. The Physical Methods section continued the development or application of newer methods, including X-ray diffraction, infra-red spectrophotometry and chromatography, to problems encountered by other sections of the Laboratory. Oil pollution of coastal waters presented new analytical problems from the early 1950s and work for the Geological Survey of Great Britain continued at South Kensington. The latter area was becoming one of the last strongholds of classical wet chemical analysis, although spectrographic and other physical methods were developed and used to advantage in several areas.

Spectrographic methods were used to determine the duty assessment for various ores and minerals, the evaluation of lead fume samples and the characterisation of quartz and asbestos varieties in industrial atmospheres. The development of field tests for toxic substances in industrial atmospheres had been a function of the laboratory since 1937 and became an important feature of the research programmes. The Government Chemist was asked to form a technical committee (which had formerly been run by the Department of Scientific and Industrial Research) in 1956 and the methods were evaluated on a collaborative basis with the HM Factory Inspectorate and industry, the method finally adopted being published in a special booklet series by the Inspectorate.

In July 1957, 10 years after the Legg Report, the Chancellor of the Exchequer appointed a committee under the chairmanship of Dr R P Linstead to review the functions and organisation of the Department of the Government Chemist, including an examination of the nature and need for work which he was asked to undertake on behalf of other government departments. The committee was also asked to consider to what extent other departments or institutions might undertake the work and to recommend what changes, if any, should be made to the organisation and staffing of the Department to ensure that essential functions were carried out as economically as possible, with maximum technical efficiency. At that time the Laboratory had an authorised complement of 62 Scientific officer staff, 122 Experimental officers, 85 assistants and 91 clerical, executive and other grades, organised into divisions as described above. Of the total scientific staff of 269, 48 were in outstations at home and overseas, together with 11½ of the 91 support staff. The committee took oral evidence from senior staff, from members of the Departmental Staff Side and representatives of the Board of Customs and the Department of Scientific and Industrial Research and written evidence from various customer government departments. It also visited both the headquarters laboratory and some of the outstations.

In its report in September the following year the Linstead Committee recommended that the Laboratory should be transferred from the Treasury and be brought under the Lord President of the Council and attached to the Department of Scientific and Industrial Research, perhaps in association with the National Chemical Laboratory at Teddington. It was felt strongly that the accommodation of the Laboratory, particularly at Clement's Inn and the Custom House was 'so far below the general standard of Government Science accommodation that it cannot but impair the morale and efficiency of staff who have to work there'. The committee felt very strongly that plans to move the Laboratory to Cornwall House (by 1961) were not sensible and that Cornwall House was totally unsuitable for the Laboratory, hence the suggested location at Teddington. It also recommended that the title 'Government Chemist' should be retained and that the Laboratory should be renamed 'Laboratory of the Government Chemist'. No change of function was suggested but a number of other detailed recommendations were made, including the rehousing of the headquarters laboratory in new, purpose-built accommodation, increase in the research programme and the fuller integration of this with the main work. It was noted that there was a certain narrowness of view amongst some staff, due to intense specialisation but that nevertheless there was a considerable amount of contact with other chemists engaged in similar

work. The good *esprit de corps* was particularly noted. Modernization of equipment and workshop facilities, the improvement of contacts in university and industry, the reduction of routine work and more internal movement of staff were also suggested. To help with this latter the reorganisation of the laboratory into four divisions with a fifth General Methods Division serving all of these was suggested. The four divisions suggested were (1) Customs, (2) Excise, (3) Drugs, Food and Water and (4) Advisory and General. This division of work fitted with the note in the report that work for Customs and Excise was still at that time the fundamental reason for the existence of the Laboratory, 85% of the samples, representing 50% of the work were for revenue purposes. Emphasis was placed on the need for long-term as well as other research, with particular regard to the use of modern instrumentation including radiochemical techniques, X-ray diffraction and fluorescence analysis, flame photometry, gas chromatography and various other chromatographic techniques.[14]

While the Linstead Committee was taking evidence the Government Chemist, Dr G M Bennett, died suddenly on 9 February 1959, from a heart attack. He had already suffered a severe heart attack in 1953, and had withdrawn from all except his official duties. His wife had died in 1957 and this may have accelerated his death. Sadly he had suggested in November 1956 that he should retire, since he was then 65, (born 25 October 1892), but was persuaded to stay on for another two years after his 65th birthday.[15] Dr Bennett's death was announced to the staff at a special gathering in the Main laboratory in the Clement's Inn building at noon the day following by Mr E H Nurse, the Deputy Government Chemist. Mr Nurse became acting Government Chemist pending a new appointment.

73   Edwin Nurse, Acting
Government Chemist, 1959–1960

The Laboratory of the Government Chemist 1959–1991

A New Beginning: DSIR

THE TWO MAIN RECOMMENDATIONS OF the Linstead Committee, that the Department of the Government Chemist should transfer to the Department of Scientific and Industrial Research and that the Laboratory should be relocated to Teddington in association with the National Chemical Laboratory, were both attractive to the Treasury but both presented difficulties. Doubt was expressed as to whether the Laboratory's research programme, which was not its dominant theme, was large enough to come within the terms of *Department of Scientific and Industrial Research Act, 1956* and whether therefore such a transfer would be legal. It was realised that most DSIR research establishments already had a 'service' element such as that which was the major feature of the Laboratory's programme, but it would clearly be desirable to encourage the Laboratory in its research activities for legal as well as other reasons As regards relocation, plans for a move to Cornwall House, then seen to be possible by 1961, were well advanced and any alternative move to Teddington would require a new building and be some three times more expensive. In these circumstances, Ministers agreed to the transfer to DSIR but the move to Teddington was abandoned in favour of Cornwall House. It was to be suggested again by the Hardman Report (see last chapter) and finally revived in a different form a quarter of a century later. In principle though the recommendations of the Linstead Committee were accepted and the new Laboratory of the Government Chemist, part of the Department of Scientific and Industrial Research, came into existence on 1 July 1959.

Results of other recommendations of the Linstead Report were an increase in modern scientific equipment for the Laboratory with special reference to radiochemistry and automation, the initiation of a course of winter lectures by visiting scientists and the reorganisation of the work into five

divisions in August 1960.[1] The new divisions, based on the Linstead Commission recommendations were (1) Customs A, Excise and Chemical Stations, (2) Customs B, Tobacco and Advisory Services to the Board of Trade, (3) Crown Contracts, Water, Physical Methods (of analysis), Forensic and Technical Services, (4) Foods, Bacteriology, Drugs and Medicines and (5) General Research and Development, Mineralogical Chemistry and Radiochemical Investigations[2]. There was no lecture or other meeting room at Clement's Inn and the winter lectures were given in a nearby room in the London School of Economics (LSE) in Houghton Street. Earlier, during the second world war, staff had the facility of using the general cafeteria on the top floor of LSE in Clare Market.

During Dr Bennett's period of office as Government Chemist he had actively promoted research allied to the Laboratory's areas of work and encouraged liaison with industry in fields where there were no conflicts of interest. This policy caused a slow change towards an independence of outlook, and indeed a further development of a mutual respect for and by individual customers which had been growing in previous years. Bennett particularly encouraged the development of new instrumental techniques, such as X-ray diffraction, X-ray fluorescent spectroscopy and gas chromatography, techniques in which the Laboratory has continued to be actively engaged.[3]

Soon after the upheaval caused by the Linstead reorganisation a new Government Chemist, Dr David Lewis, was appointed on 1 April 1960. Mr Nurse, the acting Government Chemist, retired on the previous day, after 45 years of service in the Laboratory. A few weeks later, on 9 June, Dr John Longwell became Deputy Government Chemist, having acted as such since the death of Dr Bennett. Dr Lewis was a civil servant, having served for many years (1947–1960) as Senior Superintendent of the Chemical Division of the Atomic Weapons Research Establishment at Aldermaston. Before he joined the civil service in 1941 he had been a University lecturer and teacher.

It was recognised from the start of its association with DSIR that the Laboratory was not typical of the research establishments with which it was now associated. These establishments were engaged primarily on research projects whereas the Laboratory was mainly concerned with meeting

74   Dr David Lewis, Government Chemist, 1960–1970

the needs of recognised customers, providing analytical and advisory services. Most customers were central government departments. Some of the work arose directly from statutory responsibilities and research formed no more than 10% of programme. It was for the most part applied to problems arising from samples received.[4] The numbers of samples were still high at this time, reaching the highest post-war total of 527637 in a twelve month period in 1951–52. It had climbed steadily to this after the all-time high point in 1940 of 560,354, from which it fell during the war to the low point in 1943, as already described. From this point in 1951–52 the numbers of samples declined, as it has continued to do up to the present day.[5]

The work involved in many of these samples was much as before, but some of them showed features which were to become more apparent as time went on. New analytical techniques were embraced as they became viable, often at a pre-commercial stage and more and more of the work involved arose out of the increased governmental concern for public and environmental health. These latter areas were an increasing concern of the Laboratory. Samples of national dried milk and welfare orange juice have been mentioned, others also began under the old Department and in some cases continued to the present day. One such case is fluoride in water, which has been the subject of some controversy. This work arose from studies on the prevention of dental caries through the addition of up to 1 part per million of sodium fluoride to drinking water. This amount had been added in the USA since 1945, and a mission sent there in 1952, which included Dr John Longwell, then head of the Water Division, recommended on its return that fluoridation be tried in this country. A considerable amount of work for the Laboratory ensued. Water was tested for fluoride content and the retention of fluoride in the body was also tested. Members of staff were asked to use fluoridated toothpaste, to drink tea (which contains fluoride naturally) or water for several weeks and the amount of fluoride excreted was measured.[6] Other work concerned with water, beginning in the 1950s, arose out of concern for the large amounts of detergent being found in effluent and rivers. The work continued for many years, with the focus changing to support for the search for biodegradable detergents.[7]

A similar concern for the environment is to be found in

the work on synthetic pesticides which became an important feature of the work of the Laboratory. The work initially related to the standardisation of the performance of pesticide formulations and later to pesticide residue analysis in food-stuffs and in the body, as the possible hazards to man and to animals became more obvious. Within a year of his appointment, in 1960, Dr Lewis chaired an analytical sub-committee of the Agricultural Research Council. The aim was to find a method to determine small quantities of DDT. The method adopted was the determination of chloride volumetrically. The use of organo-mercurial fungicides on apples and tomatoes, and of organochlorine compounds in sheep dips initially provided the principal challenge. Paper chromatography, and subsequently column gas liquid chromatography finally provided the key to dealing with them, the latter particularly.

75   Determination of fluoride in water, 1964

Enzymatic (Cholinesterase) and total phosphorous methods were developed initially for the organophosphorous pesticides. A total chlorine approach to the organochlorine insecticides was also developed, but this was soon abandoned in favour of the chromatographic procedures, as the limitation of the traditional techniques, in terms of sensitivity, were realised. Increasing sophistication of methods has continued to the present day, when for example High Performance Liquid Chromatography (HPLC) instruments are used with selective detectors.[8]

Much other new work was taken on in the years after the Linstead reorganisation and the increasing scope and complexity of the work was recognised in 1966 by the appointment of, in effect, a second deputy to the Government Chemist. This second Deputy Chief Scientific Officer was put in charge of an Environmental Chemistry Group, consisting of food, drugs, potable waters, pesticides, toxic gases and radiochemical work. Dr Harold Egan, until then Superintendent in charge of the Food, Drug and Agriculture division was appointed to the post.[9] The Laboratory was now divided into eight Divisions, consisting of Customs and Excise A and B, General Chemical Services, Food Drugs and Agriculture, Pesticides and Toxic Gases, Radiochemistry, Armed Services Food Supplies, and Research. In addition there were establishment, finance and library support services. The divisions thus separated the work of the Laboratory between revenue, health and environment and other subjects of public interest. The breakdown in effort in 1966 was estimated to be approximately 4:3:2.[10]

As noted above, the Laboratory has always been willing to adopt and adapt new techniques as they are developed, a process which has developed with increasing speed since the War. Some of the techniques have already virtually disappeared. Polarography for example was used from 1964 onwards for the rapid estimation of trace metals, (later using atomic absorption spectrophotometry for the same purpose), with flame photometry for alkali metals. All of these techniques replaced old and tedious 'wet chemistry' methods, although at this time (1960s) it was still necessary to retain them for reference purposes.

New methods were naturally applied to the routine areas of work where this was applicable. In 1965 for example gas chromatography was used to determine more quickly and

76 Automatic beer analyser, 1974, for the determination of alcohol, sugar and acid content

accurately glycerol and propylene glycol additives to tobacco. Another instance is the application of automation to very large numbers of samples, as in a screening process for the original gravity of beer.[11] Some of these areas of work were about to see even more radical changes however. Thus with the accession of the UK to the European Economic Community (EC) the long established proof strength system for the measurement of the ethanol content of alcoholic beverages was changed. On 1 January 1976 the basis on which duty was charged became percentage alcohol by volume. This latter was measured at 20°C, whereas the proof system had used 60°F, another fundamental change. It had been necessary to approach these major changes with great care, and from 1970 onwards a joint working party between the Laboratory and HM Customs and Excise studied the problems involved.

The new procedure caused some radical alterations in technique. Instruments such as hydrometers and thermometers had to be replaced, together with the tables of alcoholic strengths used in conjunction with them. Many of the changes to the tables were effected by simple calculation (assisted by the Laboratory's datalink with the University of London's Atlas computer), supported in some instances by experimental verification. As an interim measure a table was produced for saccharometer readings for worts to correct for temperature from 60°F to either 15°C or 20°C. It was not certain at that time on which temperature the new legal definition of gravity would be based. Another table for the conversion of the relative density of an alcoholic distillate to the degrees of gravity lost at the same temperatures was prepared. Other tables in use for temperatures down to 36°F themselves required correction before the conversion was done. For wines, the tables used by Customs officials for the estimation of strength from refractive index and density were replaced. For spirits, tables to convert reading of percentage proof spirit to percentage of alcohol at 20°C were needed, together with others for converting kilogrammes of spirits to litres (the density factors being tabulated according to the indications of Sikes' hydrometer) and for converting the volume of spirits at any temperature to the corresponding volume at 20°C. Metrication also required work in other revenue areas, notably hydrocarbon oils.[12] The end of the proof spirit system and the use of the Sikes' hydrometer, the legal standards since 1818, meant the replacement of about 2500 hydrometers used by HM Customs & Excise.

On 1 January 1978 an equally great change took place in work carried out in the Laboratory since its foundation in 1842. This was a change in the method of assessing tobacco revenue resulting from the entry of the United Kingdom into the EC five years earlier. Previously the tobacco manufacturer bore the whole duty during the entire process of conditioning, cutting, blending, cigarette making, packing and storage and it was not unusual for a factory to handle tobacco carrying a duty of one million pounds or more in a single day. On its accession to the EC in 1973 the UK had been allowed to retain this system until 1977, after which it had been required to change to the EC system based on the value of the finished product for sale (cigarettes, cigars, pipe tobacco, etc). This brought to an end an era not only in the long history of tobacco taxation (back to 1603) but also in the history of the Laboratory, the original purpose of which in 1842 had been to provide scientific assistance to the Excise in its task of collecting and protecting the revenue from tobacco. Both Customs and Excise services had relied for 135 years on the Laboratory to provide the analytical and technical advisory support for this system of revenue collection. This need was removed by the change although the Laboratory continued to provide technical advice for HM Customs and Excise on the protection of revenue, including duties derived from beer, wine, alcohol, hydrocarbon oils, import duties, betting duty and value added tax as well as assisting in the work of preventing the smuggling of restricted drugs.[13]

Tobacco work did not cease in 1978. As well as giving advice to HM Customs and Excise on technical aspects of tobacco where necessary, analyses related to the health aspects of smoking was now a major source of work. Since 1972 the Laboratory had been engaged on a survey of the tar and nicotine content of cigarette smoke for the Department of Health and Social Security (DHSS), as a result of action in Parliament arising from the Royal College of Physicians

77 Tobacco section, Cornwall House, 1966

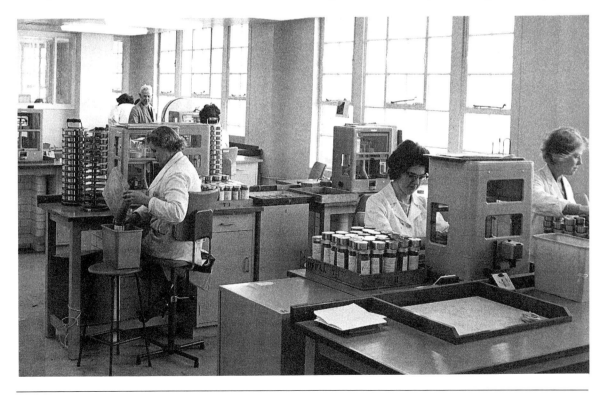

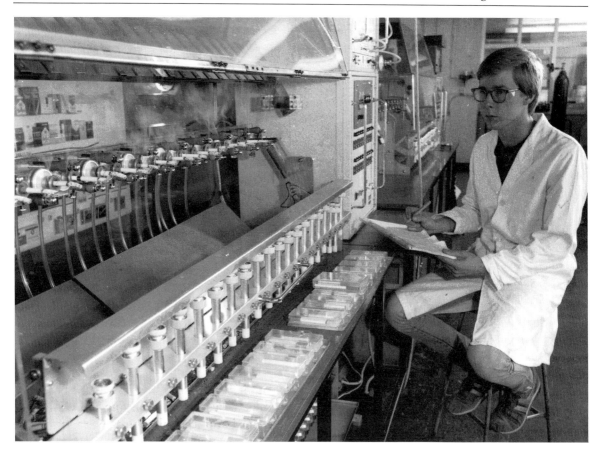

78 Smoking machine, 1989

report *Smoking and Health Now*. This report had identified a relationship between smoking and its effects on health, and gave the evidence that tar deposited in the lungs of a smoker was an important factor in the development of lung cancer, and that nicotine inhalation was possibly a cause of cardio-vascular disease. One of the recommendations in the report was that tables of the tar and nicotine yield of cigarettes should be drawn up, and packets of cigarettes labelled with the information. The Laboratory undertook the analytical work, developing the use of automatic 'smoking machines' to do so. In 1978, after a change in the basis of tax, a special health tax was laid on cigarettes with a tar yield of 20 mg/cigarette or more. This supplementary duty (later repealed) was administered by HM Customs and Excise, who then required the analysis of cigarettes to meet this purpose.[14]

Several other fields of work began in the period under

review, some of them of such as oil pollution analysis of considerable importance. Work on this had been undertaken as early as 1936, when samples of oily residues taken from the foreshore were first examined at the request of the Board of Trade. Samples continued to be analysed until the outbreak of war in 1939. In 1951, when the amount of oiling of seabirds and beaches first began to cause alarm, further samples were received. These included oiled feathers and wings of birds found dead on the foreshore, examined at the request of the Lord President of the Council for the British Section of the International Committee for Bird Preservation. The methods available in 1951 for the characterisation of oil spills were confined to physical separation into asphalt, wax and liquid oil, and determination of the congealing point of the wax and the distillation range of the oil. These procedures were used in an attempt to distinguish fuel oil, crude oil and tanker bottoms (sludge), but the effects of weathering had not then been studied and there was virtually no prospect of pinpointing individual pollution sources. It was shown, however, that much of the pollution of beaches was caused by the waxy residues remaining after tankers had discharged their cargoes, presumably due to the then prevalent practice of washing ships' tanks at sea and discharging the washings overboard. On the other hand, the oiled birds were found to be contaminated with fuel oil which was more likely to spread as a thin film on the sea than tanker residues. This concern for oil pollution led to the setting-up of the Committee on the Prevention of Pollution of the Sea by Oil by the Ministry of Transport in 1952. The Laboratory was represented on this Committee from its beginning. Work carried out by the Laboratory for this Committee included simulated weathering tests to distinguish persistent from non-persistent oils and ship trials to determine the oil content of tanker washings. The Laboratory had also been concerned in the evaluation of oily-water separators for use in ships or at shore reception facilities, and numerous samples from such separators were tested to determine their oil content.

Pollution samples continued to be examined with a view to prosecution under the *Oil in Navigable Waters Act* 1955 and 1963 or in the hope of claiming compensation, but the analytical results could still do little more than confirm that the pollutant oil was of the same general type or from the same area as a reference sample. Gas-liquid chromatography was first used in the Laboratory to examine oil pollution samples in 1964, but the relatively primitive columns then available yielded little more information than the other techniques already in use. The *Torrey Canyon* incident of 1967 aroused public concern about oil pollution, and subsequent legislation made it easier for authorities dealing with pollution at sea or on the beaches to obtain adequate compensation, provided that the source of the pollution could be identified. By 1969 it was realised that gas-liquid chromatography (GLC), particularly if used in conjunction with other techniques, offered the prospect of positive identification of crude oil samples. Subsequent work showed that GLC could not only replace distillation range as an indication of oil type, but could also supply a 'fingerprint'. Once reproducible quantitative measurements could be made, ratios of significant individual hydrocarbons could be calculated. These, together with more precise determinations of vanadium and nickel, made it possible for the first time to compensate for the effects of weathering on crude oils. An advertised service, the Oil Spillage Analytical and Identification Service (OSAIS) was set up giving advice and to some extent analytical help with oil pollution cases.[15]

A field of work in a more traditional area came about as the result of a change made by the *Finance Act* of 1960. The terms of this Act required the devising of a means to determine whether or not a gas oil was marked for use in diesel engines and intended for industrial use. This type of oil paid less duty and there was thus an incentive to use it fraudulently for ordinary road vehicles. The method finally adopted was to add not easily removed markers which were distinctive optically and chemically when tested. Furfuraldehyde and quinizarin were chosen at first, others were used later, and many samples taken from suspect vehicles were tested in the Laboratory, with a view to prosecution for illicit use, if so found. The Laboratory also devised a method for use at the road side by Customs and Excise officers in their Road Fuel Testing Units (RFTU). The system is still in operation.[16]

Not all of the work carried out in the Laboratory was as serious in intent as the examples just described. In the food divisions for example, as well as routine samples, food from the past was examined. In 1966 an emergency ration pack from the South African war of 1899–1900 was received. This

79 'Hey for the Open Road',
cartoon of 1948 from Punch

was made available through the courtesy of the relatives of
the officer who had carried it. The ration consisted of
concentrated beef, and cocoa paste. Both were found to be
sound, though unpalatable. Some deterioration had occurred
which was considerable in the case of the cocoa. The cocoa
was also found to contain a large amount of tin from the
pack. The total energy value was determined as 1100 calor-
ies. Similar historical samples had been received in 1948 and
1958.[17]

Other work out of the ordinary course of events were a
series of analyses carried out for the British Museum (National History). This work had begun in 1947 when a human
skull, femur and tibia from Broken Hill in Zimbabwe were
examined in an attempt to determine whether they came
from the same stratum. They were tested by emission spectro-

graph for lead and zinc; lead levels were not significantly different but the skull was rich in zinc, suggesting it had a different origin from the other bones. This led to about 20 more samples being examined for lead, zinc and vanadium. Dr K P Oakley at the Museum suggested that the presence of fluoride in ground waters would cause the gradual conversion of hydroxyapatite to fluorapatite, so that for a series of fossils from a given location the fluorine content would be an indication of their age. Fluorine determinations commenced in 1948 on a skeleton found at Galley Hill, near Swanscombe in Kent and showed that the skeleton was a relatively recent burial in an older deposit. In 1949 the first samples relating to the 'Piltdown Man' were received. These remains had been discovered at Piltdown in Sussex between 1908 and 1912. They had been regarded by some as a kind of 'missing link' between man and apes, but increasing doubts suggested a resort to new scientific methods of examination. In the samples from these bones phosphate was determined, and the ratio of fluorine to phosphate used as a measure of the replacement of hydroxyapatite by fluorapatite. Iron determinations were carried out in addition on some samples, as well as spectrographic determinations of chromium and manganese on some of the darkened bones.

The final results, which proved beyond all doubt that the Piltdown remains were a fraud, were published in 1955.[18] Later development of better chemical methods of determining fluoride, phosphate and iron made it possible to analyse much smaller samples and in the ten years following the Piltdown work several hundred samples of bone, teeth and ivory were examined to determine their relative ages. Samples were also examined for nitrogen content, using a micro-Kjeldahl technique on samples of milligram size. The level of nitrogen is another guide to age, the older the bone the less nitrogen is expected. Many other archaeological samples have been examined in the Laboratory, including comprehensive analyses of bronze artefacts by classical chemical methods for the British Musuem. Improved physical methods of analysis made meaningful results easier to obtain. These samples also included the tentative identification of pigments in Roman wall plaster, using X-ray fluorescence spectrometry, a technique also used for checking the composition of ancient coins.[19]

One aspect of life in the Laboratory which had remained

constant until 1960 was the disposal of surplus samples. In that year Sir John Lang, secretary to the Admiralty, chaired a small committee appointed by the Treasury to enquire into the practices used in the Laboratory for the disposal of surplus samples. An unfortunate incident involving alcohol samples had brought the practice to the public eye. Interest centred on tobacco products, tea and alcoholic beverages. Staff were allowed to consume samples of these after they had been examined and stored for long enough to ensure that no further examination would be necessary. Traders had always been able to claim the remnants of the samples, but few called to collect them. Most of the surplus was in any case destroyed, by burning or other orthodox means, but the limited distribution (originally to minimise pilfering) of some remnants had been allowed at least since early in the century, (see also chapter 13 above). Arrangements whereby rationed foods (mainly jam and canned fruit) were given to staff after testing were approved by the Ministry of Food in 1945: it was an offence at that time to destroy food. The Lang Committee concluded however that there should be no further distribution of surplus samples, and instructions governing the disposal of all samples received at the Laboratory were drawn up the following year. Such instructions are still issued to staff on joining the Laboratory. A hallowed Laboratory tradition thus came to an end: one which had perhaps been practised from its foundation.[20]

CHAPTER 20

The Changing Scene

PARALLEL WITH THE MANY CHANGES IN THE work of the Laboratory which took place in the years after 1959, there were others both in an internal sense, and in the sense of the organisation to which the Laboratory belonged. There were more such changes in the 30 years after the Linstead Report than in all of its previous history. On top of that there were two physical moves, culminating in that which took the Laboratory to the present site in Teddington. The first such move was to Cornwall House, on the south bank of the Thames, near Waterloo Station. This site was first recommended by DSIR in 1959. The actual move was precipitated by a plan to redevelop the Clement's Inn area, with consequent demolition of the Laboratory building.

As has been noticed at intervals, Government Chemists and Principal Chemists since (virtually), 1842 had complained about lack of adequate accommodation, and such had been the case with the building at Clement's Inn within 20 years or so of the Laboratory moving into it, (see above, p.xx). By 1962 the Laboratory was housed in six buildings in London, (Clement's Inn, the Custom House, 5/6 Clement's Inn, (usually known as '5 and 6') two laboratories in Cornwall House, Dudley House, (the former St Giles' Workhouse in Endell Street, WC2), and a laboratory in Barry Road, (north London). Considerable dissipation of effort was caused by this situation.[1] None of the six buildings were very suitable, all had drawbacks. The headquarters building was badly lit and parts of it were badly ventilated, and it suffered from vibration caused by the printing presses in the W H Smith building next door. At the Custom House the lift had been known to fall from the top of the building to the basement out of the control of the attendant. When it rained, buckets had to be arranged under the skylights to catch the drips. The hydraulic lift in '5 and 6' had also given cause for concern.

80 Cornwall House, Stamford Street, London

It was agreed that any new building had to be in a reasonably central London site to allow access by customer departments, particularly to allow easy delivery of the excise samples from the London docks. The building settled on was Cornwall House in Stamford Street, Waterloo. This was crown property in the Duchy of Cornwall estate, and largely occupied by the HMSO central bindery and warehouse. It had been started just before the first World War, and completed during that war. Its first use was as a reception hospital for soldiers wounded on the western front. After that war it was used as government offices, and as a stationery warehouse. It had been constructed of reinforced concrete (one of the first in London), with an exceptionally high floor loading factor with this last use in mind.

It posed many problems in conversion to a laboratory. The very high floor loading had been achieved by means of massive concrete beams arranged in honeycomb fashion under each floor. To avoid having these unsightly objects exposed, and also to conceal the service ducts it was necessary to fit false ceilings to all the rooms and corridors. This lowering of the ceiling left a head room of fractionally over eight feet which gave a rather claustrophobic effect, particularly noticeable after the Victorian grandeur of Clement's Inn. The part of the building to be occupied was on the fourth and fifth floors, with some offices and the Library on the corresponding floors of the Annex. These floors were in fact the third and fourth floors of the Annex, which sometimes caused problems for visitors, and indeed staff. The conversion of the rather unsuitable building was carried out with some difficulty by the Ministry of Works. They used the latest techniques in services provision (heating, ventilation and lighting), and equipped the laboratories with specially designed modular benches conforming to the interior structure and the many window frames. The work took three years and was completed by September 1963.

The move started in the same month, and was completed by January the following year. The new laboratory was officially opened on 21 April 1964 by the Right Honourable Quintin Hogg, MP, Secretary of State for Science and Education. The official opening was followed by a series of Open Days, the first such event since 1897. Over 1200 visitors attended, seeing exhibits of current work, unfortunately without the musical accompaniment of that previous opening.[2]

The move to new premises for the headquarters coincided with a similar move for the outstation in which the armed services food supplies work was carried out. This outstation moved to the Supply Reserve Depot at Norton Fitzwarren near Taunton in Somerset. The building was opened on 13 December 1963, and replaced earlier facilities used by the Laboratory. This new building was itself soon to close. A reorganisation within the Ministry of Defence meant that victualling control for the Admiralty, for whom the Laboratory had carried out analyses in the past, was again centralised and staff at Norton Fitzwarren were transferred to the Admiralty laboratories at the Royal Clarence Yard, Gosport.[3] Very few outstations were still in existence by 1963. Of those still open at the start of the second world war Hull, Southampton and London Dock had closed in 1940

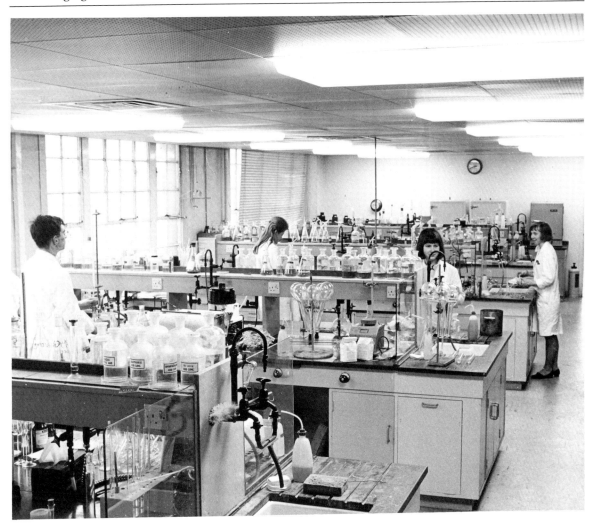

81   Food laboratory, Cornwall
House 1968

(London Dock re-opened in 1953, and closed for good in 1968). Three then remained, Liverpool, Bristol and Glasgow.

The Liverpool outstation, whose original premises in the Liverpool Custom House had been destroyed in 1941 was by far the largest of the outstations. The officer in charge supervised the other two local laboratories, which he visited from time to time. The station covered the north of England (except Tyneside), Northern Ireland and, under the Customs 'common purse' arrangement, the Isle of Man. It had several large classes of work in addition to the alcohol control functions exercised by the other outstations, including the examination of imported textiles, hydrocarbon oils, chemicals and products of the chemical industry and illicit drugs.

The work of the control laboratory at the Government Wool Disinfecting Station in the dockland area to the north of Edmund Street also came under the supervision of the Liverpool laboratory in 1964. This involved frequent visits by a senior member of staff who had first to undergo a course of inoculation against anthrax, a hazard to which all who came into contact with raw, imported wool and fleece were exposed.

Some of the investigations associated with major research projects were also carried out at the revenue outstations, as and when time permitted. Thus, at Bristol a theoretical study was made of the relationship between the refractive index of a beer, its density and its original gravity, to give a generic formula which could conveniently replace the three empirical tables used for lager, ale and stout. It was the only outstation equipped for the analysis of tobacco samples, and was thereby able to give a service to the local tobacco industry. It was also a general station for seaports in the south-west and south Wales. At Liverpool during the war an examination of spray dried egg had been carried out for the Ministry of Food, and a 'flavour score' devised. This partly involved the tasting of up to 100 samples every day, all made up without flavouring. Staff operated the tasting panel on a rota. Later work was directed at the development of a method for estimating the natural sugars present in liquorice extract in order to clarify the interpretation of the Brussels Nomenclature Tariff classification. A new set of spirit tables showing the relationship between the percentage proof spirit, percentage of alcohol by volume and by weight and the specific gravity at 20°C compared with that of water at the same temperature was compiled there. During the period 1946–1958 the examination of army food supplies formed a large part of the work of the Glasgow outstation. These came mainly from the Supply Reserve Depot at Paisley some 6 miles to the south-west of the city, where inspection facilities were provided for the analyst.[4] The Liverpool Laboratory was modernised in 1974, and the Glasgow Laboratory moved to new premises in 1978.

However, the days of revenue outstations was coming to an end. A month after the headquarters Laboratory moved in 1964 the Bristol laboratory was closed, and the work transferred partly to Liverpool and partly to Glasgow. The existence of these outstations had been questioned before, soon after the war, by a suggestion of centralisation in London, but local Chambers of Commerce had expressed opposition to such closure. The outstations were regarded as providing a useful service to local industry. Many firms established a mutual understanding (and respect) with the local Laboratory, enabling them to get early clearance of goods found not to raise problems for the Customs officers. Where controversial aspects arose, the headquarters Laboratory in London could be consulted. However, economics ruled otherwise after a few more years. The work load of the two remaining Laboratories continued to decline, in 1981 the Glasgow Laboratory closed, and the last revenue outstation at Liverpool did the same on 31 December 1983.[5]

The move of the Laboratory to Cornwall House was followed by another change of status for the organisation. This was a result of the report of the Trend enquiry into the organisation of civil science. This report (published in October 1963) recommended that DSIR be dissolved and the various component establishments be reallocated to new Research Councils. The Government of the day accepted these recommendations, although with some modifications, and on 31 March 1965 the Department of Scientific and Industrial Research was dissolved. On the next day, 1 April, the Laboratory, together with some other former DSIR research stations was transferred to the newly formed Ministry of Technology.[6] The terms of reference, statutory duties and functions of the Laboratory remained unchanged. There was little impact on its work, certainly less than had occurred in 1959.

Another such change occurred on 19 October 1970, with the dissolution of the Ministry of Technology and the formation of the Department of Trade and Industry (DTI). Once again the general character of the Laboratory's work changed very little, although perhaps it is true to say that with these successive changes the Laboratory moved a stage nearer to political matters. Since its formation the work of the Laboratory had been concerned in an advisory capacity with the work of central government departments, although it was seldom close to the main political issues, even if its scientific programme was directly related to them. Entry into the DSIR had brought closer an increased awareness of some of the political issues affecting science policy in the United Kingdom. It was probably as true to say that the government itself was becoming more aware of the importance of science to everyday life and its wider relation to policy.

Two further changes occurred in organisational status before the present parent organisation was reached, the conversion of DTI into the Department of Industry, on 5 March 1974, and the change back to DTI on 9 June 1983, following the general election of that year. One other event in 1974 was of perhaps more general interest, This was the opening in the Science Museum galleries of a permanent reconstruction of a 'typical nineteenth century laboratory', based on original furnishings taken from the 1897 Clement's Inn headquarters before their demolition in 1965 (see plate 45).[7]

During these changes in parentage the Laboratory had been guided at first by a Steering Committee, first appointed by the Council of Scientific and Industrial Research in 1962, and latterly by a Requirements Committee. The duty of the Steering Committee, of which the Government Chemist was a member, was to oversee the Laboratory and, in particular to indicate the policy which should govern the distribution of effort between the various items of its programme within the resources available. It was recognised in 1962 that the programme was largely determined by the requirements of other government departments, which accounted for about 75% of the total effort. The balance of the work was mainly on statutory duties and background research aimed at the improvement of existing methods or the development of new. It was clearly more applied in character than the general run of DSIR research stations programmes. This balance of effort changed somewhat in subsequent years.[8]

The change to a Requirements Committee was made necessary by the Government endorsement of the proposal of the Rothschild Committee that applied research and development should be organised in accordance with a customer–contractor principle (CCP). This principle was defined as the relationship between the customer (or client) who stated his needs, the contractor (the analyst in this case) carrying out the work and the customer paying. Until then the service provided by the Laboratory was an allied service to Exchequer Departments and a repayment service to the non-Exchequer customers. This was confirmed by a working party chaired by Mr R St J Walker, which also said that the financial provision for the Laboratory should be readily

82 Scanning electron microscope

variable from year to year in accordance with the needs of the work.[9]

The Rothschild Report, *Framework for Government Research and Development*, was published as a White Paper in July 1972. The Laboratory came within the terms of this Report since it was regarded as an Industrial Research Establishment of the Department of Trade and Industry, notwithstanding that its main efforts were devoted to the scientific, analytical and advisory needs of central government with only a relatively modest research programme. For the Laboratory it was recommended that its work should be commissioned through a group of Requirement Boards acting as proxy customers. This recommendation clearly did not fit the circumstances under which the Laboratory operated, so that rather than responding to the needs identified by such a group, a Requirements Committee representative of the major customers was appointed. The Steering Committee was already constituted in a fashion representative of the Laboratory's major customers and so, on 14 December 1972, it was reconstituted as a Requirements Committee, without change of membership. The terms of reference set for this committee were to advise the chairman on the programmes required to support future government policy, their relative priority, their progress and the relationship of each programme to other national and international programmes. The departments represented were Customs and Excise, Agriculture Fisheries and Food, Defence, Health and Social Security and Environment. The Laboratory continued to consult customers separately, regarding both current and future needs and the cost of projects under the various headings. The main work, which in 1971 had been formulated under 25 subject-oriented programmes, was reviewed and reorganised into eleven customer-oriented programmes. These related to the needs of the Department of Trade and Industry, the Department of Employment, the Research Councils, small groups of 'other Government Departments' and of quasi government departments and to the needs of the Laboratory itself as well as to the five 'major' customer departments. The statutory duties of the Government Chemist and the investigatory and other work in support of these duties continued to be the subject of discussion with the sponsoring department for many years.

The pattern of the Laboratory's work did not alter materially in consequence of these arrangements but discussion commenced with major customers on the transfer from allied service funding in which the cost of work was borne by the DTI vote (as had been the basis of virtually all of the Laboratory's work in earlier years) to the cost being borne on the customer's own parliamentary vote. The hope was that the new arrangements should be conducted on as simple a basis as possible, with an appropriate sum being transferred in the first place from the DTI vote to the other departments concerned. The change of the programme of work from a discipline-oriented to a customer-oriented basis was completed in 1973 and a system for the internal recording of work compatible with costing and charging work to individual customers commenced on 1 April 1974.[10]

Initially the operation of the Customer Contractor

Principle proved to contain no major problems for the Laboratory, and a simple system of identifying and charging the customer for each item of work was developed. One problem in the CCP itself lay however in the fact that as part of the Government's plan to contain public expenditure, strict limits ('cash limits') were placed in advance on all departmental expenditure. The main difficulty was the funding of work for which no specific customer department could be identified. The Laboratory had developed an ability to anticipate new analytical requirements in many of the areas of public interest and this had come to be regarded as part of the Laboratory's service, often to the great advantage of the customer, not least the the public at large. The Government Chemist also retained a number of statutory functions, some of which were created over a century earlier: for these, and for the laboratory work necessary in support of them, no customer existed other than the Department of Industry itself. This matter was eventually dealt with by the Requirements Committee itself acting, with departmental support, as a proxy customer sponsoring research in some of the fields of work.[11]

The external changes of organisation to which the Laboratory was subjected, coupled with changes in volume and type of work, caused as many if not more changes in internal organisation. These changes are too many to follow in detail, they are listed in the Annual Reports for the period, (to 1983). In general though the organisation of the Laboratory into divisions was retained, each containing a different number of sub-divisions, which were moved into different groupings as the work changed. Major areas of work arose as the customer needs changed with time and as seen in the previous chapter, some completely new programmes were added. These new programmes included what was at first the Dental Materials sub-division, initiated in 1964 at the request of the Ministry of Technology. This was very successful in developing a variety of new dental cements, and several patents resulted. It later became the Materials Technology sub-division and as such won the Queens Award for Technological Achievement in 1988 for pioneering work in the development of new materials for use in dentistry. The cements were also further developed for use as bone cements in artificial hip joints and other prosthetics. Other later results of this programme were in controlled release formula-

83   Modern direct reading
weightless balance

tions, for testing animals with nutritional deficiencies
amongst other uses. Considerably more research was undertaken at the Laboratory after it went to DSIR, following the
recommendation of the Linstead report that an increased
emphasis should be placed on long term research.[13]

All of these changes naturally had an effect on the staff
of the Laboratory. There was considerable discussion in the
1960s as to whether the dissolution of the Department of the
Government Chemist and its attachment to the Department
of Scientific and Industrial Research had been a good thing
for the Laboratory. Bearing in mind that many of the senior
staff had joined the 'old' Laboratory (the Department of the
Government Chemist) in the years immediately following the
first world war (and that retirement at the time of the change
was often nearer to age 65 than 60), there were those who
regretted the 'loss of independence'. The general balance of
opinion though was that a better understanding of the
scientific aspects of the Laboratory's purpose was possible
under the new organisation. With most of the Laboratory's
work geared to the requirements of other ministries, there

was bound to be some fragmentation of the 'Whitehall' understanding of the overall significance of the establishment. In earlier days the Laboratory had been virtually the only scientific centre available to most of the ministries and whereas the Treasury had clearly recognised at the beginning of this century that there were considerable disadvantages in having a number of smaller laboratories distributed around Whitehall they were unable to prevent their formation. Certainly the research aspects of the Laboratory's effort in terms of applied research in analytical chemistry in relation to the public need became much more clearly recognised. The Laboratory also had the resources and capability to divert effort at short notice to undertake analytical investigations into new and urgent problems which might unexpectedly arise. In addition, it was coming to be regarded as a training and recruiting ground for the smaller, specialist laboratories which were being created in other departments.

The Laboratory was never a typical 'research station' although it had contributed substantially (and continued to do so) in the field of analytical investigation and research. There was a much greater understanding in DSIR of the Laboratory's need in terms of modern, and sometimes very expensive, equipment. This situation was greatly to its advantage and widely used in terms of the judicious choice of techniques in which to invest at a time when analytical chemistry, and the important area of trace analysis in particular was fully emerging from a pre-war if not 19th century image.

Staff numbers also fluctuated, not necessarily with the changes in organisation. As seen above, in 1971, the staff numbered 383, divided into 8 Divisions, which included 325 scientific staff, and 58 support staff in technical services, Library, and Administration, some of this latter group also being scientifically qualified. The number fluctuated somewhat, but the maximum number some 464 was reached in 1977, when the number of support staff was about 80. A few years after this the numbers of staff began to decline, as the amount of routine work declined. There are at present (1991) 291 scientific staff with 84 support staff, excluding consultants and students. Other changes took place in the Directorate. Dr J K Foreman, the Deputy Government Chemist, was transferred at the end of 1976, to take up the post of Deputy Director at the National Physical Laboratory. The post of

Deputy Government Chemist was abolished and instead two Deputy Directors, of Resources and Customer Services, were appointed. One of them, Dr D C Abbott, had been appointed Deputy Director (with no subtitle) the previous year. In 1983 the Directorate was enlarged to four, with the Government Chemist, the two Deputy Directors as above, and a third Deputy, of Industrial Research, (Dr R Dietz) appointed. This structure still applies at the time of writing, although the third deputy now takes a special role in the promotion of Biotechnology.[13]

In the period under discussion there were several changes in the post of Government Chemist. Dr Lewis, appointed as the 'Linstead' changes took effect in 1960 retired on 1 October 1970. In his ten years in office he had instituted a large number of management changes to ensure that the Laboratory kept fully abreast of developments in its field, including measures necessary for a full interchange of thought and ideas. He was a greatly respected figure both within and without the Laboratory. Dr Lewis was succeeded by Dr Harold Egan. Dr Egan had been a member of the Laboratory's staff since 1943, and had taken over as Deputy Government Chemist in 1970, following the retirement of Mr E I Johnson. Dr Egan had held posts in many parts of the Laboratory. He had held every grade from Temporary Chemist and Assistant Experimental Officer, and his experience at all levels made him thus uniquely qualified to guide the Laboratory through the many changes in the years in which he was Government Chemist. He was only the second Government Chemist or Principal (the first was Dr Fox, excluding George Phillips) to rise through the ranks and reach the highest post.[14] He was succeeded by Dr Ronald Coleman, on 24 December 1981. Dr Coleman was a Deputy Director at the National Physical Laboratory when appointed Government Chemist and earlier had been Senior Superintendent of the Research and Special Services Division at the Government Laboratory in the years 1973 to 1977. Previous to this he was with the UKAEA for 20 years. In May 1987 he was promoted to be Chief Engineer and Scientist at the Department of Trade and Industry in 1987 and was succeeded by Mr Alex Williams, a physicist with considerable management experience in industry and the Civil Service, and until then under-secretary for research, technology and policy at DTI.[15]

84   Dr Harold Egan, Government Chemist, 1970–1981

Agency and after

THE 1970s AND 1980s WERE A CRITICAL time in the development of the Laboratory. Changes fundamental to the way the Laboratory was controlled and financed were being planned, as was yet another move, this time out of central London.

Cornwall House had never been a very satisfactory headquarters for the Laboratory. There never seemed to be enough room to meet the needs of the scientific programme, since the internal design of the building did not allow for an efficient use of the space. This was despite more and more of the ex-warehouse being converted for Laboratory use, particularly after the Stationery Office (HMSO) moved out in the early 1980s. By 1970 however other drawbacks were becoming obvious. The ventilation system was not adequate, despite renovation, and there was a possibility that the whole site would be redeveloped. It had become obvious that new premises would be required sooner rather than later. Other possibilities were becoming apparent by about this time however. In 1973 the *Hardman Report* was published. This dealt with the dispersal of Government establishments from London, and it recommended that the Laboratory should be moved to Teddington, to share a site and facilities with the National Physical Laboratory, a prophetic suggestion as it happened. This would avoid the need for a London out-station, which dispersal of the Laboratory too far from the centre would make necessary.[1]

Dispersal to Teddington was, after discussion, decided not to be a viable option at this time, and by agreement with the Government sites, within an hour or an hour and a half from London were examined. Three sites, Peterborough, Northampton and Swindon were chosen for examination and teams from the Laboratory visited them. Extensive discussions took place. Presentations were made to the staff by the local authority and development agency for each of the three towns, and a vote taken. The staff and management decided that they would like to be dispersed to Swindon, but a few weeks later, on 30 July 1974, the government announced in its general plan for dispersal of Government offices that the Laboratory would be sent to West Cumbria.[2]

This location did not appeal to the staff of the Laboratory, partly because of its remoteness from London, (it would have been difficult to reach likely sites in Cumbria in less than five to six hours by train), and particularly because it was remote from major customers with whom close contact was essential. It was argued that to provide the required services to most customers at such a long distance would have required more scientific staff, who would have to spend a lot of their time travelling, not a very efficient use of their time. The case against the move to Cumbria was stated in various Annual *Reports*, while discussions went ahead on how the new situation could be made to work. The government had identified phases in the main dispersal plan in which the move of the Laboratory was seen as taking place towards the end of the five year period envisaged for the whole programme. This in itself brought problems, because the Laboratory's need for additional accommodation for new and expanding programmes of work was becoming urgent, and with little prospect of new laboratories in West Cumbria before 1983 or 1984, arrangements had to be made in 1975 for 'overspill' accommodation in the ground of the National Physical Laboratory (NPL) at Teddington. Even there very little accommodation suitable for adaption to the Laboratory's need was available and plans for a suitable modern outstation Laboratory at Teddington to accommodate 40 staff, were also put in hand. This was to be located in an unused corner of the grounds, adjoining Blandford Road. It was argued that this new laboratory block would serve as a London outstation for the Laboratory (see below) when it moved north.[3]

There were already some staff on the Teddington site since arrangements had been made in 1976 to house them in refurbished buildings there. A single storey building adjoining the Gilbert Morgan Laboratories (fittingly then occupied by the NPL Division of Chemical Standards and originally home of the National Chemical Laboratory), and another out-building were fitted out for use by the Laboratory and completely occupied by 1977. The Mineralogical Chemistry

sub-division was the first to move, and in March 1977 the staff there were recognised as a distinct sub-division. The unit took over responsibility for providing an analytical service to NPL in November 1978.

A site for the new Laboratory in Cumbria had been chosen in 1977, large enough to take a building housing 360 staff. This move meant that the Laboratory would have a new purpose built building, albeit not on a site of the staff's choosing, for the first time for 80 years.[4] Many new features were planned, social and conference facilities, and a staff restaurant for example, as well as scientific facilities such as a neutron activation unit, to enable the Laboratory to use nuclear based techniques. The whole building was to be constructed on a flexible modular basis and another part of the planning involved the construction of a laboratory test module at Teddington. This was fitted out with all services to the environmental standards that were to apply in the new laboratory, and in it were tested new ideas on the design of a safely operating ventilating system as well as the ergonomic designs of furniture and fittings that were being developed by the project team. It had been decided, after extensive consultations with customers, that some work had to remain in London, and that it would be necessary to have an outstation of about 130 staff. This would have posed significant problems, both in the need to duplicate facilities and in the long distance management of such a large entity. Representations to the government to alter the decision resulted in the move to Cumbria being confirmed twice, in 1976 and 1978.[5]

A considerable amount of work was thus carried out in preparing for the move to Cumbria. Despite misgivings Dr Egan worked assiduously to support the government decision, and to find the best way to make the policy work, while also trying to have changed a decision which he regarded as likely to endanger the future viability of a Laboratory to which he had devoted most of his working life. It was probably with a great sense of relief that he announced to staff on 26 July 1979 that, following the general election, the new government had, together with several other such dispersal projects, cancelled the decision to move the Laboratory to West Cumbria. It was certainly 'widely welcomed in the Laboratory' as he said in the 1979 *Report*.[6]

The decision not to disperse the Laboratory to Cumbria was indeed welcomed, but it revived the underlying problem,

85   Dr Ronald Coleman, Government Chemist, 1981–1987

that the accommodation in Cornwall House was not adequate and little money had been spent on the building over a five year period except to keep it watertight. More and more of the building had been converted to Laboratory use, but it was still did not meet the needs of the Laboratory. It was difficult to make efficient use of the area available and it became increasingly difficult to maintain it to standards required for modern analytical work, as well as to the increased standards required by the *Health and Safety at Work Act* of 1974.[7] Work was put in hand to make the best possible use of Cornwall House, pending a decision on where a new Laboratory could be sited. The conclusions of a report drawn up in 1979 were that the Laboratory would need to

86   Cornwall House from Waterloo Bridge

use most of the existing space in the main building: Cornwall House was divided into a main building and an annex, the Laboratory already occupied a large part of the latter by 1979. A major programme of refitment and replacement of existing services would be essential if the Laboratory was to meet its future work load and to meet safety requirements. However the structure of the building, particularly in the constraints it would put on the layout of the necessary ventilation system would make the project very difficult to achieve and very costly to carry out. The building was just not flexible enough to suit modern laboratory requirements. The search for a new location thus resumed, to identify again a site within an hour or so journey from London. Sites in Milton Keynes, London Docklands and Swindon were considered. Finally though in 1982 it was decided that the Laboratory would move to the Teddington site already occupied by the National Physical Laboratory, as originally suggested by the Hardman Report in 1973. Such a location was expected to produce savings through the use of joint services, and encourage the sharing of ideas between two important national scientific establishments.[8]

The financing, planning and construction of this Laboratory took place over another six years. Construction work actually started in January 1985, and the Laboratory moved in October–December 1988. In succession to Dr Coleman, Dr Egan's successor, Mr Alex Williams, mentioned earlier, was appointed Government Chemist in 1987, an interesting moment to take over. The planning for the new Laboratory was very elaborate, as befitted the largest and most modern chemistry laboratory constructed in the United Kingdom for more than a decade. It was fully modular in plan, to enable modifications resulting from changes in work or staffing to be as easy and as cost-effective as possible. Many modern features were built in to the design, such as the recovery of energy from the air conditioning which served all the laboratory areas. Use was made again of the test module described earlier. The total cost of the new building was to be 36 million pounds. Detailed planning for the move itself began in 1987, and a very complex programme was developed to overcome the poor access to many parts of Cornwall House, and for the removal of specialised and delicate (and occasionally very large) equipment, as well as very large numbers of records and books. The amount of

87 Alex Williams, Government Chemist, 1987–1991

time allocated to planning was fully justified by the event, which went very smoothly indeed. The Laboratory was split into eleven operational areas, and each area moved on a fixed day over a twelve week period. Essential services were in place at the new site prior to the move and each operational area aimed to lose as little effective time as possible during the move. Some areas lost virtually no time at all, and the delay in reporting sample results was kept to two weeks in many others.[9] A series of open days was held following the formal opening of the Laboratory, by the Right Honourable Alan Clark, Minister for Trade, and more than 5000 customers and potential customers were received.

88 New LGC building at Teddington

Almost coinciding with the move to a new building was the most dramatic change in status undergone by the Laboratory since its foundation: the conversion into an Agency. Agencies stem from the Ibbs Report *The Next Steps* of 1988. The aim of the Ibbs Report was to create among government departments a culture in which value for money, competition and increased efficiency of operation reduced the cost to the taxpayer of government services. To this end 'Agencies' were set up. Each Agency was asked to operate under a framework document which set out its basic operating principles, supported by a Corporate Plan which identified its strategic plans for a five year period. Each Agency had a Chief Executive supported by an executive directorate. In the Laboratory the Chief Executive also bore the traditional title of Government Chemist, supported by three Deputy Directors, responsible for different aspects of the work, the provision of analytical services, the programme on food and biosciences, and resources (financial and personnel). A Steering Board was appointed to advise the DTI Chief Engineer and Scientist on the Corporate Plan of the Laboratory, on the way the Laboratory carried out its functions and to ensure that it developed in a cost and scientifically effective way. Other members of the Board represented DTI, and experts within the Laboratory's field of competence. In function and form it was similar in some ways to the old Steering and Requirements Committees.

The Laboratory became an Agency of the DTI on 30 October 1989. Agency status brought with it a number of important freedoms. It was free to compete for work within its competence, and to develop its range of skills to meet market opportunities as it perceived them. It had to recover the full economic cost of its operations through charges to all of its customers, both those in public and private sectors. The Laboratory could aim for growth, as long as that growth could be financed from increased income.[10] Mr Williams, who had successfully overseen this momentous change, as well as the very successful move to the new building, retired at the end of March 1991 after four years as Government Chemist. He was subsequently awarded the CB in the Birthday Honours list. Mr Williams was succeeded on his retirement by Dr Richard Worswick, previously a Director of AEA Industrial Technology, as Chief Executive and Government Chemist.

89   Dr Richard Worswick, Government Chemist, 1991–

This event and becoming an Agency brings the history of the Laboratory up to the present. The new Agency is still recognisably the organisation that had existed before the change, but the change brought with it different prospects for the future. In some ways the Laboratory may still be seen as that founded by George Phillips. Routine work on alcohol control goes on, as does analysis of tobacco, although the latter is not for revenue purposes but the determination of 'tar', nicotine and carbon monoxide yields of cigarettes. At the request of the Independent Committee on Smoking and Health analytical techniques had been developed for the determination of the many other harmful components of cigarette smoke. This smoke contains many thousands of separate chemical compounds. In fact cigarette tobacco itself, in an interesting example of a wheel turning full circle may now (quite legally) contain any of many hundreds of additives. The Laboratory does not routinely have to determine their presence.

Much of the Laboratory's work is now aimed at protecting the public health and consumer protection, nutritional studies on food or water quality for example, about 35% of its effort going into these fields in 1992. The Government Chemist is still referee under the *Food Act*, of 1984, as his predecessor had been under the 1875 Act, as well as referee or official analyst under 23 other Acts.[12]

Another important area of work is in the provision of an analytical service to HM Customs and Excise (HMC&E) in their work to detect and prevent the smuggling of drugs of abuse. Simple field tests and a sequential test procedure was patented by the Laboratory and licensed for manufacture as a kit. Commercial enzyme kits for detecting internally concealed drugs have been evaluated in the Laboratory and instructions given in their use to Customs staff. Another branch of forensic work which has grown in importance is the examination of questioned documents. This work has been carried out since the early part of the century. It now involves such varied work as the giving of evidence in massive VAT frauds to giving opinions in such cases as that of Clive Ponting and on the Rudolf Hess suicide note. The customer base for the forensic services has expanded considerably and now also includes police forces, banks and building societies.

Revenue work for HMC&E also continues, providing them with scientific support, thus giving a continuing connec-

90   Modern optical microscope

tion with the Excise. Work connected with the Methylated Spirit Regulations is still carried out, determining the identity of denaturants in spirits, and checking for the presence of markers in hydrocarbon oils intended for industrial use, and indeed for the illegal removal of such markers, now using total luminescence spectroscopy. Much of the work for the Customs is in the form of advice. The increasing sophistication of industrial products (paint, food, toiletries and chemicals for example), and the more complex tariff classifications that follow means that there is an increasing need for technical advice to the Board of Customs. The entry of the United Kingdom into the EC in 1973 increased the complexity of the tariffs, the classification frequently depends on the composition, which increases the need for analytical examination. Advice is also given to the Department of Transport on the classification of goods under the European Agreement concerning the International Carriage of Dangerous Goods by Road, and the International Regulations concerning the carriage of dangerous goods by rail. In addition advice is also given in the carriage of dangerous goods by air and to the Home Office on the Hazchem code used in the UK to inform the emergency services of hazards from chemicals carried by road. Environmental work, for example in the analysis of water, air and asbestos samples, remains an important field in which the Laboratory is involved, building on experience built up over many years. Much of the work is in traditional fields in which important analytical advances have been made, for example pollutants in water. In one important case this has involved pollutants in the North Sea. The recent 'Red List' drawn up by the Department of the Environment laid down levels of these lower in some cases than previously detectable limits. New and more sensitive methods for various pesticides and herbicides residues have been developed to meet this problem. Pollution in another form, that of *Legionella pneumophilla*, has also been detected by new methods.[13]

Other fields of work are entirely new. From the 1970s fostering the use of Biotechnology was supported by the Laboratory on behalf of DTI, as part of the DTI effort to support advanced technology. A Biotransformation 'club' grew from this work, one of a number of such 'clubs' supported by the Laboratory. More important work was done in chemical sensors, and in a Chemical Sensor club. Another field of work developed in the late 1980s and now a large and important part of the Laboratory effort is the Validity of Analytical Measurement (VAM) programme. This promotes quality assurance within the UK and Europe, particularly promoting an awareness of the importance of valid results, the use of valid methods and the use of reference materials. It also conducts research into new analytical techniques, encouraging their adoption. This development of reliable methods giving consistent results was something

91 Entrance to LGC building at Teddington

which concerned Phillips in his work on tobacco and the considerable furore over methods of food analysis and their results in 1875 was described above, in chapter 7. The Laboratory has been generally concerned with developing new standard (and valid) methods for the whole of its history, a particularly important activity now given the importance of chemical analysis to modern life. The Laboratory continues to act as the focus for analytical chemistry within the UK, a fitting position for a Laboratory with 150 years of experience in the field and its customers are drawn from industry and commerce, no longer from solely within the Government domain.

By 1991 the Laboratory had thus come a long way from the methods and techniques used by its founder, George

Phillips. The enormous importance of 'black box' instruments, the great importance of automated methods of chemical analysis, would hardly be recognised by him as chemistry. Even the optical microscope, which he used extensively, had changed out of all recognition. He might also have noted with surprise that there were 375 permanent scientific and support staff against the most he had which was 24 (permanent and temporary staff).[14] He would have found that opposition over the necessity for the use of science had certainly disappeared, to be replaced by a widespread acknowledgement that for 150 years his Laboratory had played a major part in the development of analytical chemistry both to the nation and to the international scientific community. George Phillips would no doubt be well satisfied with the results of his work.

Appendices

APPENDIX 1

The Laboratory of the Government Chemist Today: a summary

The Laboratory of the Government Chemist (LGC) is an Executive Agency of the Department of Trade and Industry. It provides comprehensive analytical and advisory services based on chemistry to a wide range of customers in both the public and private sectors. It also provides consultancy and support in the areas of biotechnology and biomaterials.

As has been described, LGC has experience and expertise in most areas of analytical chemistry and plays an important part in the enforcement of law and the protection of public health, the consumer and the environment. Both human and animal health and nutrition are important areas of work for the Laboratory. Food analysis is carried out for a variety of customers for many different purposes. These include nutritional value and compliance with the Food Safety Act, tariff classification and quality assurance of food supplied to the armed services. Animal feedstuffs are examined for the presence of medicinal additives and contaminants.

Involvement in law enforcement and the protection of Government revenue have enabled the Laboratory to gain considerable expertise in the examination of goods for tariff and excise duty purposes and the forensic examination of materials, particularly drugs of abuse and questioned documents. LGC's experts often attend courts and tribunals as expert witnesses in these areas.

Environmental services include sampling and analytical services for asbestos and other airborne pollutants, waters and advice on the treatment of contaminated land. A comprehensive range of analyses of pesticide, fungicide and herbicide products are carried out as well as of tobacco products and of substances for consumer hazards.

The Laboratory is the centre for analytical chemistry in Government and is the focus for chemical aspects of the National measurement system. It works to improve the quality of analytical measurement in the UK by providing advice on analytical quality systems and marketing certified reference materials, in the VAM programme described in the last chapter.

The Biotechnology Unit and Sensors Technology Unit are the focus of the Government's efforts in these technologies. They provide advisory services and support to industry for collaborative research with the science base.

APPENDIX 2

Principals, Government Chemists, etc

Excise/Inland Revenue

George Phillips (Principal of Laboratory, June 1859)	October 1842 – March 1874
Dr James Bell	April 1874–January 1894
Sir (Thomas) Edward Thorpe (Principal of the Government Laboratory)	February 1894–September 1909

Customs

James L Johnstone (Principal Inspector of Gaugers and Head of the Wine Testing Laboratories)	1860–1872
Alfred Baker	1872–1879
James B Keene (Analyst for Tea from 1875)	1879–1886
George Excell	1886–1890
W Cobden Samuel (Principal of Laboratory, after 1894 under Sir Edward Thorpe)	1890–1901

Department of the Government Chemist

Sir James Dobbie (first Government Chemist)	December 1909–July 1920
J Connah (Acting)	August 1920–March 1921
Sir Robert Robertson	March 1921–April 1936
Sir (John) Jacob Fox (died in office)	April 1936–November 1944
Dr A G Francis (Acting)	November 1944–September 1945

Dr G M Bennett (died in office)	September 1945–February 1959
E H Nurse (Acting)	February 1959–March 1960
Dr David T Lewis	April 1960–October 1970
Dr Harold Egan	October 1970–December 1981
Dr Ronald Coleman	December 1981–May 1987
Alex Williams	May 1987–March 1991
Dr Richard Worswick	April 1991–

APPENDIX 3

Principal Dates in the Development of the Laboratory

1842	October	Excise Laboratory
1849	27 February	Inland Revenue formed
1853		Laboratory moved to temporary premises in Arundel Street
1858	31 January	Laboratory recognised as a sub-department of the Inland Revenue
1859	June	Laboratory moved to Somerset House
1861	1 January	Customs wine testing offices
1871		Inland Revenue provincial chemical stations
1875		Customs tea laboratory
1894	5 March	Customs laboratories and Inland Revenue Laboratories brought together under joint management as the Government Laboratory
1897	October	Laboratory moved from Somerset House to Clement's Inn Passage
1911	1 April	Laboratory given independent status as the Department of the Government Chemist
1959	1 July	Laboratory becomes part of the Department of Scientific and Industrial Research
1964	January	Clement's Inn Passage and Customs House laboratories closed; moved to Cornwall House
1965	1 April	DSIR dissolved; Laboratory becomes part of the Ministry of Technology
1970	20 October	Laboratory part of the Department of Trade and Industry
1974	5 March	Laboratory part of the Department of Industry
1983	11 June	Laboratory part of the new Department of Trade and Industry
1989	30 October	Laboratory becomes an executive Agency within DTI

APPENDIX 4

Reports of the Government Chemist

Reports of the Principals of the Laboratory, and subsequently of the Government Chemists were published continuously from 1857 to 1981, with breaks in 1916 and 1940–1949. The first series were published as appendices to the Reports of the Commissions of Inland Revenue, and cover the years 1 April to 31 March.

1857	1st Report. Appendix 7. pp.xii–xix
1858	2nd Report. Appendix pp.xvii–xxi, xxx–xxxiii (a summary of samples analysed appears in the *Report* proper, pp.12–16)
1859	3rd Report. Appendix pp.xii–xix
1860	4th Report. Appendix pp.xiii–xxii
1861	5th Report. Appendix pp.xiii–xxvi
1862	6th Report. Appendix pp.xiii–xxii
1863	7th Report. Appendix pp.xiv–xxvi
1864	8th Report. Appendix pp.xii–xxiv
1865	9th Report. Appendix pp.xiii–xxvi
1866	10th Report. Appendix pp.v–xix
1867	11th Report. Appendix pp.xii–xxiv
1869	12th Report. Appendix pp.xiv–xxii
1870	No 13th *Report* was published. Instead in this year was published *Report of the Commissioners of Inland Revenue* the Duties under their management for the years 1856–1869 inclusive, etc., in two volumes. Reference to the work of the Laboratory appears on pp.153–159 of volume 1.
1870/	14th Report. Appendix pp.xiv–xxix
1871	14th Report. Supplement Appendix, pp.xxiv–xxix
	This *Report*, so far as the Laboratory is concerned, only covers the year 1869–1870, the Supplement covers the year 1870–1871.
1872	15th Report. Appendix pp.xiv–xxi
1873	16th Report. Appendix pp.xv–xxii

1874	17th Report.	Appendix	pp.xv–xxiii
1875	18th Report.	Appendix	pp.xv–xxvii
1876	19th Report.	Appendix	pp.xxii–xxxiv
1877	20th Report.	Appendix	pp.xxiii–xxxiii
1878	21st Report.	Appendix	pp.xxii–xxxii
1879	22nd Report.	Appendix	pp.xxii–xxxii
1880	23rd Report.	Appendix	pp.xxii–xxxi
1881	24th Report.	Appendix	pp.xxi–xxix
1882	25th Report.	Appendix	pp.xxv–xxxv
1883	26th Report.	Appendix	pp.xxvi–xxxv
1884	27th Report.	Appendix	pp.xxvi–xxxv
1885	28th Report.	Appendix	pp.126–129
1886	29th Report.	Appendix	pp.xxviii–xxxvii
1887	30th Report.	Appendix	pp.xxviii–xxxvii
1888	31st Report.	Appendix	pp.xxviii–xxxviii
1889	32nd Report.	Appendix	pp.xxxiv–xliv
1890	33rd Report.	Appendix	pp.xxxvi–xlvi
1891	34th Report.	Appendix	pp.xxxvi–xlvi
1892	35th Report.	Appendix	pp.xxxvi–xlvii
1893	36th Report.	Appendix	pp.xl–l

Throughout this series scattered references to the work of the Laboratory occur in the main body of the *Reports*.

In 1894 the Government Laboratory was formed, and this was noted in the 37th Report of Inland Revenue (p.44). No Laboratory report appeared, but tables of the number of revenue samples analysed appeared in an appendix, pp.xl–xliv. This occurred also in 1895 in the 38th Report, (pp.xliv–xlvii), in the 39th (pp.148–149), and the 40th (pp.162–163). The full Report was sent to the Treasury, manuscript copies of these years, 1894–1897, are to be found in DSIR 26/139, pp.1– 33.

In the following year, (1898), the first of the next series of Annual *Reports* was issued. *Report of the Principal*

Chemist upon the Inland Revenue Branch of the Government Laboratory for the year ended 31 March 1897. The second of these *Reports* was also issued in 1898, for the year ended 31 March 1898. In the following year was issued the *Report of the Principal Chemist, Government Laboratory*, upon the work of the Laboratory, for the year ended 31 March 1899. In this *Report* Customs samples (see below) were included for the first time. These *Reports* were issued covering subsequent years from 1899 to 1911.

The third series of *Reports* began in 1912, with the *Report of the Government Chemist* upon the work of the Government Laboratory for the year ended 31 March 1912. Similar *Reports* appeared in 1913 and 1914. In 1915, to save money a number of reports were not published, among them that of the Government Chemist. It was prepared however, for presentation to the Treasury, and exists in a typed copy. The *Report* for 1916 was once again printed, as were those for 1917–1939. In 1922 they ceased to be Command Papers as they had been since 1898.

No *Report* for 1940 was issued, nor apparently pre-pared, although samples returns were kept throughout the war, and those for one division (Crown Contracts) exist in manuscript (DSIR 26/– [Box61], as does a letter asking for a return from the Park Royal Laboratory, DSIR 26/– [Letters]). Total sample figures for the whole of the Laboratory were published in 1950, the first *Report* after the war period. It also included a brief account of the Laboratory's work during 1940–49. The title was changed to *Report upon the work of the Department of the Government Chemist* for the year ending 31 March 1950. These were issued until 1954, in 1955 it became *Report of the Government Chemist* upon the work of this Department for the year ending 31 March 1955, repeated in 1956–1958.

In 1959, with the change of status from full Department to part of DSIR was issued the first of the *Report of the Government Chemist*. In the case of this first one it went from '1st April 1959 to 31st December 1959'. Those from 1960–1981 thus reported on a calendar year. That published in 1981 was the last full *Report*. It is now issued annually in a glossy format.

Customs Reports

No full reports on the Customs Laboratories were ever issued. The annual report of the Commissioners contain scattered references to the Laboratories and to their staff commencing with the 4th Report in 1860. In the Report for the year ending 1879, and with increasing regularity in subsequent years an appendix gave the number of tea samples received. In the same Report another appendix set out the numbers of samples of wines and of British and foreign spirits tested, and this too appeared with some regularity from then. Occasionally a comment on one or other of these appendices was included in the main text of the Report. There were never any systematic reports of the work of the Custom House Laboratory as such. The sample summaries ceased in 1909 with the formation of the department of Customs and Excise, the references to samples under the Food and Drugs Act continued until 1917.

APPENDIX 5

The New Government Laboratory

The following description of the new, 1897, laboratory at Clement's Inn was published in *Nature* (October 7, 1897, vol 56, pp.553–554). It was written by James Woodward, at that time a First Class Analyst and later Superintendent in charge of the Tobacco Division. It is published here by permission of *Nature* as a full description of a typical nineteenth century laboratory.

The New Government Laboratories

The new Government Laboratory is built on a rectangular plot of land, 120 feet long by 65 feet wide, in Clement's Inn Passage, adjacent to King's College Hospital.

The exterior of the building is faced with red bricks with bands, corners and windows of Portland stone, and consists of four floors surrounded by an area whose retaining wall is faced with white glazed bricks. Central corridors run from end to end of the building on the basement and ground floors; a staircase at each end and a hydraulic lift give access to the various floors. The main entrance faces the gateway leading into Clement's Inn, and at the opposite end are two entrances for service purposes.

The architectural treatment of the first and second floors differs wholly from that of the ground-floor and basement: the entire central portion of the building forms one large room, 49 feet long by 43 feet wide, lighted by eight lofty mullioned windows and a flat-roofed dormer lantern, the open roof being carried on light iron principals. The floor of this room is about five feet above the ceiling level of the ground-floor rooms, and the space thus gained is utilised in raising the height of the principal rooms on the ground-floor, and as a duct, seven feet wide, below the floor of the central room, for holding the heating appliances, and water, gas and drainage pipes. The remainder of the building is divided into two sections by this room: each section consists of two floors with flat asphalted roofs, one roof carrying the water cisterns, and the other affording space for operations which it is desirable should be performed in the open air, a spiral iron staircase affording the necessary access.

The ground-floor corridor has a mosaic pavement, and with the exception of a few rooms in the basement, which, as well as the other corridors, are "granolithic", all the rooms have pitch-pine parqueterie flooring. The interior walls of all the laboratories, store rooms, and corridors, are faced with white glazed brick relieved by an ornamental dado of coloured glazed bricks; the only rooms with plastered walls being those intended for office purposes.

The basement floor contains a boiler house, engineer's workshop, store rooms, a mechanical laboratory, and laboratories for bacteriological work, water analysis, standardising scientific instruments, and verifying the hydrometers and saccharometers used in the Revenue Service. The mains for gas, water, and steam are carried along the corridor immediately below the ceiling, and are supported on light iron girders, every pipe being in view throughout its entire length. Underneath the corridor floor is the main ventilation shaft, a long chamber seven feet square, with which the several ventilating shafts and fume flues are connected. A powerful fan, worked by a silent one-horse engine, keeps up the air circulation and discharges the foul air into an upcast shaft surrounding the boiler furnace flue. A "return clean water main" also runs under this corridor floor, and after picking up branch mains from all the working laboratories, ends in a concrete tank of 7000 gallons capacity. Stores for house and steam coal, and a room for refrigerating machinery, have been constructed outside the main building, the former under the street pavement and the latter also partly in the area, which is here roofed in with Hayward's lights.

The main entrance leads into the ground floor, which contains on the left a waiting-room, the principal's private office, the reference library, and the research laboratory (a room 34 × 17 feet); on the right are the Crown contracts laboratories, a suite of three rooms having a total length of 69 feet by 17 feet, the private office of the deputy principal, and the reference sample laboratory, which is 28 feet long by 20 feet wide.

The chief feature of the first floor is the main laboratory, the central room already mentioned, adjoining which is a dark room for polarimetric work and a refrigerated room for storing samples. A short corridor leading to the main staircase gives access to two rooms for the superintending analysts and to the two tobacco laboratories.

The second floors contain photographic rooms, typewriter's office, museum, and four laboratories.

The building is lighted throughout by electricity obtained from the Strand Corporation, whose continuous 100-volt current is also employed for working various motors.

Rooms intended for offices have open fireplaces fitted with Teale's slow combustion stoves; the remaining rooms are heated by passing steam through iron radiators. In the main laboratory the radiators are below the floor in the central duct, and are connected with the external atmosphere by air channels covered with slate slabs, and the warm air enters the room through iron gratings which cover the duct. To prevent down draught a copper steam pipe runs all round the base of the dormer lantern; in all the other laboratories the radiators are on the slate slabs covering the air channels, usually in the centre of the room.

For ventilation, four large air shafts run from the upper corners of the main laboratory down to the basement, where they connect with the main shaft already mentioned, and in every room through which they pass there is an opening controlled by a "hit and miss" grating. The mouthpieces at the back of all the evaporation and draught closets are contained by downward flues into the same main shaft.

The water supply is from the New River Company's high-pressure main, branches from which run throughout the building direct to the various tables for working filter-pumps, turbines, and similar contrivances. For other purposes the water is stored in three cisterns on the roof, having a total capacity of 7000 gallons, from which it is distributed for boiler feed and ordinary laboratory work. To economise water, all the working tables are provided with special drainage outlets, which are connected by a system of iron pipes to the "return clean water main". The water discharged through this main into the concrete tank is puimped up into the service cisterns on the roof; the only water run to the drains is that used for cleansing purposes.

For ice making and refrigerating, one of Messrs. J. and E. Hall's carbonic anhydride refrigerating machines is employed, in which "brine" is cooled by the evaporation of liquid carbonic anhydride in copper coils surrounded by the brine, the cooled brine being used for making ice, cooling water, and for maintaining a low temperature in the sample store adjoining the main laboratory. This store is an insulated chamber with hollow walls, made of steel plates placed immediately in front of the insulation, through which the cooled brine circulates. The main laboratory has been specially designed for the evaluation of spirituous liquors, in connection with which a great desideratum is a supply of water fairly uniform in temperature all the year round. In the summer months the temperature of the ordinary water is lowered by passing it from the cisterns on the roof down to the refrigerating machine-room, where it runs through a cooler fitted with coils through which cold brine circulates. From the cooler the water is pumped by a centrifugal pump up to a special insulated cistern holding 1000 gallons, from which all the tables in the main laboratory are served.

The working tables have mahogany tops 1½ inches thick, with fronts and ends of varnished Riga wainscot. In all rooms, except the main laboratory, the tables are placed against the outer walls immediately underneath the windows; they stand on a 3-inch plinth, which is protected by a recessed toe space and by making the table-top overhang 3 inches. They are uniformly 37 inches from floor to top of table, with a row of cupboards above the plinth topped by a single row of drawers. A space between the removable backs of the cupboards and the walls serves for carrying the water-pipes and draining troughs.

A white ware sink (12 × 9 × 4½ inches) is provided for each pair of workers, and behind it is a water standard fitted with Kelvin tap delivering into the sink, and side pipes with lever cocks for condensed water. The outlet of the sink connects through a wooden pipe with a V-shaped wooden trough lined with lead and pitched, which, after picking up from all the sinks in the table, discharges into a galvanised iron funnel, also coated with pitch; a continuing pipe conveys the dirty water into drains outside the building.

Fischer's brass filter pumps with vacuum gauge attached are fixed to the tables, and the water passing through them is conveyed by a system of pipes into the clean water return-main, as is also the water drawn from the side tubes of the water standard.

Sets of shelves for reagent bottles, consisting of three plates of glass supported on gun-metal brackets, are fixed on the walls at the back or ends of the tables.

In the main laboratory there are eight tables arranged in two rows, four tables being on each side of a wide central gangway, with a clear space of five feet between each table.

The tops are twelve feet long by five wide, and each table affords working space for four operators: a white ware sink (20 × 9 × 4½ inches) is placed at each end of the table, and the standards delivering water into the sinks serve as pillars carrying a shelf nine feet long by one foot wide, which runs down the centre of the table between the two sinks at a height of one foot above the table top. Along the under surface the wires for the electric light are carried, and a plug is fixed on each side for motor attachments.

The two water standards are connected by a water pipe running underneath the shelf, and from this pipe four branches are taken on each side for supplying water to the still condensers. This water comes from the cooled water cistern, and after doing its work is passed through nozzles fixed on the table top into a pipe running underneath along the middle of the table, which finally connects with the return clean water main.

The tables stand on a plinth with recessed toe space, the top overhangs three inches all round; the cupboards and drawers are similar to those already described. Between the backs of the opposite cupboards is a space similar to that between the backs of the cupboards and the walls in the other laboratories, and this space is utilised for holding the water pipes and draining trough. The gas supply pipes are carried along the fronts of the tables in all the laboratories immediately below the over-hanging portion, with off-takes leading to nozzles fixed on the top of the tables at the back. These off-takes are copper tubes which pass through the framing of the drawers, the control cocks being in front of the table.

All the principal laboratories are provided with evaporation closets, steam sand trays, steam drying ovens, distilled water apparatus, and cabinets for holding and using standard volumetric solutions.

The evaporation closets are very similar in character to those already in use at the Yorkshire College and elsewhere. They consist of a slate slab placed in front of a flue mouthpiece; a copper conical vessel is bolted to the under surface of the slab, which is perforated with a large bevelled hole in which is fitted a white stoneware collar. This collar, together with the copper portion, forms a hollow inverted cone, passing through the slate with a base 12 inches in diameter, on which is placed a thin steel plate coated on both sides with a rubber composition called "woodite", and perforated with holes of various sizes for holding basins, capsules and similar vessels. The upper portion is enclosed in a glass case resting on the slate slab, the front being a glass door sliding up and down by means of a counterbalancing weight working over pulleys. The roof is a plate of glass sloping down from front to back, with its back edge placed just above the top of the mouthpiece. A valve placed underneath the slab admits steam into the copper under portion of the cone, and any accumulation of condensed water flows away by a pipe fixed at a level slightly lower than the steam inlet. This pipe connects with a cubical cistern of brass with plate-glass front, arranged to act as a constant level apparatus in the event of steam not being available, in which case the bath is heated by a safety Bunsen burner placed immediately below the inverted copper apex.

The drying ovens, steam sand trays, and distilled water apparatus are all constructed as constituent parts of one appliance, through which steam from a single inlet circulates. The sand tray is a shallow copper vessel 30×12 inches; below it is a copper jacket lined with tin, through which the steam passes. It is well insulated and lagged round the sides and bottom, and forms the top of an enclosed oak cabinet fitted with wooden rails for holding dusters and towels, which are dried by the waste heat. Reduced steam first passes through a steam trap which automatically discharges the accumulation of condensed water, the outlet from the sand tray being so arranged that there is always available a supply of hot distilled water which can be drawn off as required through a Kelvin bib cock placed immediately over the cabinet doors.

From the sand tray the steam passes into the drying oven, which is fixed on the wall immediately above the sand tray. This oven is a stout copper-jacketed vessel insulated and lagged, the doors being fitted with plate-glass panels. Air for ventilation is admitted at the bottom and passes through a copper coil in the steam jacket, so that on entering the bath it is heated up to the temperature of the steam, and escapes through a similar opening at the top.

From the top of the oven a copper pipe leads the excess steam into the distilled water apparatus. This is an iron cylinder supported on brackets, and contains a block-tin worm, the upper end of which is connected with the pipe from the oven. The lower end delivers distilled water into a large earthenware jar standing on a wooden pedestal. A glass cock passes through a tubulure at the bottom of the jar in front, and through a similar tubulure on the right-hand side of the jar is fitted a glass water gauge, which also serves as an automatic overflow by being bent over into a funnel placed behind the jar; this funnel also receives water from the cylinder containing the block-tin worm, and by suitable connections delivers the water into the tank under the basement.

The appliance for holding the standard solutions is a shallow cabinet of Riga wainscot fixed against the walls, with polished plate-glass top and four doors; the panels of the upper doors are of glass, and the plinth is protected by a countersunk band of brass. The bottles containing the standard solutions stand on a shelf immediately behind the glass doors, and are fitted with two-holed rubber stoppers, through which pass a soda-lime guard tube, and a glass tube dipping down to the bottom of the bottle. This glass tube is connected, by india-rubber tubing which passes through a bevelled hole in the plate-glass top, with the stoppered side tube of a burette. Each burette is held in position by a pair of small clips fixed on two parallel brass bars, the bars being supported between a pair of brackets fixed to the ends of the cabinet on the plate-glass top. These clips ensure a perfectly rigid perpendicular position, and at the same time allow the burette to be easily raised or lowered. The burettes are filled with the standard solutions by suction through the top end of the burette, which is fitted with guard-tube continued by a depending piece of india rubber to a glass mouth-piece.

The tobacco laboratory is provided also with special drying ovens and furnaces for incinerating vegetable substances. The drying ovens, three in number, are placed on the wall one above the other, and steam for heating them is generated in a special boiler standing close by, the condensed water flows back into the boiler, which is also connected with an independent water supply, having a valve and ball-cock for keeping constant level.

The carbonising and muffle furnaces are arranged in two chambers of white glazed bricks, supported on arches which spring from a large York flagstone; apertures in the roof communicating with a flue for carrying off the fumes and heated air. For carbonising the tobacco a special furnace has been designed. From the gas main in front of the chamber five branch pipes connectr with long rectangular tubes. On each tube are screwed eight brass boxes, with a lever gas-cock between each box and the tube. Near each corner of the box a small Bunsen burner is fixed, and by this means a small flame plays uniformly over the under surface of the platinum dishes, which are supported on a light wrought-iron nickel-plated grid. The furnace is capable of holding forty dishes at one time. The front of the chamber, in which the furnace stands, is closed by a counterpoised glass door sliding up and down.

The incineration of the samples is completed in three muffle furnaces, of special design, heated by gas.

The whole of the work in connection with the building and fittings has been carried out under the immediate super-vision of H.M. Office of Works, from designs supplied by Dr. Thorpe, and the manner in which the work has been executed reflects the highest credit on that department.

APPENDIX 6

Examination Papers

Examples of the papers set for candidates 'for attendance of the Laboratory' (and subsequent attendance at the Royal College of Science) by Revenue Assistants in 1881 and 1908 (and Customs Assistants in the latter year), are given below. A different paper was set each year, but the pattern of questions, four on 'elementary' inorganic chemistry, four on 'elementary' organic chemistry 'with special reference to the chemistry of substances subject to Revenue control', three on 'elementary' physics, three on 'algebra up to and including Quadratic Equations' and one on 'Euclid (Books I and II)' was the same. The time allowed to answer 14 of these questions varied from five hours initially to seven hours later. The spaces at the top of each paper for 'name' and 'collection' (changed in 1901 to 'collection or port') were presumably filled in by the invigilator before the paper was handed out, The papers were taken locally, not centrally in London. The number of those sitting was quite high (100 in 1897, see above p.160), and candidates sometimes sat more than once, some apparently many times, (see DSIR 26/–, [Examination papers]).

Candidate for Attendance at the Laboratory

Name .... *Collection* ....

EXAMINATION PAPER, 1881

TIME ALLOWED – FIVE HOURS

Three questions *only* to be answered in each of the groups marked *A* and *B*, and two in each of the groups marked *C* and *D*. The last question on the paper to be answered, and the working of the answers to all the mathematical questions to be given in full.

The number given at the end of each question indicates the value attached to the correct answer.

Group A

1. Give an example of the operation of each of the chief natural forces, and describe the characteristics by which chemical affinity is distinguished.   (10)

2. Define empirical, molecular, and constitutional formulae. Illustrate your definitions by examples.   (10)

3. What are crystalloids and colloids? Describe the construction and use of the dialyser, and state what you know of the diffusion of gases. Describe experiments illustrating your answer. (10)

4. Describe the manufacture of ordinary and amorphous phosphorus. Give sketches of the apparatus employed and equations showing the chemical reactions. (10)

Group B

1. What are the fundamental facts on which spectrum analysis is based? Explain how you would observe the spectrum of sodium. (8)

2. What is meant by the term "specific gravity"? If a body *A* weighs 7.55 grammes in air, 5.17 grammes in water, and 6.35 grammes in another liquid B, what would be the density of the body A, and that of the liquid B? (8)

3. What instruments are necessary for determining quantitatively atmospheric pressure, temperature, and weight? (8)

4. Explain the causes of atmospheric electricity, and describe how you would illustrate the existence of such electricity. (8)

Group C

1. Given $3x - \dfrac{3x - 10}{9 - 2x} = 2 + \dfrac{6x^2 - 40}{2x - 1}$

   to find the value of x. (5)

2. Given $\left. \begin{array}{l} x^2 - xy = 48y \\ \text{and} \quad xy - y^2 = 3x \end{array} \right\}$ to find the value of x and y. (7)

3. Given $\left. \begin{array}{l} x^2y + xy^2 = 180 \\ \text{and} \quad x^3 + y^3 = 189 \end{array} \right\}$ to find the value of x and y. (7)

Group D

1. The sum of £210 was divided amongst three persons in geometrical progression, and the first had £90 more than the last. How much had each? (7)

2. What two numbers are those which being both multiplied by 27 the first product is a square, and the second the root of that square; but being both multiplied by 3,

the first product is a cube, and the second the root of that cube? (7)

3. What fraction is that to the numerator of which if 4 be added the value is one-half, but if 7 be added to the denominator its value is one-fifth? (7)

If a straight line be divided into any two parts, the square of the whole line is equal to the squares of the two parts, together with twice the rectangle contained by the parts. (6)

Candidate for Attendance at the Government Laboratory

Name .... *Collection or Port* ....

EXAMINATION PAPER, 1908

TIME ALLOWED – SEVEN HOURS

Three questions *only* to be answered in each of the Groups A and B, and two in each of the Groups C and D. Question E must be attempted, and the working of the answers to all the mathematical questions must be given in full.

Answers to be written on one side only of the paper.

The Group and Number of the question should be given in each case.

Candidates may retain their copy of the questions.

The number at the end of each question indicates the value assigned to the correct answer.

Group A

1. Describe experiments which show that ammonia gas contains hydrogen and nitrogen. (10)

2. What is silica? How is it found in nature? Describe its appearance and mention its chief chemical properties. (10)

3. Describe the mode of manufacture of oil of vitriol; state its chemical composition and give some account of its chief commercial applications. (8)

4. Give some account of the methods of extracting silver from its ores. (10)

Group B

1. How is pure methyl alcohol prepared? By what tests would you distinguish it from ethyl alcohol?   (10)

2. State what happens when (*a*) sodium oxalate, (*b*) potassium acetate, (*c*) ethyl alcohol is heated with strong sulphuric acid, and reprsent the reactions by means of equations.   (10)

3. How would you prove that carbon and hydrogen are present in alcohol, sugar and starch?   (8)

4. What is "invert" sugar; how is it prepared and to what property has its name reference?   (10)

Group C

1. Give some account of the undulatory theory of light, and explain the terms Double refraction, Interference, and Polarisation.   (8)

2. Give a general description of the construction and use of the Polarimeter.   (9)

3. What is meant by the term Electrolysis? Give some examples of electrolytic action.   (8)

Group D

1. Simplify –

$$\frac{3xyz}{xy + yz - xz} - \frac{\dfrac{x-1}{x} + \dfrac{y+1}{y} + \dfrac{z-1}{z}}{\dfrac{1}{x} - \dfrac{1}{y} + \dfrac{1}{z}}$$ (7)

2. £1,000 is divided between A, B, C and D. B gets half as much as A; the excess of C's share over D's share is equal to one-third of A's share, and if B's share were increased by £100, he would have as much as C and D have between them. How much does each get?   (8)

3. Find the two times between 7 and 8 o'clock when the hands of a watch are separated by 15 minutes.   (7)

Group E

Trisect a given straight line and prove that the sections are equal to each other.   (7)

Select Bibliography

MANY BOOKS, ARTICLES AND REPORTS from Select Committees and Royal Commissions have been used in this book. All of them are given in full in the appropriate reference. This bibliography is therefore restricted to books etc. of more general interest. The classes of the original manuscripts used are indicated in Sources and Abbreviations. All books were published in London unless otherwise shown.

Accum, Frederick, *Treatise on the Adulteration of Food*, 1820

Alton, H. and Holland, H. H., *The King's Customs*, 1908

Anon, 'The Government Chemist', *Chemist and Druggist*, 1935, pp.787–798

Bateman, Joseph, *The Excise Officers' Manual*, ed. by James Bell, 1865

Bell, James, *Compendium of General Orders from the Year 1700 to the Year 1857 inclusive*, 1858

Bell, James, *The Analysis and Adulteration of Food*, 2 vols., 1881, 1883

Bell, James, *The Chemistry of Tobacco*, 1887

Bellot, H. Hale, *University College London 1826–1926*, 1929

Bud, Robert, *The Discipine of Chemistry, the Origins and Early Years of the Chemical Society of London*, PhD thesis, University of Pennsylvania, 1960.

Carson, Edward, *The Ancient and Rightful Customs*, 1972

R C Chirnside and J H Hamence, *The Practising Chemists: A History of the Society for Analytical Chemistry, 1874–1974*, 1974

Cholmondeley, John St. Clair, *The Government Laboratory*, 1902

Church, Richard, *The Golden Sovereign*, 1957

Church, Richard, *The Porch*, 1937

Church, Richard, *The Stronghold*, 1939

Dickens, Charles, 'Weighed in the Balances', *All the Year Round*, 1877, vol. 18, pp.12–17

Dyer, Bernard, *The Society of Public Analysts: Some Reminiscences of its First Fifty Years*, 1932

Fox, J. J. and Bowles, T. H., *The Analysis of Pigments, Paints and Varnishes*, 1926

Ham, G. D., *Customs Year Book*, 1891 (and other years)

Ham, G. D., *Ham's Revenue Officers Vade Mecum*, 1876 (and other years)

Hamlin, Christopher, *A Science of Impurity: Water Analysis in Nineteenth Century Britain*, 1990

Hassall, A. H., *Adulterations Detected*, 2nd edition, 1861

Hassall, A. H., *Food and its Adulterations*, 1855

Highmore, Nathaniel J., *The Excise Laws*, 1st edition, 2 vols., 1899

Highmore, Nathaniel J., *The Excise Laws*, 2nd edition, 2 vols., 1923

Johnston, W. E., *Loftus' Inland Revenue Officers' Manual*, 1865

Keene, James, *A Handbook of Hydrometry*, 1875

Keene, James, *A Handbook of Practical Gauging*, 1861

Lambert, B., *History of London*, 1806

Leftwich, B. R., *A History of the Excise*, 1908

Loftus, W. R., *Handbook for Oficers of Excise*, 1857.

Loftus, W. R., *Inland Revenue Almanack and Official Directory*, 1856, 1857, 1858

Luke, Claude, 'Safeguarding the Public: Problems of all kinds solved by the Government Laboratory', *Strand Magazine*, 1939, vol. 96, pp.674–681

McCoy, C., *Dictionary of Customs and Excise*, 1938

Mesley, R. J., *Liquor Duties and the Government Laboratory*, unpublished manuscript, 1974, (DSIR 26/–)

Mills, John, 'The Government Laboratory', *Strand Magazine*, 1902, vol. 21, pp.561–571

Owens, John, *Plain Papers relating to the Excise*, Linlithgow, 1879

Pilcher, Richard, *The Institute of Chemistry of Great Britain and Northern Ireland: History of the Institute, 1877–1914*, 1914

Pocock, Tom, 'Chemists Behind the Law', *Leader*, 1946, vol. 3, pp.10–11

Rayment, Albert M., *Memories of an Excise Officer*, Barrow in Furness, 1975

Report (Second) from the Select Committee on Adulteration of Food, 1855, Parliamentary Paper 480

Report from the Select Committee on Tobacco Trade, 1844, Parliamentary Paper 565

Routh, Guy, *Occupation and Pay in Great Britain 1906–1979*, 1980

Russell, Colin, *Science and Social Change 1750–1900*, 1983

Simmonds, Charles, *Alcohol*, 1919

Tanner, Arthur, *The Tobacco Laws and their Administration*, Stroud, 1898

Tate, F. G. H., *Alcoholometry*, 1930.

Taylor, Geo., *The General Letters and Orders Issued by the Hon. Board of Excise*, 1837

Thorpe, T. E. and Horace Brown, *Reports on the Determination of the Original Gravity of Beers by the Distillation Process*, Institute of Brewing, 1915

Ure, Andrew, *The Revenue in Jeopardy from Spurious Chemistry*, 1843

Wanklyn, J. Alfred, *Milk Analysis, A Practical Treatise*, 1874

Woodward, J., 'The New Government Laboratories', *Nature*, 1897, vol. 56, pp.553–554

Notes

Notes to Introduction

1   Although James Marsh (inventor of the Marsh test for Arsenic) had been appointed chemist at the Royal Arsenal, Woolwich before 1829, no proper laboratory in a continuing sense seems to have existed. Much has been written lately on the climate of opinion and the scientific situation in which the Laboratory was founded, as well as the government attitude towards science in the nineteenth century. A good survey, which refers to previous work is Colin Russell, *Science and Social Change 1750–1900*, 1983. See also Robert Bud, *The Discipline of Chemistry, The origins and early years of the Chemical Society of London*, PhD thesis, University of Pennsylvania, 1980, and Bud, 'The Chemical Society – a glimpse at the foundation', *Chemistry in Britain*, vol 27, 1991, pp.230–232.

2   C. McCoy, *Dictionary of Customs and Excise*, (1938), p.249.

3   Act 'William and Mary' *cap*.24. Despite the legal use of the term 'proof spirit', probably intended originally to be a mixture of equal parts by weight of the strongest known alcohol and water, it was not legally defined until 1816. In that year in the Act 56 George III, *cap*.140, it was defined in terms of weight relative to that of an equal volume of water, ('spirits which at a temperature of 51 degrees F weigh twelve thirteenth parts of an equal bulk of water'). In 1908 the strength of proof spirit was determined at the Government Laboratory to be 49.28% by weight of absolute alcohol, (F.G.H. Tate, *Alcoholometry*, (1930), p63; R J Mesley, *The Liquor Duties and the Government Laboratory*, (1974), p.46. (unpublished ms), see also pp.152–155 below. It was determined as 49.24% in 1846 by Joseph Drinkwater, an Excise Officer associated with the Laboratory (see chapter 4 below), 'On the Preparation of Absolute Alcohol', *Memoirs and Proceedings of the Chemical Society*, vol. 3, 1845/46 – 1847/48, pp.447–454.

4   Tate, p.xi

5   Tate, pp xi–xii.

6   Tate, pp.xiv–xv. In an extract quoted by Tate from Clarkes book *The Hydrometer, or Brandy Prover*, of 1746, Clarke cites an experiment carried out at the Custom House in 1746, and also his contacts with 'principal officers of the Excise', p.xv, xvii–xviii.

7   *Exports Act*, 27 George III, *cap*. 31.

8   W R Loftus, *Handbook for officers of Excise*, 1857, pp 85–86.

9 C Blagden. 'Report on the best method of proportioning the Excise on Spirituous liquors', *Philosophical Transactions*, vol 80, 1790, p.321; G Gilpin, 'Tables for reducing the Quantities by weight in any mixture of pure spirit and water, to those by measure, etc', *Philosophical Transactions*, vol 84, 1794, part 2 pp.1–110; Tate pp 1–10, Sikes work, pp 11–32. A pyknometer is a container whose exact volume is known and is used to determine the density of a liquid. It is filled with the liquid and then weighed.

10 *Sikes Hydrometer Act*, 56 George III, *cap*.140. This Act provisionally sanctioned the hydrometers and tables and remained in operation until 1 August 1818. In May 1818 the Spirits Strength Ascertainment Act. 58 George III, C.28 gave permanent authority to Sikes' system.

11 *Duties on Spirits Act*, 6 George IV, *cap*.80.

12 Bate's Saccharometer was directed to be used under the above Act, by Order of the Commissioners of Excise on 27 June 1829. Loftus pp 161–162; above Act, Section liii.

13 See section lx and lxii of above Act.

14 McCoy, p.73. McCoy gives no evidence for this assertion. By choosing attenuation as the basis of duty the illegal removal of spirit was avoided.

15 CUST 121 TE 1258, 7 July 1829. These papers deal with some 'naphthalene' seized as being, or containing, spirit. It was sent to the Commissioners by the Port Surveyor, who sent it to the Surveying General Examiners (of Excise) to be examined. Since these were unable to come to a conclusion, they suggested that it should be referred to 'Mr Faraday'. Mr Faraday (Michael Faraday, who at that time acted as a private analytical chemist) said that it did not contain any spirits of wine. Other similar cases are found in TE 1419, 1829 (a case of artificial colouring in beer, the chemist was Faraday again), and TE 20576, also of 1829, (the chemist was Richard Phillips). There was an Excise committee of the Royal Society, CUST 121 TE 4524 of 1834, but it is unlikely that they were involved with this work. For the still at St. Katherine's Dock, Owens, *Plain Papers*, pp.498–499; B R Leftwich, *History of the Excise*, 1908, p.141.

16 The adulteration had been very sophisticated. One snuff, popular at court, was made largely of ground bitter almonds, ambergris and attergul (a scented wood). See an amusing account of the early history of tobacco smuggling and adulteration in Arthur Tanner. *The Tobacco Laws and their Administration* (London 1898), pp 9–16. Arthur Tanner joined the Inland Revenue in 1882 and published (privately) his book on the tobacco laws while an officer at Kingsbridge (Devon) Station. He had worked at the Laboratory 1891–1896, joining as a student. The book is dedicated to his 'valued friend' H J Helm, a senior analyst, and he speaks of his time in the 'Tobacco Room' of the Laboratory.

17 Tanner, p.20.

18 *The Tobacco Act*, 3 and 4 Victoria, *cap*.18 s.11.

19 Tanner, p.23. The situation concerning adulteration and the final results in 1842/44 is discussed from the point of view of the Excise in CUST 121 TE 2360 new series.

20 *The Tobacco Act 5* and 6 Victoria, *cap*.93.

21 *The General Letters and Orders issued by the Hon. Board of Excise*, edited by Geo. Taylor, 1837, pp.10–11. Interestingly another General Order of 2 November 1837 ordered that samples of substances suspected of adulteration should be divided into three portions, of which one should be kept for reference and two sent to the Senior General Examiners at the 'Chief Office' (*Compendium of General Orders from the Year 1700 to the Year 1857 inclusive*, by James Bell, 1858, p.95). This practice of dividing samples is identical to the later (and indeed current) practice.

22 *Dictionary of National Biography*, *sub* Ure; *Report from the Select Committee of the Tobacco Trade*, (1844), HC, paper 565, p.459, (subsequently referred to as *Report 1840*).

23 CUST 121 TE 2360 new series, 1840–1849, report (headed TE 9692) of 24 January 1842. In this report 'Mr Phillips' is credited with the analysis of a tobacco adulterant in 1841/42 but this person is almost certainly Richard Phillips, for whom see next chapter.

Notes to Chapter 1

1 The Board of Inland Revenue was formed from the amalgamation of the Boards of Excise and of Stamps and Taxes on 27 February 1849. The Excise function was combined with Customs when the Board of Customs and Excise was formed in 1909; *Second Report from the Select Committee on Adulteration of Food*, 1855, Parl. Paper 480, p.50; General Order 31 October 1842, see plate 4.

2 *Report from the Select Committee on Tobacco Trade*, (*Report 1844*), p.405. It has sometimes been said that Phillips studied chemistry under Richard Phillips, apparently no relation, but in view of George Phillips' own statement this seems unlikely. Richard Phillips was Chemist and Curator of the Museum of Practical Geology 1839–1851, and previously Professor of Chemistry at the Royal Military College, Sandhurst. He had turned down the Presidency of the new Chemical Society in 1841 in favour of Thomas Graham. See *Dictionary of National Biography*, under Phillips.

3 B. Lambert, *History of London*, vol. 2, 1806, p.421. The building stood on the site of the ten alms houses founded by Sir Thomas Gresham in 1575, hence the name.

4 Tanner, p.24

5 *Report*, (1857), p.xii; Letter *Daily News*, 11 October 1849, cited by Michael Stanley, 'Thomas Graham 1805–1869', *Chemistry in Britain*, vol 27, 1991, pp.239–242.

6 *Report*, (1859), p.xiii; *Report 1844*, p.414.

7   *Report 1844*, pp. 364, 367, 390, 411 *et seq.*. For Richard Phillips see note 3 above. For an example of Richard Phillips' work for the Board of Excise, see CUST 121, TE 1469, Phillips' report of August 1847.

8   *Report 1844*, pp.405, 413, (and many other references), 589. For 1855 Committee, see note 14 below.

9   *Report 1844*, p.408

10 *Report 1844*, pp.413, 415–416.

11 In this attachment to the Board of Customs he received 2 guineas for each analysis performed, see Andrew Ure, 'The Revenue in Jeopardy from Spurious Chemistry demonstrated in research upon Wood Spirit and Vinous Spirit', London, 1843, p.iii. Ure was here in dispute with Graham and Thomas Brande, also a consulting chemist. Because of his appointment to the Board of Customs Ure has been described as 'in effect the first "Government Chemist", perhaps rather overstating the situation, see W V Farrar, 'Andrew Ure, FRS and the Philosophy of Manufactures', *Notes and Queries of the Royal Society*, vol 27, 1972, p.320.

12 *Report 1844*, pp.466–467

13 *Report 1844*, pp.414–416; *Second Report from the Select Committee on Adulteration of Food* 1855, p.63

14 *Report 1844*, p.418, and see Graham's answers in particular, pp.364, 367.

15 Tanner, pp.27–28. See also CUST 121 TE 2360 new series for petitions from manufacturers pressing on general grounds for a reduction in duty in 1843, 1844 and 1849, and the Excise response.

16 *Report*, 1844, p.368. Phillips too was quite sure that this was so, pp.416, 421

17 Tanner, p.28.

18 CUST 121 TE 180 (new series) 1846; Tanner, p.28

19 CUST 121 TE 105 (new series), 1845–1846. Another file, CUST 121 TE 242 (new series) of September 1845 deals with a case of snuff said to be adulterated with sugar. Phillips and John McCulloch (another Excise Officer) analysed this and Graham was also asked to do so. The results were ambiguous in this case and so it too was abandoned in May 1846. See also CUST 121 TE 11716, a tobacco case in Bristol in which Dr Ure was also involved.

20 *Report*, (1857), p.xiii. By 1864, and presumably earlier, an allowance was made for natural sugar, DSIR 26/200 p.103; DSIR 26/200, a work book of 1860–1864, pp.103–103a. The method used in the Laboratory in 1887 was described by Bell: James Bell, *The Chemistry of Tobacco*, 1887, pp.23–28. For 1874 case DSIR 26/– [Tobacco Correspondence 1868–1883, Box 55] pp.83–88.

21 In one adulteration case the Excise Board Solicitor commented in a letter that Phillips 'only' had analysed a sample, i.e. his result in this

case was not backed up by results from Professor Graham, CUST 121 TE 180, 1846. For the University College students see chap. 4 below.

22 *Report*, (1857), pp.xiii–xiv; A H Hassall, *Food and its Adulterations*, (1855), pp.595–596

23 *The Tobacco Act*, 5 & 6 Victoria, *cap*.93, s.4; *Report*, (1857), p.xiv; Tanner, pp.34–35

24 *Report*, (1857), p.xiv, and subsequent *Reports*; DSIR 26/202, Reference Experiment Book no. 13, p.82; DSIR 26/216, Reference Experiment Book 18, pp.115–122; DSIR 26/202, Reference Experiment Book 13, pp.151–152; The Revenue Act, 30 & 31 Victoria, *cap*.90, s.19

25 *Report*, (1869), p.xvii

26 *Customs and Inland Revenue Act*, 41 & 42 Victoria, *cap*.15, s.25; *Report*, 1874, p.xviii; *Report*, 1861, p.xvii; DSIR 26/201, p.215, April 1867. The snuff, 'Murphy's Genuine Irish High Toast Snuff' contained wheat starch, yellow ochre, clay and water.

27 *Report*, 1859, p.x111; 1860, p.xiv; 1861, p.xvi; CUST 121 TE 4553. The tobacco in 1852 had been amalysed by Phillips and Adam Young. Young had attended University College with Phillips in 1844, (see chap XX below), as had Edward Dodd who as Secretary to Inland Revenue Board sanctioned Phillips' expenses in 1852. Another visit by Phillips in 1852 to give evidence in case of pepper adulteration involved a four day absence and a visit to Leeds. He claimed a total of £6.14s.6d (£6.72p) in travel and subsistence (CUST 121 TE 4485).

28 *Report*, 1857, p.xv; 1861, p.xvi

29 *Report*, 1860, xiv; 1864, p.xvi; 1865, p.xvi; 1866, p.viii; CUST 121 TE 14431 of 1872. No details of the method for separating the liquorice sugar can be found.

30 *Report*, 1874, p.xvii; Tanner, p.60; DSIR 26/201, Reference Experiment Book no. 13 pp.186–195.

31 General Order, 12 August 1871; Tanner, p.36

32 DSIR 26/133 ff 6b–15b (Laboratory Correspondence, 1869–1901) gives copies of minutes and reports. *Reports* from 1872 give returns from the out-stations.

33 *Report*, 1872, p.xv; 1877, p.xxiii; 1882, p.xxix. For adulteration in Ireland. *Report*, 1863, p.xvi, 'The adulteration of tobacco appears to be practised more at the present time in Ireland than in Great Britain.'

34 Tobacco Act, 26&27 Victoria, cap 7, s.1; *Report*, 1864, p.xii, xiii.

35 DSIR 26/– [Tobacco Correspondence 1868–1883 Box 55], memo. signed by Bell, March 15 1879, pp.115–118. Pages 119–120 contain a table of results of 'moisture in manufactured tobacco'. The matter was reconsidered in 1883 after a petition from manufacturers to be allowed to add more moisture to tobacco. Bell considered that up to 15% would not cause any analytical or financial problems. This figure

had originally been recommended by Phillips in 1863, see pp.226–255. For Bell's recommendation, see pp.158–162.

36   Tanner, pp.38–46 (the whole affair was very complicated); *Report*, 1881, p.xxvi; *Customs Consolidation Act*, 50 and 51 Victoria, *cap*.15. For the clarification of moisture statement, see Tanner pp.44–45.

37   *Report*, 1860, p.xiv; 1886, p.xxxii; 1889, p,.xxxviii; 1877, p.xxvi.

Notes and References to Chapter 2

1   CUST 119/333. The list of samples is given on a stray sheet in a file dealing with Excise students at University College. These are dealt with in chapter 4. The sample year runs from 1 June 1849 to 1 June 1850; CUST 47/695, a minute of 20 Februry 1854; An Act to Replace the Duties etc. in Soap, 1853, 16 and 17 Victoria, c.39; CUST 121 TE 1206 new series, 1846, a file of documents and analyses trying to discover whether or not unadulterated soap ever naturally exceeded the 'Excise limit' of a specific gravity of 1.05 (it did sometimes). The analyst concerned was Richard Phillips.

2   A. L. Hassall, *Food and Its Adulteration*, 1855, p.589; Patent 13747, 18 September 1851, 'Preventing the injurious effects arising from smoking tobacco'.

3   *Report*, 1857, p.xiv; 1862, p.xvi; 1865, p.xx.

4   *Report*, 1860, p.xvi; 1861, p.xviii; 1865, p.xix. Hassall was not popular among the 'Excise Chemists' as late as 1865, Bateman, *The Excise Officers Manual*, edited James Bell, 1865, p.316.

5   *Report*, 1859, p.xv; Hassall. *Food and its Adulterations*, pp.xxxii–xxxiii, xxxix; *Adulterations Detected*, 2nd edition, 1861, p.36. There are frequent references to microscopes and their use in Phillips' reports in fact, and in 1861 (*Report*, 1861, p.xv) he refers to drawings he had commissioned by one Rochfort Connor, a member of his staff in order to use them as aids when teaching the use of the microscope in identifying adulterants. Most of these remarkable drawings are now lost, but the two remaining are reproduced in plate 10.

6   *Report from the Select Committee on Tobacco Trade*, 1844, pp.465–466. Pepper adulteration of the 1880s, DSIR 26/203, pp.18, 87. In the latter case the pepper was mostly ground rice.

7   *Report*, 1857, p.xv; H. Alton and H. H. Holland, *The King's Customs*, (1908), p.218; CUST 121 TE 4721.

8   For a discussion and description of substances found in tea, see Hassall, pp.273–320. See pp.278–9, 288, 301–303 for a description of some teas analysed by George Phillips, and pp.288–289 for a fascinating account of a spurious tea factory in London in 1851. For 'Lie tea' see Hassall, Adulterations Detected, (1861), pp.77–81. Some of the spurious tea manufactured in this country was far from sophisticated, one examined in 1870 contained iron oxide, (13.8%), wood (12.2%), other inorganic matter (12.4%), dead insects, and

appeared to be in a state of decomposition, 'emitting a foetid odour', (*Report*, 1870, p.70). See also *Report*, 1862, pp.11–12.

9   Highmore, *Excise Laws*, vol 1, pp.771–780; Owens, *Plain Papers*, p.524.

10   *Report*, 1857, tables of samples received to date; *Report*, 1860, p.xxii. For some unexplained reason Phillips gives numbers of samples received from 1844, not 1842.

11   CUST 33, Excise Lease Book 1786–1886, entry 153. The lease cost £52.10s (£52.50), taken out 24 June 1853, and apparently ended 13 March 1858. Phillips was still signing 'Laboratory, Inland Revenue, Old Broad Street' on 10 June 1853, (CUST TE 4721). Leases were also taken out on houses in Norfolk Street and Surrey Street nearby at the same time. *Report*, 1860, p.xiii; *Report*, 1861, p.xiv; J Mills, 'The Government Laboratory', *Strand*, vol. 21, 1902, p.562. *The Civilian*, in 1897 describes the original accommodation as 'one room', perhaps poetic exaggeration, 2 October 1897, p.383, see also 9 October p.393. For Maclean, DSIR 26/– [Historical file box 55]; DSIR 26/139, p.167. For a list of rooms occupied in the early 1890s see page 334 below.

12   CUST 47/718, minute of 3 March 1858; Loftus, *Inland Revenue Almanack and Official Directory*, 1856, 1857 and 1858. Prior to the recognition of the Laboratory as a sub-department of the Inland Revenue Phillips appeared in the Imperial Calendar as a First Class Clerk in the Surveying–General Examiners department. His Inland Revenue rank was Senior General Examiner. The following year in the Imperial Calendar, 1859, there was an entry for the Chemical Laboratory with Phillips as Principal. For the position of George Kay and the growth of the staff see chapter 4.

13   *Report*, 1859, p.xii; *Report*, 1860, p.xiii; Inland Revenue Establishment, 1867. DSIR 26/216, Reference Experiment Book no. 17 kept by Richard Bannister is a good example of a well kept work book in a an extraordinarily clear hand. There are no examples of 'ruled ledgers' extant, the existing books are ordinary lined note books, albeit mostly bound in leather.

14   *Report*, 1861, p.xiv.

15   *Inland Revenue Establishment*, 1867; *Government Estimates* 1910–1911.

16   *Report*, 1861, pp.xv–xvi. This description is borne out by the procedure described in the 'Reference Experiment Books', for example DSIR 26/201, 202.

17   CUST 119/426, a group of letters, memos and copies of the various reports. The work cost the Board more than £1000 in fees. Bound copies of the reports still exist in the Laboratory Library. Bell's involvement with the coffee and chicory work at University College is shown by one of the monthly student reports, noting that he was absent four times in the period 15 November–31 December 1852 due to being detained at the coffee experiments, CUST 121, TE 4385.

18   *Report*, 1857, p.xvi; *Report* 1860, p.xvii. Even Phillips seems to have found the situation confusing, since in his evidence in 1855 to the *Select Committee on the Adulteration of Food*, he said (untruly) that the regulations allowed traders to sell coffee and chicory when coffee was asked for, provided it was labelled as a mixture, contrary to statements in his Reports, (*Second Report of the Select Committee*, 1855, Parl. Paper 480, p.53). The report of one case from 1853 exists, where a mixture was sold *as* a mixture and found to be one, but was not labelled as such 'as required by the General Order of 25 February 1853'. The report was signed *inter alia* by James Bell, CUST 121 TE 5025. For the case of a sample sold in 1859 as pure coffee but labelled as mixture, see *Report*, 1859, p.17.

19   DSIR 26/200, pp.42–51; DSIR 26/206, pp.30–61; *Report*, 1861, pp.xix–xx; *Report*, 1863, p.xix.

20   *Report*; 1857, p.xvi; *Report*, 1864, p.xix.

21   *Report*, 1870/71, p.xviii; *Report*, 1874, p.xix; *Report*, 1875, p.xxi; *Report*, 1878, pp.xxvii–xxix; DSIR 26/206, pp.168–173.

22   *Report*, 1870, p.136.

23   *Report*, 1872, p.xiv; *Report*, 1874, pp.xv, xvi; *Report*, 1877, p.xxiv; *Report*, 1883, p.xxvii.

24   DSIR 26/202, Reference Experiment Book no.13, p.110; *Report*, 1875, p.xxiv; DSIR 26/– [Historical Documents, no.14, Box 55]; *Civilian*, 29 May 1897, p.1; *Civilan*, 1897, p.345; Christopher Hamlin, 'Edward Frankland's Early career as London's Official Water Analyst; the Context of Previous Sewage Contamination', *Bulletin of the History of Medicine*, 1982, vol. 56, pp. 56–76, particularly pp.64 and 66; Christopher Hamlin. *A Science of Inpurity: Water Analysis in Nineteenth Britain*, Berkeley, California, 1990, particularly pp.152–298; *Report*, 1888, pp.xxix; DSIR 26/254, Book of Signatures, pp.198, 204.

25   *Report*, 1878, p.xxii.

26   PRO IR 28/10.

27   *Merchant Shipping Act* 1867, 30&31 Victoria *cap*. 124, s 4(3). The board of Trade and Inland Revenue letters are in CUST 121 TE 1034/68, numbers of samples are in a memo by Phillips dated 1 February 1869, Board of Trade regulations with their letter of 9 March 1868. The work created a great deal of correspondence. *Report*, 1975, p.115 for a reference to the end of the work.

28   CUST 121 TE 1034/68, letters of November–December 1879 concerning a case in Glasgow; *Report*, 1874, p.xv; *Report*, 1875, p.xv; *Report*, 1877, p.xxv; *Report*, 1878, p.xxv; A Harden and S S Zilva, 'The Antiscorbutic Factor in Lemon Juice', *Biochemical Journal*, 1918, vol. 12, pp.259–269.

29   DSIR 26/216, Reference Experiment Book no. 18, (of Richard Bannister), pp.212, 217; DSIR 26/133, Laboratory Correspondence, ff.16–17.

Notes and References to Chapter 3

1   *Report*, 1857, p.xv.

2   *Report*, 1857, p.4. This question had arisen before, at least twice. Once at the end of the eighteenth century, (A Clow and N L Clow, *The Chemical Revolution: A Contribution to Social Technology*, 1952, p.549) and in the early 1830s when the question had been considered by a Parliamentary Committee on the use of molasses in Brewing, see CUST 121 TE 2578.

3   *Parl. Paper 26*, 28 January 1847, p.3.

4   *Parl. Paper 26*, pp 2, 6.

5   *Parl. Paper 426*, 18 May 1847; *Parl. Paper 529*, 7 June 1847; George Fownes, 'On the value in Absolute Alcohol of Spirits of different specific gravities', *Philosophical Transactions of the Royal Society*, 1847, vol.137, pp.249–251.

6   *Sugar and Brewing Act*, 10 and 11 Vict, *cap*.5; *Distillation of Spirits from Sugar Act*, 10 and 11 Vict, *cap*.6, ; *Distillation of Spirits from Sugar etc. Act*, 11 and 12 Vict, *cap*.100, (use of mixtures in distilling).

7   The method was based on the assumption that if an alcoholic liquor was boiled until the alcohol had all been removed and then restored to its original volume with water, the difference in specific gravity before and after evaporation would be equal to the difference between the specific gravity of the spirit removed, made up to the same volume with water, and that of water itself. Having thus obtained a measure of the spirit content, a factor was used to calculate the attenuation or 'gravity lost' and this figure was added to the specific gravity of the bulked-up residue after the evaporation (the 'extract gravity') to give the original gravity. See below (p.56), for evidence that Phillips and Dobson worked in their own time.

8   Phillips and Dobson, *Report*, of 2 August 1847 (quotation from p.3), CUST 119/449. A full description of their proposed method is given on p.23. The later reports are also found in CUST 119/449. The Phillips and Dobson report is also copied in CUST 121 TE 14431 of 1872.

9   10 and 11 Vict, C.5, sect.10.

10 *Report*, 1857, p.xviii; *Report*, 1862, p.xxii.

11 Graham, *et al*, *Report*, reprinted in *Journal of the Chemical Society*, (1853), vol.5, pp.229–256. See pages 252–253 for comments on the Phillips and Dobson table. I am grateful to Dr Colin Richards for this reference. The Report was also produced as a pamphlet in 1852 by the Inland Revenue Board. The case of 1853 is found in CUST 121 TE 4979.

12 Graham must have made a considerable part of his income from such work. He sometimes received large sums (£210 for his work on chicory for example (see above), CUST 119/426, and regularly received fees of 2 guineas (£2.10), with a discount for ten or more

samples. He charged 3 guineas for court attendances in town and 5 for country courts, plus expenses. The Excise had a discount as frequent users, CUST TE 6636 of 1846. According to his own evidence he examined 100 samples in the 15 months following the *Tobacco Act* of 1842, *Daily News*, 11 October 1849.

13   CUST 121 TE 1005 new series, June 30 1847; *Excise Act*, 17 and 18 Vict, *cap*.27, (1854), sect.8; *Excise Act* 1856, 19 and 20 Vict, *cap*.34, sect.xvi. The wording of this method was as follows:–
'a definite Quantity by Measure of such Beer shall be distilled, and the Distillate and the Spent Beer respectively shall be made up with distilled Water each of them to the original Measure of the Beer before Distillation, and the Specific Gravities thereof respectively shall be then ascertained, and the Number of Degrees and Parts of a Degree of Specific Gravity by which such Distillate shall be less than the Specific Gravity of distilled Water shall be deemed to be the Degree of Spirit Indication of such Distillate, and the actual Specific Gravity of such Spent Beer added to the Degrees of original Gravity set forth in the Table contained in the schedule annexed to this Act opposite to the Degree of Spirit Indication of the Distillate contained in the same Table shall be deemed to be the original Specific Gravity of the Worts from which such Beer was made before Fermentation; and for the Purposes aforesaid all the Weighings and Measurings that may be necessary to be made of the respective Liquids shall be made when the same are at the Temperature of Sixty Degrees of Fahrenheit's Thermo-meter, and at that Temperature distilled Water shall be considered as One thousand.'

Appended to the Act was a Schedule 'Containing a Table to be used in determining the original Specific Gravity of Worts of Beer'. This was in two columns, 'Degrees of Spirit Indication' corresponding to 'Degrees of original Specific Gravity', from 1003 to 1076.5. The interpretation of this method by the Inland Revenue Laboratory is given in Joseph Bateman, *The Excise Officers Manual*, ed. James Bell, 1865, pp.265–267. Bell was later deputy Principal of the Laboratory, and then Principal. A 'reference' method for the distillation method of determining original gravity based on that used in the Government Laboratory was published as late as 1975, J R Hudson, 'The Institute of Brewing Analysis Committee Estimation of the Original Gravity of Beer', *Journal of the Institute of Brewing*, (1975), vol.81, pp.318–321.

14   *Report*, 1857, p.xix; *Report*, 1859, p.xviii; *Report*, 1863, p.xxii, see also *Report*, 1864, p.xxi. One Brewer, writing in 1852 was certainly not convinced, the acrimonious correspondence between him, the Board and Graham *et al*. is printed in the Board's pamphlet edition of the Graham report. This may perhaps be the brewer referred to by Phillips.

15   *Report*, 1864, p.xxi.

16   CUST 121 TE 9958, 1863, copied into CUST 121 TE 14431 of 1872 with the Dobson and Phillips report; *Report*, 1875, p.xxvii.

17   Not only for work in the fields hitherto described: in 1854 and 1855 Phillips was asked to examine samples of fibre claimed to make better

paper than fibres previously used. As part of his investigation in the second case he actually made several sheets of paper, CUST 121 TE 5789, and 6030. This work arose because of the need to find substitutes for the rags etc. hitherto used to make paper and then in short supply because of greatly increased demand for paper, *Report* 1857, Appendix 6, pp.x–xii.

18 *Report*, 1857, pp.11 (quoting the definition of malted barley grain in 19 and 20 Vict, *cap*.34), p.xiii.

19 *Report*, 1857, p.xvii; *Report*, 1881, pp.xxvi, xxiii; *Report*, 1861, p.xxii–xxiii.

20 *Plain Papers*, p.331; Inland Revenue Act, 1880, 43 and 44 Vict, *cap*.20, Sect.13; *Report*, 1881, p.xxxii; *Report*, 1891, p.xliii. For a few years from 1889 the figures relating to beer duty were always expressed in terms of barrels (standard barrels) in the *Report*, corresponding to the standard gravity of 1055 degrees.

21 DSIR 26/202, Reference Experiment Book no. 13, pp. 149–150, (another group of new standard hydrometers were checked two years later, pp. 171–172); *Report*, 1889, p.xxxiv; *Report*, 1885, p.127. The increasing importance of original gravity estimations prompted J A Nettleton, a member of the Laboratory staff, (and one of the Temporary Assistants), to write a series of articles on Original Gravity in the *Brewers Guardian* in 1881, published as a booklet in the same year. *Inter alia* this describes the apparatus and methods used in the Inland Revenue Laboratory to determine original gravity.

22 Cholmondeley, p.34; Mesley, p.34.

23 *Report*, 1859, p.xiii; Report 1860, p.xiv; Report, 1861, p.xvi; Tate, pp.54–55; Mesley, pp.32–33. DSIR 26/93 is part of Phillips original draft of his report on this work, in his own hand and dated 31 December 1860. It is in a very fragmentary and fragile condition due to its being 'saved from the dustbin many years ago by Mr O'Loghlen and handed to H J Helm 16/3/97' according to a note on the front in Helm's hand.

24 *Report*, 1861, p.xxii; DSIR 26/216, Reference Experiment Book no. 18, used between 1865 and 1869 contains several samples adulterated with *cocculus indicus* (e.g pp.246, 251, 252); DSIR 26/203, a work book of H J Helm contains results of over 200 such samples between June 1882 and February 1888, pp.52–63, 93–103, 114–117. Other substances used for the adulteration of beers were ground ginger, liquorice, coriander and carroway seeds and capsicum pods.

25 *Report*, 1887, pp.xxviii, xxxvi; *Intoxicating Liquors Act*, 35 and 36 Vict, *cap*.94, sect.19; *Customs and Inland Revenue Act*, 48 and 49 Vict, *cap*.51, sect.8(2).

26 *Report*, 1881 (for 1881), p.xxi; *Report*, 1882, p.xxxii.

27 *Report*, 1882, p.xxxii; *Report* 1881 (for 1880), p.xxviii; *Report*, 1881 (for 1881), p.xxi, xxvi.

28 *Report*, 1886, p.xxxiv.

29   *Report*, 1894, p.xl.

30   *Report*, 1873, p.xx; *Report*, 1878, p.xxiii. See also chapter 5 below. A similar case occurred in the following year, DSIR 26/118, pp.30–33.

31   The Graham, Hoffman and Redwood report was originally published as *Parl. Paper* no.201 of 30 April 1855. It was reprinted in 1856 with the 'testimonial' letters added as *Reports on the Successful Application of Methylated Spirits* by HMSO. The quotation is from p.30. The authorising Act was 18 and 19 Vict., *cap*.38. See also *Report of the Commissioners of Inland Revenue*, 1857, p.7. Mr Hutchinsons' letter appears on p.41 of the Methylated Spirits Report.

32   *Report*, 1857, p.xix. Regulations made later (in 1899) said that 'The wood naphtha must be sufficiently impure to import to the methylated spirits . . . such an amount of nauseousness as would, in the opinion of the Principal of the Government Laboratory, render such spirits incapable of being used as a beverage, or of being mixed with potable spirits of any kind without rendering them unfit for human consumption', (Nathaniel J Highmore. *The Excise Laws*, vol.1, 2nd edition, (1899), p.322).

33   *Report* 1858, pp.xvii–xxi. Phillips described his experiments in detail. The *Pharmaceutical Journal* report had been referred to Phillips by Thomas Dobson. The 1858 *Report* contains the analytical part in the main body of the text. Phillips' appendix consists entirely of the report on methylated spirit and on copper arsenite in wallpaper (see previous chapter).

34   DSIR 26/216, Reference Experiment Book no. 18 (of Richard Bannister), particularly pp.7–8 *et seq*. *Report*, 1863, pp.xxiv–xxv; Patent 11695 of 16 November 1847, *Purification of certain oils and spirits*.

35   Hyponitrous ether is ethyl nitrate.

36   *Report*, 1864, p.xxiii; *Report*, 1865, p.xxv; *Report*, 1866, pp.xvii–xviii. Many samples of 'whiskee' and 'brandee' appear in the workbooks of the 1860s and 1870s, e.g. DSIR 26/200 and DSIR 26/201. 'Ginee' is found in DSIR 26/216, August 9, 1866, p.62. The sample of 'Pure Islay' etc. is in DSIR 26/202 p. 41, in 1864. The analyst, Mr Helm, remarks that it had a 'peculiar taste'.

37   *Report*, 1874, p.xxi.

38   These two colour tests are referred to in use in several places in the workbooks of this period, see for example DSIR 26/216, p.62, DSIR 26/203, p.85. The Riche and Bardy test produced a purple aniline dye with methyl alcohol, but not with ethyl alcohol, (Charles Simmonds, *Alcohol*, 1919, pp.182–184). Reduction of Potassium Permanganate was frequently used earlier, (DSIR 26/206), or a 'mercury' test, possibly the reduction of mercuric nitrate by ethyl alcohol but not methyl alcohol.

39   *Report*, 1886, p.xxxi, xxxv, xxxv–vi.

40   *Customs and Inland Revenue Act*, 53 and 54 Vict, *cap*.8, sect 32(i)

had allowed any combination of substance approved by the Commissioners to be mixed with the spirits, and by regulations dated 15 June 1891 they allowed ⅜ of 1% of mixed naphtha to be added to methylated spirit which was to be sold retail (see Highmore, (2nd edition, 1899) vol.1, p.365).

41 Highmore, 3rd edition, 1923, vol. 1, p.553; *Report of the Departmental Committee on Industrial Alcohol*, 1905, Cd 2472, p.10 (para 35); The Methylated Spirits (Other than Power methylated Spirits) Regulations, 1925, sect.3; T E Thorpe, *Dictionary of Applied Chemistry*, 1937, vol. 1, p.183. I am grateful to Dr Colin Richards for his help with the complicated subject of methylated spirit.

Notes and References to Chapter 4

1 For example Edward Carson, *The Ancient and Rightful Customs*, London, (1972), p.294.

2 Dobson signed first, as Surveying General Examiner, the joint report (see above) on sugar and molasses in brewing that he wrote with Phillips in 1846, Phillips signing as Supervisor. Phillips was not promoted to an SGE (fourth Class) until 1847. Dobson became Assistant Secretary to the Inland Revenue Board in 1849 and Secretary in 1860. In 1863 he retired on a pension equal to his salary.

3 For work suggested by Dobson see note 8, chapter 3 above; staff and salaries, *Chemical Establishment, Excise Department*, Parl. Paper 509, 11 August 1855. McCulloch, obviously highly regarded as a chemist was suggested as one of those suitable to work on the constitution of tobacco in 1845/46.

4 CUST 47/718, Minutes of 3 March and 11 March; *Report*, 1867, p.xvi; Kay was in the Laboratory by 1856, Loftus, *Inland Revenue Almanack*, 1856, p.64. He published the Laboratory's method for determining the original gravity of beer in 1857, partially reprinted in W E Johnston, *Loftus' Inland Revenue Officers Manual*, 1865, p.304, from, according to Johnston, the *Engineer*.

5 CUST 121 TE 11029; CUST 121 TE 10191; CUST 121 TE 10142 of 1863; Fletcher Norton, staff member in 1859 (*Report* 1859 p.17) trained at University College 1852/54, see below; DSIR 26/201, work book of H J Helm, pp.12–13 showing O'Loghlen's work, 1863; DSIR 26/136, Record of Hearings, p.3.

6 CUST 47/662, minute of 2 October 1845; University College Register of Students 1840–41, 1845–46. The other officers were Joseph Drinkwater, John McCulloch, Robert Bainbridge, Edward Dodd, Andrew Cornwall, Thomas Rodway, Andrew Young, Charles Oding, and Charles Forsey. The additional officers were Thomas Brook, Charles Boys, Henry Prescott, George Pierce and John Sibthorpe. Thomas Allanson registered in February. Forsey had attended University College previously as a private student in 1840–41. Matriculation at University College at this time appears to have meant proving

one's ability to study. The tobacco work is described in chapter 2, the Graham/Drinkwater report was dated 21 August 1846 and represented several months work. Drinkwater's paper, 'On the Preparation of Absolute Alcohol and the Composition of "Proof Spirit"', *Memoirs and Proceedings of the Chemical Society*, 1845/46 – 1847–48, vol. 3, pp.447–454. Cust 121 TE 4104, 1846–1854, contains reports on the progress of the students as well as Drinkwater's report to the Board of 1846, letters to the Treasury and monthly reports on the students. Owens, *Plain Papers* (p.543) gives examples of examination papers which he says were prepared for candidates in 1857. They are very elementary and may be papers for students intending to matriculate.

7 H Hale Bellot, *University College London, 1826–1926*, 1929, pp.249–252; CUST 121 TE 4104, the expenses for outside expert witnesses are given as £282.7s.8d (£282.38) in 1843, £335.1s.2d (£335.6) in 1844 and £408.7s.11d (£408.40) in 1845; University College archives, John Wood correspondence, letter to Charles Atkinson, July 1846; CUST 119/333, TE 512 (new series September 1846. One of the papers by Richard Railton who studied 1851–1855 was 'On the Use of Hydrogen in determining Vapour Densities and on the acidification of alcohols by Oxygen gas or Atmospheric Air', *Journal of the Chemical Society*, vol. 6, 1854, pp.205–209. Another paper of his was sent to the Royal Society. The researches carried out by the Excise students were typically on 'derivatives from hydrates of Phenyle' or on new derivatives of chloroform, the latter by George Kay, whose article on this appeared in the *Journal of the Chemical Society*, vol. 7, 1855, pp.224–231, in the same volume as another of Railton's papers and papers by his University colleagues Henry Scrughan, James Fairlie and James Spencer.

8 As well as CUST 121 TE 4104, CUST 121 TE 4385 of 1854/55, CUST 119/333 and CUST 121 TE 512 (new series) of September 1846 all contain monthly students' reports. CUST 121 TE 4104 contains a joint letter from five students, dated 14 July 1854 asking for the 'usual' leave of absence for one month in August.

9 CUST 121 TE 4104; *Civil Service Gazette*, 2 September 1854 p.558.

10 *Civil Service Gazette*, 12 August 1854, p.509; 19 August 1854 p.526; 26 August 1854 p.541–542; 2 September 1854 p.558; 16 September 1854 p.590; 22 March 1855 p.315; 29 March 1855 p.332

11 CUST 121 TE 10142 of 1863, (in a minute of George Phillips; *Report 1860*, p.24; *Chemical Establishment Excise Department*; *Report 1860*, p.xiii; *Report 1861*, pp.24–25; Owen, *Plain Papers*, p.299. The prize of a microscope was awarded at least from 1872–1899, DSIR 26/205, under 'Students'; CUST 121 TE 8745 of 1859 (5 December), in DSIR 26/– [Historical File box 55].

12 *Report 1867*, p.xvi; *Report 1876*, pp.xvii–xviii; DSIR 26/205, under 'Bond'.

13 DSIR 26/133, Laboratory Correspondence, ff.3a–6

14 *Report 1863*, p.xv; *Report 1865*, pp.xiv–xv; DSIR 26/– [Historical

File, item 4, Box 55], a draft letter of Phillips dated March 1857 giving the names of six officers he had just examined, and praising their qualities.

15   *Report* 1864, p.xiii; CUST 121 TE 15291 of 1877 (copied into DSIR 26/– [Historical File, items 10–12, Box 55] and DSIR 26/133, ff.39a–40a, 55a–65). The duties of each officer, Keeper of Chemicals, Deputy Principal and Principal are laid down in DSIR 26/133 f.39a. Bell's salary rose in 1877 to £900, that of his Deputy (Richard Bannister) to £550, First Class Analysts to £370 (an increase of £20), and Second Class Analysts to £190 (an increase of £30).

16   DSIR 26/– [Historical File item 9, Box 55]. The two lists of books and journals are not the same and very few books are common to both. Since some of the books in the first list but not in the second were bought, (they are still held in the Library), it must have been intended to buy both sets. As a matter of comparative costs *Nature* then cost 4d per week, (ie 17s 4d or c.86p) per year. It now costs £136. The chemistry books included Ure's *Dictionary of Arts, Manufactures and Mines*, George Fownes' *Manual of Chemistry*, and Gmelin's *Inorganic Chemistry*, volumes 11–17 only. The Laboratory authorities obviously thought again about this, and bought the complete set.

17   *Inland Revenue Establishment*, 1867, 1868, 1872, 1873 and 1894; CUST 121 TE 14251 of 1868; CUST TE 121 14251 of 1871 (copied into DSIR 26/133 ff.19b – 20), on the Keeper and Deputy Keeper; CUST 121 TE 14896, resignation of Terry in 1875. Terry had been promoted from Messenger to Porter in 1863 on the grounds that he was carrying out more responsible duties than his grade or salary made equitable, CUST 121, TE 10191.

18   Phillips' official service record said that his first certificate was dated 17 February 1826 and that he was Expectant of Excise at 17 April 1826. This record (Service Book 12955) no longer exists but was cited by B F Leftwich, an accurate historian of the Excise. The same service record gave his length of service as 47 years and eleven months in April 1874. This fits with the above dates. However the first certificate, which still exists, is dated 23 April 1827, and Phillips became Expectant 6 August 1827 (CUST 119/62). The discrepancy in dates seems impossible to reconcile.

19   *Report* 1875, p.xxvii; *Civilian*, 2 October 1897, p.383; CUST 121 TE 10142 of 1863; CUST 121 TE 10191 of 1863: letter of Phillips dated 16 May 1859; DSIR 26/216, Reference Experiment Book no. 18, pp.219–220. Membership of the Chemical Society, Robert Bud, *The Discipline of Chemistry*, PhD thesis, 1980, p.399; *Report*, 1858, pp.30, xxx–xxxiii; first hand testimonial to Phillips' abilities, letter of Thomas Cheater (joined the Laboratory in 1884) to A G Francis, 12 December 1942, DSIR 26/– [Letters].

20   *Inland Revenue Establishment*, 1868; CUST 121 TE 4104. The *Dictionary of National Biography* claims that he went to the Laboratory in 1846. There is no evidence for this.

21 CUST 121 TE 16773 (copied into DSIR 26/133, ff.48–55 and DSIR 26/134, Private Reports and Special Subjects, under 'E'. The staff numbers are found in the *Inland Revenue Establishment*; DSIR 26/136, Record of Hearings, p.4; DSIR 26/133, f.71a.

22 DSIR 26/133, ff.65a–76a, 110a–116a; CUST 121 TE 16773; DSIR 26/– [Historical File, item 16, Box 55].

23 Treasury argument in a letter of 4 May 1877, CUST 121 TE 15291 of 1878 and DSIR 26/– [Historical File item no.12, Box 55]; Guy Routh, 'Civil Service Pay 1875–1950', *Economica*, 1954, vol. 21, pp.201–223, particularly pp.206–207.

24 *Spectator*, 30 November 1882, p.1518; W A Mackenzie, 'Changes in the Standard of Living in the United Kingdom 1860–1914', *Economica*, 1921, vol. 1, pp.211–230. See also George Wood, 'Real Wages and the Standard of Comfort since 1850', *Journal of the Royal Statistical Society*, 1909, vol.72, pp.90–103.

Notes and References to Chapter 5

1 *Report*, 1862, p.xiii

2 F. Accum, *Treatise on Adulteration of Food*, 1820. 'Dutch pink' was ground bark of the oak *Quercus tinctoria*, (ie Quercitron), absorbed on baryta or alumina. See pp.230–23, 290–297.

3 A. Hassall, *Food and Its Adulterations*, 1855. Hassall had a very low opinion of 'Excise Chemists' (as already described), and offered several plans for better carrying out their work of preventing adulteration in this book, (pp.xxxii–xxxiii, xxxix–xi); *First Report of the Select Committee on the Adulteration of Food*, Parl. Paper 432, July 1855, pp.28, 30–31, *Second Report*, Parl Paper 480, August 1855, pp.49–70 (Phillips' evidence), see particularly pp. 50 and 62. Phillips' summarised remarks were to some extent a riposte to Hassall. Tanner, pp.31, 31–32.

4 *Report from the Select Committee or Adulteration of Food Act (1872)*, 1874, Parl. Paper, 262, pp.91, 288; *Report*, 1874, p.xxii.

5 *Report from the Select Committee*, p.328

6 Bernard Dyer, *The Society of Public Analysts: some Reminiscences of its First Fifty Years*, (1932), pp.12–13; *Analyst*, vol.1., 1876, pp.137–139, 157. The Laboratory subscribed to the *Analyst* from its first appearance and to its one volume predecessor the *Proceedings of the Society of Public Analysts*. Both these and other early volumes of the *Analyst* contain marginal notes, showing close reading.

7 *Chemical News*, vol 30, 1874, p.11, vol 31, p.66.

8 *Chemical News*, vol 30, 1874, pp.39, 45.

9 *Report from the Select Committee*, p.vi.

10 *Chemical News*, vol.30, (1874), pp.73–74, 97, vol.31, (1874), p.92; *Proceedings of the Society of Public Analysts*, vol 1, (1876), p.127.

11 Guidance note issued by the Local Government Board, September 1875, *Proceedings*, vol.1, pp.231.

12 *Proceedings*, vol.1, p.199; *Chemical News*, vol.31, p.111. The section of the Act dealing with the Laboratory (the *Sale of Food and Drugs Act* 1875, 38 and 39 Vict, *cap*.63, sect.22) said "The justices before whom any complaint may be made or the court before whom any appeal may be heard, under this Act may, upon the request of either party, in their discretion cause any article of food or drug to be sent to the Commissioners of Inland Revenue, who shall thereupon direct the chemical officers of their department at Somerset House to make the analysis, and give a certificate to such justices of the result of the analysis; and the expense of such analysis shall be paid by the complainant or the defendant as the justices may by order direct." Sections 13 and 14 dealt with the taking and division of a sample. Regulations on the size of samples issued by the Local Government Board, 26 February 1894, copy in DSIR 26/252.

13 *Report from the Select Committee*, pp.95, 306, 326; *Report 1876*, p.xxii–xiii; DSIR 26/189 Register of Analyses of Butter etc. Other butter samples were received from Ewell, and milk samples too, perhaps all inspired by Bell; DSIR 26/134, Private Reports on Special Subjects, under G, dated 17 February 1876; butter results DSIR 26/134 under B, published in Parl. Paper 293, 15 June 1876. The milk results are in the same register, starting at M.

14 For acceptance of Laboratory point of view, see address by the President of the Society of Public Analysts to their annual meeting, 1878, *Analyst*, vol.2, p.192; DSIR 26/205, Index book of H J Helm, last page; DSIR 26/215, Index book of Charles Burge; DSIR 26/133, Laboratory Correspondence, p.86; for not joining the Society see Bernard Dyer, *The Society of Public Analysts*, 1932, p.5. That the *Food and Drugs Act* gave the Laboratory no power to fix any particular standards was noted by Bell in a letter to the Torquay magistrates in 1882, DSIR 26/118, p.56.

15 DSIR 26/118, *Food and Drugs Act* correspondance, contains many letters from both analysts and from magistrates' courts asking for guidance on standards, see pp.12–18 for help on standards; *Analyst* vol. 2, 1878, p.128 and *Analyst* vol. 8, 1880, pp.217–219 for decomposition of milk allowance. The 'Somerset House' method was severely criticised in the *Analyst* later, vol. 8, 1883, pp.252–253; *Analyst*, vol. 4, 1879, pp.216, 217 for fat in milk; DSIR 26/120 for milk certification; *Analyst*, 1879, vol. 4, p.12 for butter standard worked to by the Laboratory. As late as 1893 George Baden Powell, an MP, wrote to Dr Bell asking for the standard he adopted for milk. The reply was 2.75% fat and 8.5% non fatty solids, but that the Laboratory did not 'necessarily adhere in all cases to [these] limits', DSIR 26/122, p.263.

16 Dyer, p.5 for attainments of the Public Analysts in the early days. The Somerset House analysts were painstaking and accurate workers as may be seen from their workbooks.

17 *Analyst*, vol. 7, 1882, pp.60–64; *Analyst*, vol. 8, 1883, pp.253–254 (articles by Otto Hehner on Wanklyn's method; Dyer, p.15; DSIR 26/118 p.59, Bell's comments on Hehner's article. The slander against the 'Somerset House Chemists' is reluctant to disappear, being repeated as recently as 1977 in Colin Russell, *Chemists by Profession*, p.106. A very fair summary of the controversy is given in R C Chirnside and J H Hamence, *The Practising Analysts*, 1974, p.73.

18 The Select Committee of 1872 noted that some analysts were too rigid in their standards, *Report etc.* 1872, p.iv; J Alfred Wanklyn, *Milk Analysis, A Practical Treatise*, 1874, p.7 (repeated in the second edition, 1886); DSIR 26/118 pp.46–59, letters dealing with the milk and butter cases. Bell's final letter ends with the deleted remark 'I cannot find any justification for Mr Blyth writing such a letter to the justices and I can discover no other object for his doing so than a desire to shift his troubles from his own shoulders on to those of others'. One wonders if this or a similar paragraph appeared in the copy sent. See *Report* 1881 for further remarks by Bell on milk analysis.

19 DSIR 26/120, Sale of Food and Drugs Samples No. 1, ff.32a–33; *Analyst*, vol. 2, 1877–78, pp.121, 127–128, 130–133; *Analyst*, vol. 8, 1883, pp.39–41.

20 *Analyst*, vol. 2, 1877, p.204, for courteous suggestion by Bell to an analyst that he come to London to discuss milk analysis; *Analyst*, vol. 2, p.223 for correspondence between Bell and the Society of Public Analysts (also in DSIR 26/118 pp.21, 25, 27–30); DSIR 26/133 Laboratory Correspondence, p.85a for visits by Public Analysts. This correspondence was submitted to the *Final Report from the Select Committee on Food Product Adulteration*, 1896, pp.143–145, 152–155 by Otto Hehner, a Public Analyst; James Bell, *The Analysis and Adulteration of Food*, two volumes 1881, 1883; correspondence between Bannister and Allen, *Analyst*, vol. 19, 1894, pp.231–240. Bannister who was rather obstructive in this correspondence was a senior analyst of considerable standing, being asked to give the Cantor lectures to the Royal Society of Arts in 1888. He spoke on 'Our Milk, Butter and Cream Supply'. DSIR 26/– [Spirits of Nitre] is a series of letters between Bell and Allen in 1885, perfectly courteous in tone, which is interesting in view of the exchange between Allen and Bannister.

21 Submission by the Society of Public Analysts, DSIR 26/133, ff.79–82, Bell's reply, ff.83a–89; Dyer, p.40. A conference on food adulteration in 1884, hosted by the Institute of Chemistry, was attended by many Public Analysts and by members of the Laboratory staff. Some bad feeling is revealed in the exchanges between them, *Analyst*, vol. 9, 1884, pp.133–150, 154–163.

22 'Report from Dr Bell . . ., Principal of the Laboratory of the Commissioners of Inland Revenue, August 15th, 1893'. An annotated copy is to be found in DSIR 26/252. It was published in the *Analyst*, vol. 20, 1895, pp.163–168.

23 Joseph Bateman, *The Excise Officers' Manual*, third edition, revised and enlarged by James Bell, Surveying General Examiner, 1865. See Preface, pp.241, and 251–252 for evidence of Bell keping in touch with chemistry and with his old colleagues; James Bell, 'On Fungi and Fermentation', *Journal of the Chemical Society*, Vol. 8, 1870, pp.387–400. This was published after he became Deputy to Phillips. Bell also published 'On Fungi and Fermentation' in the *Transactions of the Royal Microscopical Society*, vol 4, 1870, pp.1–14 and 'Microscopical Examination of Water for Domestic Use', *op. cit.*, vol. 5, 1871, pp.163–68.

24 James Bell, *The Chemistry of Tobacco*, 1887.

25 T E T[horpe], 'Dr James Bell, CB, FRS', *Nature*, vol. 77, 1908, pp.540–541; 'personification of kindliness', a remark by Dr Bernard Dyer in his obituary in *Analyst*, vol. 33, 1908., pp.157–159; DSIR 26/– [Historical File, nos 3b, 3c, box 55]; Richard Pilcher, *The Institute of Chemistry of Great Britain and Northern Ireland: History of the Institute, 1877–1914*, 1914, p.90; CUST 121, TE 12698, permission date 23 June 1868.

26 DSIR 26/216, Reference Experiment Book no. 18, p.196; DSIR 26/203, 'Helm no. 5', p.18.

27 CUST 121 TE 16827, July 1884.

28 *Inland Revenue Establishment* April 1893; Board Minute of 22 December 1890 in DSIR 26/252, p.23.

29 DSIR 26/– [Dinner programmes]; DSIR 26/– [Lab–Gas]; *Civilian*, 3 February 1894, p.235, an advance notice for the 1894 'Cinderella Dance'.

Notes to Chapter 6

1 The *Customs Tariff Amendment Act*, 1860, 23 and 24 Victoria, *cap*.22; Customs *Report*, 1860, p.20

2 The method and the equipment used, one aspect of which, the connection between the distillation flask and the tube to the conden-ser, was 'devised by the Principal Inspector of Gaugers', i.e. James Johnstone, is described in the first edition of James Keene's *A Handbook of Practical Gauging*, (pp 69–75), published in 1861, just as the wine laboratories started work. For the 'Phillips' still used, see plate 35. Modifications to the equipment are described in subsequent editions of the work, i.e. 3rd, (1868), 4th, (n.d.) and 5th (1883). The still was that invented by George Phillips in the 1850s (W H Johnston, *Loftus' Inland Revenue Officers' Manual*, 1865, pp.306–307). The condensing power of Phillips' condenser was tested in the Laboratory in 1886, DSIR 26/203, p.65.

3 Customs *Report*, p.20; Customs *Report*, 1861, p.29. The Act of 1860 actually gave 1 February for the start of the new duties. The Customs Report for 1861 says 1 January. Concerning outstation Laboratories, the implication of the wording in the 1860 Customs Report, 'In the

course of our inquiries into the accuracy of the test, [for trying the strength of wines], applied by our gaugers we found that the means resorted to by them were altogether unfit for and inadequate to the purpose' (p.20), is that laboratories had existed before 1859. There is certainly evidence that laboratories of some kind existed in certain (unknown) ports for the determination of the original gravities for beer samples by 1857, and probably earlier. Since this evidence occurs in a book written for use by Excise Officers, (*Handbook for Officers of Excise*, by W.R. Loftus, (1857) pp149–50) these were presumably Inland Revenue Laboratories (and thus existing well before Phillips set up his outstation Laboratories in 1871) but could have been used jointly by Customs and Excise Officers.

The designated testing ports were Bristol, Folkestone, Gloucester, Hull, Liverpool, Newcastle, Plymouth, Southampton, Glasgow, Leith and Dublin. The testing station in Victoria Dock closed in 1867, but the others continued for several decades, and two further stations were opened in Cork and Newhaven later. This list was very similar to the ports chosen by Phillips.

4   Customs *Report*, 1860, p.20; Customs General Order, 115/1860; Customs *Report*, 1861, pp.28–30. The original forty ports were:– London, Barnstaple, Bristol, Chepstow, Dover, Exeter, Falmouth, Folkestone, Gloucester, Goole, Grimsby, Hartlepool, Hull, Ipswich, Liverpool, Lynn, Newcastle, Newhaven, Plymouth, Poole, Portsmouth, Rochester, Shields, Shoreham, Southampton, Stockton, Sunderland, Swansea, Truro, Weymouth, Whitehaven, Yarmouth; Aberdeen, Glasgow, Greenock, Leith; Belfast, Cork, Dublin and Waterford. Chester, Berwick, Penzance, Dundee and Grangemouth were added within a few months and others at intervals over the next four years. Staffing of the London Laboratories were London docks, 5 gaugers and 3 weighers; St Katharine's docks, 1 and 1; Victoria docks, 1 and none; Customs House 2 and 2, see Customs *Report*, 1862, p.28

5   The sample bottles and test notes were described by Dr Egan in an unreferenced note.

6   Copy of letter to Treasury from the Customs Board, requesting permission for extra expenditure, 22 April 1861 (Customs 30/481); copy of Board Minute noting permission received, dated 1 May 1861 (Customs 28/244). The cost of the laboratories at Victoria, St Katharines and London docks were £86, £87 and £304 respectively, although more was anticipated on the latter.

7   CUST 28/255, Minute of 22 January 1864; Customs *Report*, 1863, p.39. Concerning the question of overall control of the testing stations, Albert Baker, Johnstone's successor admitted he was 'virtually' head of the Gauging Department in 1879: *Report from the Select Committee on Wine Duties*, 1879, p.11.

8   Customs *Report*, 1863, pp 33–37.

9   Customs *Report*, 1864, pp 36,38; *Report*, 1864, ppxii, xiii; DSIR 26/206, Reference Experiment Book no. 18, pp.67, 77 *et seq.*; DSIR 26/200, p.83 (H J Helm's remarks). Phillips also comments on this

work in 1865, (*Report*, pp xiii–xiv), as do the Commissioners of Customs, noting that the 'wine' was now 'decidedly improved in character', were fewer in number, and express their 'obligations to the great assistance . . . received from Mr Phillips of the Inland Revenue Department', Customs *Report*, 1865, p.42.

10 Customs *Report*, 1867, pp 52–53; CUST 28/270, Minute of 22 November 1870, for closing of Victoria Dock laboratory; retirement of Mr Johnstone, 26 June 1872, due to 'reduction of staff', information from Dr Mesley; *Report from the Select Committee on Wine Duties*, p.11. Baker had been in the Customs Department itself since 1836.

11 Customs *Report*, 1875, pp 74–75. In the same *Report* the Commissioners noted with some satisfaction their introduction for the first time of steam launches for waterguard purposes (p.75).

12 *Customs Consolidation Act*, 1876, 39 and 40 Vict, *cap.*35 Schedule; Customs *Report*, 1880, p.54; Customs *Report*, 1882, pp 44–45; Customs *Report*, 1895, p.34, table xvi.

13 DSIR 26/200, Reference Experiment Book no. 12, pp.58, 74. Reference to Mr Perkins who wished to import Fusel oil, p.28. This method for determining the ethanol content of fusel oil was described as 'the erroneous method employed by the Excise' by T.E. Thorpe (later Principal of the Laboratory), in the 1893 edition of his *Dictionary of Applied Chemistry* (vol 2, p.171). Some doubts as to the method were undoubtedly felt and experiments on fusel oil samples continued at intervals, see for example DSIR 26/202, Reference Experiment Book no.13, pp.166, 262–265, 1867 and 1868. A (presumably) unofficial method of unknown date, using a saturated solution of calcium chloride and extraction of the higher alcohols with petroleum ether, is described DSIR 26/215, under 'F'. For the new method DSIR 26/137, Letter Book, 1885–1911, under 'F'.

14 DSIR 26/133, Laboratory Correspondence, pp.20b–28; *Report from the Select Committee on British and Foreign Spirits*, 1890, Parl paper 316, pp 2–3; 1891, Parl Paper 210, pp. 38–59. Samuel also consumed ether (apparently taken in Ireland in a solution of alcohol) to test the intoxicating effects. He reported that it did cause intoxication. It was apparently taken neat in 1867, (DSIR 26/202, p.204). Ether was made a poison in 1891 under the Sale of Poisons (Ireland) Act of 1870 in an attempt to reduce ether drinking, Inland Revenue General Order, No. 2 of 1891, DSIR 26/252. Appendix One of the second *Report from the Select Committee*, 1890 , Parliamentary Paper 316, pp.125–129 is a 'Precis of Physiological Effects of Alcohol and Fusel Oil' a 'Paper handed in by the Excise' and Appendix Four (pp.132–133) a 'Paper handed in by Dr Bell' an 'Estimation of the Free Acids in Spirits'. It was reprinted in the *Analyst*, vol. 16, 1891, pp.171–174. Bell's original work book for this is DSIR 26/– [Box 64]. Work on Fusel oil was continued regularly in the following century, first of all in connection with work for the Royal Commission on Whiskey and Other Potable Spirits in 1908–1909, vol. 1, 1908, Cd 4182, vol. 2, 1909, Cd 4796. For subsequent work see *Report*, 1965, p.20.

15 Customs *Report*, 1880, pp.55, 57; *Report from the Select Committee on Wine Duties*, p.36 (Keene's evidence); Customs and Inland Revenue Act, 1881, 44 and 45 Vict, *cap*.12, sect. 8.

16 *Spirits Act*, 1880, 43 and 44 Vict, *cap*.24 sect 95 (8). This clause allowed a deduction of 5 degrees from the proof strength found, a deduction repealed in 1889 by the *Revenue Act* of that year (52 and 53 Victoria, *cap*.42, 5, 25)

17 Customs *Report*, 1882, pp 55–57; Customs *Report*, 1886, p.23, Table xvi. The figures given in this latter table do not entirely tally with those given in 1882.

18 Tate, *Alcoholometry*, pp.77–85; Mesley, p.36. Tate explains subsequent attempts to overcome the loss of spirit in a simple distillation, particularly at high proof strengths, including suggestions by James Keene in his book *Handbook of Hydrometry*, 1875.

19 The *Sale of Food and Drugs Act*, 1875, 38 and 39 Vict, *cap*.63, s.30

20 *Report from the Select Committee on Adulteration of Food Act* (1872), 1874, Parl Paper 262, pp.iv, 84–85, 85–86, 88, 93, 96–97.

21 Customs 28/299, Board Minute, 16 March 1875; Customs 28/301 Board Minute, 20 September 1875; CUST 87/73, Out Port Records, letter to the Collector of Customs at South Shields, 3 September 1891, (and possibly to other Collectors as well); G D Ham, *Ham's Revenue and Mercantile Vade Mecum*, (1876), where it is said (p.667) samples should be sent to 'James B Keene, Esq., Analyst for Tea, Customs House, London'. Customs *Report*, 1881, p.79; Customs *Report* 1885, p.16. The distinction in 1891 between a laboratory and a testing station is all the more odd in that in 1861 a Board Minute (see note 6) setting up the dock laboratories referred to it as a 'Laboratory', as did occasional subsequent references.

22 The tea examining process, as it took place at the Custom House for many years is described accurately by Richard Church in his novel *The Porch*, (edition of 1937, p.127). For reference to Church see below, Chapter 13. Lecture notes and an examination paper from about 1952 still exist.

23 Customs *Report*, 1882, p.43. Green tea is unfermented tea, a face tea was coated with such substances as gypsum, prussian blue, and orpiment (arsenic sulphide), to enhance the green colour. Congou is a large leaf black (fermented), tea, fannings were siftings of tea, (or sometimes specifically fragments of Congou apparently).

24 The *Customs Consolidation Act*, 1876, 39 and 40 Vict., *cap*.36; CUST 28/330, Minute of 1 November 1882; Customs *Report*, 1890, p.12; In the *Report from the Select Committee on British and Foreign Spirits*, 1891. Parl Paper 210, Samuel referred to himself as 'Analyst Principal of the Customs Laboratory', (p.55). Keene was described as of the 'Hydrometer Office' on the title page of his book *A Handbook of Hydrometry*, 1875.

25 Keene, *Hydrometry*, pp.37–42, *Customs Report*, 1887, pp.16–17, 20; Customs *Report* 1888, pp19–20; Customs *Report*, 1893, pp34. Acetic, butyric and sulphuric ethers are ethyl acetate, ethyl butyrate and ethyl hydrogen sulphate respectively. For Keene's condenser DSIR 26/133, Laboratory Correspondence, pp.104–105a. It was withdrawn on the grounds of inefficiency and replaced by the 'Thorpe' condenser which had a spiral condensing coil.

26 *Third Report of the Royal Commission Appointed to Inquire into the Civil Establishments of the Different Offices of State ...*, 1889, C.5748, pp91–95. Gladstone said that in 1860 the Customs department were very reluctant to take on the new work relating to Wine duties, and that the 'alcoholic test' (ie distillation), was suggested to him by a private firm who used it in their business, not by the Customs department, (p.34). This is almost certainly a mistake, but is a measure of Gladstone's opinion.

27 Customs *Report*, 1890, p.20; *Ham's Customs Year Book*, 1891, pp.72, 128 (citing General Order 37/1886 and General Order 33/1890). In one issue of 'Ham' (1882) the Outdoor department is credited with a 'Laboratory' staffed by George Miller, but never before or subsequently. The attitude of the Board of Customs towards its own laboratory is certainly puzzling. As described, in 1891 they refused to accept that the wine samples were tested in a 'laboratory', but for some years (since at least 1886) asked for these samples to be sent to the 'Laboratory Surveyor'. The laboratory staff retained their Customs grades in the official staff lists, as part of the 'Outdoor Department' and no 'Laboratory' was ever admitted as existing in these official lists. Even Cobden Samuel when head of the Laboratory seemed unsure of his office, referring to himself as Analyst Principal, Principal of the Customs Laboratory, and Principal Chemical Officer. His official grade from 1892 until he retired was Surveyor First Class. Another measure of the value that the Customs Board placed on their Laboratory was the sporadic way that they reported its work. In terms of salary the head of the Customs laboratory did not compare well with the Principal of the Inland Revenue Laboratory. In 1865 for example James Johnstone received £500 a year when George Phillips received £750.

28 CUST 121 TE 1034/68, memo of Phillips dated 1 February 1869; *Report ... Adulteration of Food Act*, 1874, p.87, (evidence of Frederick Goulburn, Chairman of Customs), p.97, (evidence of Robert Ogilvie, Surveyor General of Customs).

29 CUST 121 TE 15220, July 1876; DSIR 26/133 Laboratory Correspondence, ff.31–36; *Report*, 1858, p.16.

30 Customs *Report*, 1890, p.16

31 DSIR 26/254, Book of Signatures. The two memoranda are now filed together. There is no internal evidence as to the actual date of the first (on the Civil Establishments evidence). In this first document Cobden Samuel mentions a Mr Sheridan, his chief assistant, not mentioned elsewhere.

Notes and References to Chapter 7

1   56 and 57 Victoria *cap*.56, section 4(1); DSIR 26/125, *Fertilisers and Feeding Stuffs Act*, Correspondence, pp.1–3.

2   The Treasury Report is given in a letter to Professor Thorpe, dated 2 March 1894. A copy of this letter is now DSIR 26/– [Historical File no. 19 Box 55]. The amalgamation of the two laboratories and the setting up of the Committee is announced to Dr Bell in a letter dated 18 January 1893 [Historical File no. 18]

3   Treasury letter, see above.

4   Cholmondeley, pp.49 and 50.

5   This Dictionary was to a large extent the successor to the last 1888–98 edition of the *Dictionary of Chemistry* by Henry Watts.

6   See "Sir Edward Thorpe", by "PPB", Obituary notice *Journal of the Chemical Society*, Vol 127, 1926 pp 1031–1050, and "Sir Thomas Edward Thorpe", by A E H Tutton, *Proceedings of the Royal Society A*, vol 109A, 1925, pp. xviii–xxiv. Thorpe was in fact something of a modern polymath.

7   R C Chirnside and J H Hamence, *The Practising Chemists*, 1974, p.72. Cholmondeley attributes to this defeat the subsequent hostility which he says Thorpe showed to the "Excise Chemists", (p.51). Bannister had made a case to the Treasury to succeed Bell so that some bad feeling between himself and Thorpe might have been anticipated.

8   *Report* 1884 p.xxvii. This milk case was something of a *cause célèbre*, see above, chapter 5. It was reported extensively in the *Analyst*, vol. 8, 1883, pp. 185–239, 243–260 and *Analyst*, vol. 9, 1884, pp.1–9.

9   These arrangements are described in the Treasury letter referred to above, (note 2). The salaries were increased by £50 *per annum* for First Class Analysts, from £500 to £550, and by the same amount for Second Class Analysts, from £300 to £350. These increases were sanctioned in the Treasury letter.

10 DSIR 26/1, Review of the Customs Branch of the Government Laboratory.

Notes and References to Chapter 8

1   CUST 121 TE 16773 (of 1884) requests permission for extra staff and notes that four rooms were occupied in 1873 and eight by 1884; John Mills, "The Government Laboratory", *Strand Magazine* 1902, Vol.21, p.562. A list of the rooms occupied in the 1880s is given in a work book (no. 5 DSIR 26/203, front end paper) of H J Helm. They were as follows:

    108   Public. Beers (?)
    109   GL (?)
    110   Lemon Juice Composition
    110a Deputy Principal

128   Hydrometry
129   WH
130   Worts
131   Water
132   India
133   Tobacco
134   Export Beer
135   Students
136   Tinctures
137   GNS
138   Analysts
139   Messengers
140   Principal

These rooms are given exactly as in Helm's book. The description of conditions in Somerset House is from the *Civilian*, 29 May 1897, p.1. It may owe something to journalistic licence.

2   Tutton, *op.cit* pp xxi–xxii; *Civilian*, 2 October 1897, p.383; *Report*, 1898, p.3. For the Inland Revenue 'General Order' regarding the new address, see plate 42.

3   Programme of Conversazione, DSIR 26/–.

4   DSIR 26/252 contains a copy of the Board Minute on signing on.

5   Typists had first been employed at Somerset House in 1894, (the term then used was "Typewriters"). Their wages in 1899 were from 14 to 15 shillings per week. By March 1901 this had risen to 17 shillings per week, paid for by the Inland Revenue, DSIR 26/205 under T.

6   In 1897 the rooms were used as follows:
        1,2   Boiler house
        3   Store (old samples)
        4,5   Crown Contracts, India
        6   Hydrometers etc.
        7   Housemaids
        8   Porters
        9   Workshop
        10   Reagents
        11,12   Storekeeper
        13   Sample reception
        14–17   Crown Contracts
        18   Deputy Principal
        19   Reference sample laboratory
        20   Door attendant
        21   Principal
        22   Library
        23   Research
        24   Lavatory
        25   Cold Chamber
        26   Polariscope
        27   Main laboratory
        28   Superintending Analyst I

29	Tobacco
30	Tobacco furnaces
31	Superintending Analyst II
32	Lavatory (analysts)
33	Photographic room
34	Museum
35	Vacant [later water analysis]
36	Typewriting
37	Supervisors
38	Vacant

Rooms 1–13 were in the basement, 14–24 on the ground floor, 25–32 on the first floor, and the rest on the second floors. This list of rooms also comes from Helm's book, (p.256), note 1 above. A plan of room 35 as it was in about 1930 is found in DSIR 26/231, p.14, a note book describing Infra Red experiments carried out by Dr Fox. Room 9, which apparently had a stone table was eventually used. The numbers given to these rooms have been disputed by Dr Griffiths, who remembers three rooms and not four in the Crown Contract laboratories and also points out that no number has been allocated for the office which was between the Principal's room and the Library.

Full descriptions of the new building are given in J. Woodward, "The New Government Laboratories", *Nature*, 1897, Vol.56 pp 553–554, and Charles J Meads "The Government Laboratory", *The Chemical News,* 1897, vol 76, pp 173–174, see Appendix 5. Woodward was a First Class Analyst at the Laboratory in 1897 and Meads a Temporary Assistant. The descriptions given in the text also derive from personal knowledge, both of the present authors worked at "Clement's Inn" as it was colloquially known.

7   *Report,* 1898, p.3

8   DSIR 26/133 Laboratory Correspondence, ff.117–125; DSIR 26/205, under Engineer, Stoker and Watchman. The Stoker received 1 shilling (5p) extra per week for attendance not later (underlined) than 6.30. Fire instructions in 1890 are given in DSIR 26/– [Historical File, no. 23, Box 55].

9   DSIR 26/133, ff.123–123a for salaries; DSIR 26/252, no 120, 'Arrangement of Work, Engineer, Storekeeper, Porters and Charwomen; change of hours in 1901 DSIR 26/215 under Laboratory Hours and DSIR 26/205 under Typewriter etc. Note concerning attendance, DSIR 26/234, no. 71, 27 April 1911.

Notes and References to Chapter 9

1   *Report,* 1909, p.29

2   Letter of Thorpe to Sir Alfred Milner (Chairman of the Board of Inland Revenue), 24 August 1896, and Treasury reply, 8 September, 1896, DSIR 26/– [Historical File nos. 20 and 20b, Box 55]. The correspondence is also given in DSIR 26/137. The previous entry on the 'War Office' shows that the work was done from 1893.

3   *Report*, 1898, p.5. The amount per head had increased to over 2 lbs by 1900,with a marked increase in the popularity of cigarettes.

4   *Report*, 1897, p.5

5   *Report*, 1899, p.5

6   Barm is the froth that forms on the top of fermenting malt liquors.

7   T E Thorpe and John Holmes, 'The Estimation of Ethyl Alcohol in Essential Oils and Medical Preparations', *Journal of the Chemical Society*, vol. 83, 1903, pp.314–317; *Report*, 1903, pp.10, 11; Highmore, *Excise Laws*, 1st edition, 1899, p.322

8   T.E. Thorpe and G. Stubbs, 'Taxine', *Journal of the Chemical Society*, 1902, vol, 81, pp 874– 883.

9   *Report*, 1897, p.7.

10 *Report*, 1900, p. 14

11 *Report*, 1897 p. 8, 1899, pp 9–10.

12 DSIR 26/119 *Food and Drugs Act* 1899, Correspondence, ff.20, 20–22, 24b, 23b–24, 26b, 28, 32, 35; *Butter and Margarine Act* 1907, 7 Edward VII *cap*.21; *The Margarine Act* 1887, 50&51 Victoria *cap*.29 s.3.

13 *Report*, 1901, p. 18.

14 DSIR 26/125, Fertilisers and Feeding Stuffs Correspondence etc, pp.33–4, 36, 48–51, 54–55, 57, 63–64; *Report*, 1898, p.9

15 Cholmondeley, p.65

16 *Report*, 1901, p. 13; 1902, p. 15; DSIR 26/–, letters between Thorpe and Scott, and others, particularly of 26 February 1901, 18 March, 29 March, (see plate 48), 11 April, 27 May, 1903.

17 *Report*, 1898. p.4.

18 *Report*, 1899 p. 3.

19 DSIR 26/133, ff.89–90, 91–93a. Mr Cobden Samuel had five members of staff in 1894; *Inland Revenue staff Establishment* 1903, 1904; DSIR 26/1, sheet headed 'Comparison of existing staff', apparently for the year end March 1903, and memo on the re-organisation of the staff of the Custom House laboratory, p.2.

20 *Report*, 1902 pp. 6, 7.

21 *Report*, 1902, p.5.

22 See for example *Report*, 1901, p.4. The denaturing was done under supervision in bond.

Notes and References to Chapter 10

1   *Report from the Select Committee on the Adulteration of Food Act*, 1872, 1874, (C 262). Bell also gave evidence to various other committees, see above.

2   *Report of the Departmental Committee appointed by the Board of Agriculture to-Inquire and Report upon the desirability of Regulations under . . . Food and Drugs Act 1899 for Milk and Cream,* 1901 (Cd 491), and see notes 6,7 and 8 below.

3   A.E.H. Tutton, *op.cit.,* p. xxi

4   *Report,* 1899, p.12; Cholmondeley, p.64; *Reports – on the use of Phosphorus in the Manufacture of Lucifer Matches,* 1899 (C 9188).

5   *Report,* 1908, p.24; T.E. Thorpe 'Note on the detection of white or ordinary phosphorus in the igniting composition of Lucifer matches', *Journal of the Chemical Society,* 1909, vol, 95 pp 440–441; DSIR 26/– Boxes 25 and 26, 'Matches'].

6   Simmonds later wrote a standard work on alcohol, *Alcohol, Its Production, Properties, Chemistry and Industrial Applications,* (London 1919). He and Thorpe published two papers on 'Lead Silicates in relations to pottery manufacture', *J-.Chem.Soc,* 1901 vol. 79. pp 791–807 *Journal of the Chemical Society,* 1910, vol. 97 pp.2282–2287. The Command papers were *Report . . . on the Employment of Compounds of Lead in the Manufacture of Pottery,* etc, 1899, (C 9207), for the thanks to Simmonds, Templeman and More see p.41; *Report . . . on The Work of the Government Laboratory on the Question of the Employment of Lead Compounds in Pottery,* 1901 (Cd 679); *The Use of Lead in the Manufacture of Pottery, Reports from Professor T E Thorpe,* 1901 (Cd 527); *Report . . . upon the Pottery Industry in France* by Thomas Oliver, 1899, (C.9526). See also *Report,* 1900, p.13, *Report,* 1902, p.18, and DSIR 26/– (Thorpe 'Pottery Inquiry 1901–1909' file). Resistance by manufacturers, see for example *Staffordshire Sentinel,* January 17, 1901, p.2 (copy in pottery enquiry file.

7   *Report of the Departmental Committee . . . Use of Preservatives and Colouring Matters in the Preservation and Colouring of Food,* 1901 (Cd 833). The most alarming colouring matters, such as Prussian blue (potassium ferrocyanide) and copper sulphate, were being used.

8   *Report,* 1902, p.11, 1903, p10; *First Report of the Royal Commission appointed to Inquire into Arsenic Poisoning from the consumption of Beer* etc. Part I, 1901 (Cd 692); Part II, 1903, (Cd 1845); *Final Report,* Part I, 1903, (Cd 1848); Part II, 1903 (Cd 1869). *Report of the Committee Appointed by the Commissioners of Inland Revenue to specify the Ingredients of Beer . . . which are liable to be contaminated by Arsenic and to prescribe tests* etc. (HMSO, 1903). For the thanks to Cheater and Stubbs, p.17. The papers of Cheater as Secretary, DSIR 26/69, Arsenical poisoning of Beer.

9   See for example 'Report of the International Committee on Atomic Weights', *Journal of the American Chemical Society,* 1903, vol 25 pp 1–5 and 'Reports of the International Committee on Atomic Weights for 1921–22', *Journal of the American Chemical Society,* 1921, vol. 43, pp 1–3.

10 T.E.Thorpe, 'On the Atomic Weight of Radium',*Proceedings of the*

Royal Society 1908 vol 80A, pp.298–309; E.Thorpe and A.G.Francis, 'The Atomic Weight of Strontium',*Proceedings of the Royal Society,* 1910, vol *83A,* pp.277–289; In a staff list of 10 May 1909 (see note 3, chapter 13 below), Mr Francis was noted as being employed on 'special research work under Dr Thorpe's direction'; Thorpe/Francis correspondence DSIR 26/254, 1908–1910, no.24, and folder; DSIR 26/– [1909–1910, box 64]; Thorpe/Rutherford correspondence, 1907, DSIR 26/254, (folder); other correspondence on Radium between Thorpe and Professor Livering at Cambridge, DSIR 26/254, 1907, nos. 23, 26, between Thorpe and James Dewar, 1907, no. 25a, from United States Geological Survey, 1908, no.25.

11 Edward Thorpe and Horace Brown, *Reports on the Determination of the Original Gravity of Beers by the Distillation Process* (Institute of Brewing, 1915), pp 606–607; Mesley, *The Liquor Duties and the Government Laboratory,* pp 51–56. The original gravity of beer is the specific gravity of the unfermented wort.

12 Francis Tate. *Alcoholometry,* (HMSO, 1930), pp 52, 53–54, 56–57, 68; Simmonds, *op.cit,* p.466; Mesley *op.cit.,* pp 44–50; DSIR 26/200, pp.101a–102, (work carried out by H J Helm).

13 *Report of the Committee Appointed by the Treasury to consider the desirability of establishing a National Physical Laboratory,* 1898, (C 8926), pp iii, 5.

14 See chapter 12.

Notes and References to Chapter 11

1 The *Sale of Food and Drugs Act* 1875, 38 and 39 Victoria, *Cap.* 63, section 22.

2 In the period 1875–1894 there were disagreements with Public Analysts' certificates in 188 cases out of the 678 disputed samples referred to the Laboratory, 96 and 411 respectively related to milk samples; see *Final Report from the Select Committee on Food Products Adulteration,* etc. 1896 (Cd 288), p.xviii.

3 For Bannister's remarks, *Final Report,* pp 124–130. Both Hehner and Bannister handed in papers to the Committee which were intended to reflect badly on the other party, *Final Report,* pp.143–145, 150–155, 159–160, 172–173.

4 *Report of the Select Committee on Food Products Adulteration,* 1894 (Cd 253), pp. 88–89.

5 *Final Report,* pp xviii–xx, xli–xlii. For this controversy see also Cholmondeley, pp. 52–53, 54, 55–56.

6 T E Thorpe, 'The Analysis of samples of milk referred to the Government Laboratory in connection with the *Sale of Food and Drugs Act*', *Journal of the Chemical Society,* 1905, vol 87, pp. 201–225. This method continued to be used during the next 70 years whenever it was necessary to examine referee samples of sour milk.

7   *The Analyst*, 1896, vol 21, pp. 34, 42–43.

8   *The Analyst*, 1900, vol 25, pp. 63, 309–313. The joint report was also summarised in the Laboratory *Report* for 1900 (p. 9). There was a terse exchange of correspondence between Thorpe and the Public Analysts in 1900 (over milk standards) but this may well have been the last such exchange, DSIR 26/256, no. 35.

9   Cholmondeley, p. 51; *Final Report from the Select Committee on Food Products Adulteration*, p. 123; DSIR 26/2, Committee of Inquiry into the organisation of the Government Laboratory, Memorandum of Dr Thorpe, 1909; DSIR 26/1, draft letter to the Treasury, 57882/07, p.1

10   DSIR 26/133, ff.97a–98, 99–99a; DSIR 26/256, no. 38 and letter in attached folder.

11   *Report*, 1897, p.3.

12   Cholmondeley, pp.56–57; DSIR 26/205, under Assistants and Students; DSIR 26/215, under the same.

13   Cholmondeley, pp.79–81. See note 14 below. Cholmondeley has six months' service before application as Laboratory Students, but Charles Burge, Superintending Analyst 1902/1904, Deputy Principal 1904/1906,, gives twelve months for Excise Students and none for Customs applicants, DSIR 26/215 under Assistants.

14   This proof, together with examples of the papers in this period, and of some of the certificates signed by students in 1901 (including Arthur Francis, who later became Deputy Government Chemist), are now in the PRO (DSIR 26/–, [Examination papers]. The elaborate procedure gone through each year to hold this examination is given briefly in DSIR 26/215 under Laboratory Students.

15   DSIR 26/133, Laboratory Correspondence, ff.108a–109a; DSIR 26/2, Dr Thorpe Memorandum of 1909 for full discussion of allowances etc.; DSIR 26/215 under Laboratory Students for Bond; DSIR 26/133, f.43 (letter of 1887); DSIR 26/252, no. 131, Treasury letter allowing bond for Established staff to be dispensed with, and DSIR 26205, DSIR 26/215 for refences to Bond (under Bond and Students). The bond originally asked for is described above in Chapter 4.

16   DSIR 26/215, under Assistants; Cholmondeley, p. 76; *Report*, 1902. p. 3.

17   *Report*, 1905, p. 10.

18   See note 14 above

19   Cholmondeley, pp. 69–70, 75, 76.

20   *Report*, 1900, p. 6.

21   DSIR 26/215, under Assistants. It appeared to be paid to all Assistants. See interesting and slightly hostile comments in the *Civilian*, Saturday May 29, 1897, p.1. For a full account of these allowances see DSIR 26/2 (see above, note 9).

Notes and References to Chapter 12

1   *Civilian*, 1909

2   A E W Tutton, 'Sir Edward Thorpe, 1845–1925', *Proceedings of the Royal Society A*, 1925, vol.109A, p.xviii

3   J.W. 'Sir James Johnstone Dobbie', *Proceedings of the Royal Society*, 1925, vol 107A, pp vi–viii.

4   *Report*, 1910, p.7. Samples were received under this Act until it was repealed in 1953, but no white phosphorus was detected after 1918.

5   See for example *Report*, 1910, p. 21; *Report*, 1911, p.22; DSIR 26/254, Book of Signatures, p.266. The reference in the 1910 *Report* is the earliest to this type of work.

6   *Report*, 1911, p.22.

7   DSIR 26/42, Paris Conference: correspondence.

8   'International Conference on Food Analysis', *Analyst*, 1911, vol 36, pp 536–539.

9   DSIR 26/1, Review of the Customs Branch of the Government Laboratory.

10 DSIR 26/2, Organisation of the Government Laboratory: Committee of Inquiry, Dr Thorpe memorandum, (also DSIR 26/1, draft letter to Dr Thorpe, 15221/07, p.2); File relating to the reorganisation, held by HM Customs Library, (25000 series) numbered on cover 11557.

11 DSIR 26/6, Government Laboratory: Reorganisation of Staff; DSIR 26/5, Government Laboratory: Establishment for 1911/12; Excise Transfer Order, Statutory Rules and Orders 1909, no. 197, clause 18; *Report*, 1912, p.5.

Notes and References to Chapter 13

1   *Report*, 1912, pp 4–5. Dobbie in fact said (erroneously) that the Inland Revenue Laboratory was founded in 1843. See also HM Customs document cited in note 10 above.

2   List of the Officers and Clerks of the Department of the Government Chemist, 1911; *Report*, 1909. p 29.

3   DSIR 26/1, 'Government Laboratory Staff', 10 May 1909.

4   According to a sheet advertising vacancies for Temporary Assistants, published by the Laboratory in February 1911.

5   Albert M Rayment, *Memories of an Excise Officer*, Barrow in Furness, 1975, p 39.

6   Rayment, pp 40, 43.

7   Rayment, pp 43, 44. This is an interesting confirmation of the evidence given to the 1910 enquiry, see above, p.167.

8   Rayment, p 43. Rayment, when describing Church implies that he was in the same room as himself. However, since at that time Church, from his description was undoubtedly in the Export Sugar Laboratory, and Rayment was in the 'Import Section', the explanation is probably that Rayment's expression 'in the Laboratory rooms' is a loose one for 'in adjoining rooms': these two laboratories were next door to each other. See list of staff and the work they did in DSIR 26/1.

9   Richard Church, *The Porch*, (London 1937), sequel. *The Stronghold*, (London, 1939).

10 H. Egan, *Chemistry in Britain*, 1983, vol 19, p 906.

11 This preoccupation with punctuality is also described by Rayment (p 40). The description of Cheater has been described as quite wrong by Dr J G A Griffiths who met him in the final years of Cheater's service. He remembers him as a keen faced individual who supported his junior staff, the latter fact brought out by Church.

12 It now seems unlikely that 'Glasgow' was A. Cameron, a Temporary Revenue Assistant (Egan, note 10 above), since 'Glasgow is described by Church as a Customs Watcher, employed as a porter. We know that there were five such porters at the Customs House at this time.

13 Church, *The Porch*, p 124. The room numbers, staff occupying them, and Halpin's promotion are inferred from the documents cited in notes 2 and 3 above. 'Jack' Sheppard was described by Dr Geoffrey Smith in a letter dated 7 December 1982 (DSIR 26/– [Letters]) as 'tall, gloomy and balding, with large glasses, perpetually worried', but this is a memory of much later. Dr Griffiths remembers Mr Halpin as having red cheeks but not Mr Sheppard.

14 Church, *The Porch*, pp 99–100. Dr Griffiths remembers only one adding machine in the Laboratory in the 1930s, for adding pounds shillings and pence.

15 *Report*, 1911, p 6.

16 *Report*, 1911, p 11; *Report*, 1914, p 11.

17 DSIR 26/3, DSIR 26/4, correspondence files between the Treasury, the Local Government Board and HMSO of November and December 1910; *Report*, 1913, p 4.

18 *Report*, 1913, p 4; *Report*, 1914, p 4; information concerning F R Ennos, letter by him to Dr Egan, 12 July 1965, DSIR 26/– [Letters].

19 *Report*, 1914, p 20; James Johnston Dobbie and John Jacob Fox, 'The Composition of Some Mediaeval Wax Seals', *Transactions of the Chemical Society*, 1914, vol. 105, pp 795–800. No conclusions as to the 'better preservation' of the seals was reached.

20 *Report*, 1914, p 5; F. Baines Westminster Hall, *Report to the First Commissioner of H M Works etc, on the Condition of the Roof Timbers of Westminster Hall, etc*, 1914, (Cd 7436), pp.12– 15, 55–61. Nothing appears to have been done until 1920, and the final (chemical) solution owed little to the work of the Laboratory or Dr Westergaard.

21 *Report*, 1914, p 17.

22 DSIR 26/– [item 1, Box 64], DSIR 26/– [items 5, 6, 7, Box 62]

Notes and References to Chapter 14

1   Richard Church, *The Stronghold*, Chap. 7. Church describes her being a 'bachelor of science', and as having 'a most attractive figure, a wild head of hair and a dangerous button of a nose'. This may possibly have been Miss Chatt, who certainly had a degree. See also Church's autobiography *The Golden Sovereign*, 1957, p.228, where his description of the arrival of this woman is even more alarming (for the woman).

2   *List of the Officers and Clerks of the Department of the Government Chemist*, 1st April 1920. In addition there were four women replacing Temporary Revenue Assistants at the London Docks and Liverpool outstations, and one Temporary Woman Clerk. These numbers had reduced to 11 in scientific grades by 1925, and two by 1930.

3   DSIR 26/–, Government Laboratory staff; Staff Book, Tobacco Department, pp.10, 16, held at LGC. 'Sanitary premises' for women were constructed in 1915 in new temporary accommodation being arranged at Clement's Inn, DSIR 26/254, no. 79. Information on coffee breaks etc from a letter of T W Harrison, 2 July 1965, DSIR 26/– [Letters].

4   WNH[artley] and A.L.[ander], 'James Johnstone Dobbie', *Journal of the Chemical Society*, vol. 125, 1924, p.2689.

5   *Reports*, 1915, p.4; l916, p.4; 1917, p.4; 1918, p.6; 1919, p.4. The *Report* for 1915 was only issued in typescript form, (signed personally by Dobbie), as a war economy measure, but reverted to the normal printed form in 1916, although the type was considerably smaller in 1918 and 1919 presumably to save paper.

6   *Report*, 1917, p.15.

7   *Report*, 1915, p.54. At the conclusion of the war this Department was merged with the Board of Trade.

8   *Report*, 1918, p.11.

9   *Reports*, 1917, p.4; 1918, p.11; 1919, p.12; 1920, p.11. This 'prize' work is found in a work book of Dr Fox, DSIR 26/–, ['special work 1040'], started in 1916.

10 *Report*, 1915, p.43; DSIR 26/– [files 14, 15, Box 52].

11 *Reports*, 1917, p.4; 1915, p.37; 1919, p.11; *Journal of the Board of Agriculture*, vol. 23, 1916–17, pp.1087–1091.

12 *Reports*, 1919, pp.4, 8; 1920, p.4.

13 Hartley and Lander, *op. cit.*, p. 2687; J.J. Dobbie and J. J. Fox, 'The constitution of sulphur vapour', *Proceedings of the Royal Society*, 1919, vol. 95A, pp. 484–92; J.J. Dobbie and J.J. Fox, 'The absorption

of light by elements in the state of vapour', *Proceedings of the Royal Society*, 1920, Vol. 98A pp.147–153; *Proceedings of the Royal Society*, 1921, vol. 99A, pp. 456–461; The earlier papers, with Fox and (twice) also with Arthur Gauge appeared in the *Transactions* and *Journal* of the Chemical Society, 1911–1914.

14 Hartley and Lander, *op. cit.*, p. 2681. Letter from Dobbie to Connah, DSIR 26/254, no.22

15 R.C. Farmer, 'Robert Robertson, 1869–1949', *Obituary Notices of Fellows of the Royal Society*, vol. 6, 1949, pp. 539–550. Connah retired as Deputy Government Chemist, in August 1921. He had joined the Civil Service in 1881.

Notes and References to Chapter 15

1 *Report*, 1921, p.36; *Report*, 1939, p.49; DSIR 26/10, Reorganisation of Technical Staff 1921–1926, letter of Sir Robert Robertson, 13 June 1921. The full sample figures are 204653 in 1921 (with another 104022 from the Chemical stations) and 430885 in 1939, with another 124160 from the Chemical stations.

2 Anon, 'The Government Chemist', *Chemist and Druggist*, June 29, 1935, p.796. This is a long and well illustrated article and contains a very useful summary of work in the Laboratory in the early 1930s; F Aumonier, 'A Simple Mercury Cathode for Arsenic Determination', *Journal of the Society of Chemical Industry*, 1927, vol. 46, pp.341T–345T. This Foods and Drugs reference work had in fact reached its peak at the turn of the century, as may be seen below. The first effective *Food and Drugs Act* was passed in 1875.

Period	Samples Analysed
1875–1884	227
1885–1894	466
1895–1904	1203
1905–1914	931
1915–1924	707
1925–1934	287
1935–1944	161
1945–1954	143
1955–1964	66
1965–1974	31

3 *Report*, 1921, p.5

4 *Finance Act* 1928, 18 & 19 George 5, *cap.* 17; *Report*, 1929, pp18–19. Letter calling Sir Robert to the first meeting of the 'Application Committee' to discuss the definition of Motor spirit in 1926, DSIR 26/77. Various other relevant papers are scattered throughout this file.

5 See for example *Report*, 1921, p.12. Home grown tobacco samples continued to appear in the Annual Reports until 1938.

6 *Report*, 1929, pp.17, 32; 1932, p.17; the *Dangerous Drugs Act*, 1920,

10 & 11 George 5, *cap.* 46; the *Dangerous Drugs and Poisons (Amendment) Act* 1923, 13 & 14 George 5, *cap.5*; the *Dangerous Drugs Act* 1925, 15 & 16 George 5, *cap.* 74, (quotation from clause 1(i); the *Dangerous Drugs Act* 1932, 22 & 23 George 5, *cap.* 15.

7 For example J R Nicholls, 'The Estimation of Morphine' *Analyst*, 1923, vol 47 pp 506–510, (given also as a paper at an internal Laboratory meeting); 'The Determination of cocain alkaloids in mixtures with other alkaloids and local anaesthetics', *Analyst*, 1936, vol 61 pp 155–159; J King, 'The identification of dicodide, eucodal and dilaudide', *Analyst*, 1931, vol 56 pp 498–503 (dicodide is dihydrocodeine bitartrate and dilaudide is dihydromorphine dihydrochloride).

8 *Chemist and Druggist*, 1935, p.10. The micro-determination of drugs was described to me by Dr Griffiths.

9 *Dyestuffs (Inport Regulations) Act* 1920, !0 & 11 George 5, *cap.* 77; *Safeguarding of Industries Act* 1921, 11 & 12 George 5, *cap.* 47; *Merchandise Marks Act* 1911, 1 & 2 George 5, *cap.* 31; *Report*, 1922, p.9

10 *Merchandise and Marks Act* 1926, 16 & 17 George 5, *cap.* 53; *Import Duties Act* 1931, 22 & 23 George 5, *cap.* 8; *Ottawa Agreement* 1931, 22 & 23 George 5, *cap.* 53; *Report*, 1922, pp.4–5; *Agricultural Produce (Grading and Marking) Act* 1928, 18 & 19 George V, *cap.* 19 and various regulations laid down under this Act, including the *Agricultural Produce (Grading and Marking) (Wheat Flour) Regulations* S R & O 592, 1933. There were very many other such regulations covering most kinds of food; J R Nicholls, 'Water and Fat in English Cheeses', *Analyst*, 1941, vol. 66, pp.276–286; J R Nicholls, *Aids to the Analysis of Foods and Drugs*, 6th edition, 1942, p.115; *Chemist and Druggist*, 1935, p.794.

11 *Report*, 1933, p.20.

12 *Chemist and Druggist*, pp.8–9; DSIR 26/77, letter of 7 January 1935 from the Secretary to the Customs Board, and of 15 January, letter of Sir Robert; DSIR 26/– [KID file].

13 *Report*, 1926, pp.17–18. Dr Fox was working on such artificial silks early in 1925, DSIR 26/– [Special Work 1040]; *Chemist and Druggist*, 1935, p.791

14 S A Ashmore, C E Barton, J A Heald, John King, J Pooley and R Sutcliffe (DSIR 26/10). Pooley had been an unsuccessful candidate in what may have been the last competition held under the older regulations for three Second Class Analysts in August 1920. The three successful candidates in this competition 'of candidates nominated to compete for three situations as Second Class Analysts in the Government Laboratory' were B A Ellis, R A Fitzsimons and A F Weiss, one of whom (Ellis) also went on to become a Superintending Analyst (source, copy of Civil Service Commission Report, held at LGC).

15 Staff list for 1922; DSIR 26/25, Committee on Scientific Staffs in Government Departments, Minutes, 5th meeting.

16   DSIR 26/9, Customs and Excise, substitution of unattached officers by temporary clerks; 26/10, substitution of Customs and Excise officers by Laboratory Assistants; 26/11; 26/27, Proposals for an increase in staff, 1925.

17   DSIR 26/25–30, Committee on Scientific staff in Government Departments, Minutes and Documents; *Report of the Committee on the Staffs of Government Scientific Establishments*, HM Treasury, 1930. The quotation is from DSIR 26/25, First set of Minutes, 27 May 1929. In 1930 the staff consisted of 47 established staff, that Government Chemist and Deputy, 6 Superintending Chemists, 12 Chemists Class I, 27 Chemists Class II, 35 Temporary Assistant Chemists, and 90 Laboratory Assistants in two grades. The number of officers of Customs and Excise seconded to the Department for service and training had dropped to 36. By 1932, after the Superintending Chemists there were 11 Senior Chemists, 4 Higher Grade Chemists, 33 Chemists, and 6 Assistants III, a total of 62, together with 23 Temporary Assistant Chemists, 91 Laboratory Assistants, (both said to be obsolete grades, although the latter still appear in the 1939 *Report*), and 36 Customs and Excise officers. By 1936 the latter had all gone and there was only one Temporary Assistant Chemist. The Assistant grades were in state of confusion throughout these years. (*Reports*, 1930, p.39; 1932, p.39; 1936, p.40, 1939, p.42).

18   *Report*, 1939, p.40

19   *Imperial Calendar*, 1931; Guy Routh, 'Civil Service Pay, 1875–1950', *Economica*, vol. 21, 1954, pp.206, 216; Institute of Chemistry, *Journal and Proceedings*, Part 1, 1931, pp.202–203. It is difficult to make comparisons between grades at this period, see also Guy Routh, *Occupation and Pay in Great Britain 1906–1979*, 1980, p.143.

20   DSIR 26/256, nos 5, 79. This use of the building had been suggested by Dobbie. There was also the 'Vestry Hall', in Clare Market, just to the north of the Laboratory. This was apparently occupied by part of the Tobacco section in 1918 (p.16, Tobacco Staff Book, held in LGC).

21   Bankruptcy Building: the use of this does not appear to be documented and depends on surviving staff memories; addition to headquarters building, memorandum by Dr Bennett, then Government Chemist, in 1955, says this occurred in 1935, DSIR 26/– [Cornwall House]. This date is confirmed by the anonymous *Chemist and Druggist* article, published in 1935, which says the addition was then recent. The luncheon club is remembered by Dr Griffiths who served on the committee as the Laboratory representative.

22   This does not appear to be documented. This date depends on the memory of surviving staff of this period.

23   The headquarters building, the Bankruptcy Court building and the legal chambers on the other side of Clement's Inn were demolished during the period 1965–69 and replaced by an extension of the London School of Economics (on the Laboratory site) and by modern offices.

24 The Park Royal laboratory is mentioned in the *Report* for 1936, p.5, its subsequent fate in a letter of Norman Squance, January 1983, DSIR26/– [Letters]. Gilbert appeared in the Laboratory staff lists as the sole 'Technical Chemist'.

25 *Reports*, 1917 *et seq*; 1933, p.14; 1934, p.17; DSIR 26/9.

Notes and References to Chapter 16

1 *Report*, 1930, p.39. There is no reference to the Metropolitan Police laboratory in the *Report*.

2 Eg. *Report*, 1929, p.33. These were two chemists 'holding official posts in India' who had received training in the methods used at the Laboratory.

3 The Department of Scientific and Industrial Research (DSIR) was set up to promote and organise scientific research with particular reference to its application to trade and industry, work the laboratory was increasingly involved in.

4 *Report*, 1921, p.9; 1931, p.39; 1932, p.6; 1933, p.20. Advice was also frequently sought in the case of the *ad valorem* duty samples (see chapter 9 p.138) asking for the distinction between various grades of articles rather than sampling. Presumably this was the reason for the committees of the Board of Customs and Excise referred to above.

5 DSIR 26/76, meeting of the Power Alcohol Committee, 17 October 1921. The recommendation was incorporated in the Power Methylated Spirits Regulations, SR & O 1921, no.1318.

6 DSIR 26/76, meeting of the Power Alcohol Committee, 25 May 1923.

7 The specification, one for motor spirit and two for aviation fuel, was published as British Standards no.121, 1923, and covered distillation characteristics, acidity, specific gravity, residue on evaporation, aromatic hydrocarbons and sulphur content. The work originated in that of the Laboratory Power Alcohol Committee, see DSIR 26/76.

8 DSIR 26/57–60.

9 *Report*, 1927, p.12; 1929, pp.10, 34, 36; 1930, pp.38–39

10 *Report*, 1930, p.38

11 *Report*, 1935, pp.35, *et al*

12 *Report*, 1931, p.14. A similar remark was made in the 1922 *Report*, p.15.

13 Some of the internal papers exist, eg 'Pollution of Rivers' by A J Gauge, 1923, DSIR 26/78. Examples of both Papers and Reports are found in DSIR 26/77 (Miscellaneous Correspondence and Reports), a file of Dr Fox, all dated 1921 or 1922. Possibly all later ones became 'colloquia' papers. These series may have been instituted in preparation for methods to insert into the 'Departmental Process Book' of which, unless it is the methods master file remembered by Dr Griffiths,

there is no other trace than its name. In the opinion of Dr Griffiths the meetings were a waste of private time and valueless to the Department.

14 J J Fox and T H Bowles, *The Analysis of Pigments, Paints and Varnishes*, 1926; F G H Tate, *Alcoholometry*, 1930. Fox was a Superintending Chemist in 1926, Bowles was a Chemist Class I (the grade below Fox) as was Tate.

15 R C Farmer, *Journal of the Chemical Society*, 1950, pp.434–440; R C Farmer, *Obituary Notices of Fellows of the Royal Society*, 1959, vol.6, pp.539–561. Information concerning carrying out research in staff's own time, letter from F R Ennos 12 July 1965, DSIR 26/– [Letters]. Thanks are due to Dr Griffiths for comments on the character of Sir Robert Robertson.

16 J J Fox and A E Martin, *Nature*, 1937, Vol.139, p.507. Fox's new infra red instrument DSIR 26/– ['Papers for attention', Fox, Box 64]. The plate is from the maker's leaflet with the papers.

17 *Report*, 1937, p.19

18 A G Francis, 'The recovery of Radium from luminous paint', *Journal of the Society of Chemical Industry*, vol. 41, 1923, pp.94T–96T; DSIR 26/77, Miscellaneous Correspondence.

19 *Report*, 1955, p.14.

20 DSIR 26/– [Annual Dinners] and information from Dr Griffiths.

21 Claude Luke, 'Safeguarding the Public: Problems of all kinds solved by the Government Laboratory', *Strand Magazine*, 1939, vol. 96, pp.674–681, (this also appeared in a condensed form as 'Which Chemical Laboratory?', *Armchair Science*, 1939, vol. 12, pp.8–15); Charles Dickens, 'Weighed in the Balances', *All the Year Round*, 1877, vol. 18, pp.12–17, (the article from which this book obtained its title); Tom Pocock, 'Chemists Behind the Law', *Leader*, 1946, vol. 3, pp.10–11; DSIR 26/– [Press Cuttings on the Annual Report, 1916–1929, Box 63]

References to Chapter 17

1 *Report*, 1950, p.3; CUST 49/4208, Customs and Excise, Registered Papers, letter dated 11 November 1939. Thanks are due to Mr Gilbert Denton for drawing my attention to this letter; DSIR 26/– [Letters], F R Ennos to J G N Gaskin, 12 July 1965, A T Parsons to J G N Gaskin, 3 July 1965, J G N Gaskin to H Egan, 21 December 1982, E I Johnson to H Egan 8 December 1982. Much of the information in this chapter owes very much to the memory of Dr J G A Griffiths. He is the source where no other reference is given. Dr H G Smith, an overseas analyst during the war, was also of considerable help.

2 DSIR 26/– [Letters], B A Rose to H Egan, 17 December 1982, J Buchan to H Egan, 30 August 1982.

3 DSIR 26/– [Production of Toluene, Box 59]; DSIR 26/– [Fox papers,

envelope, Box 64]. There is also a file marked 'Confidential Laboratory war work'; DSIR 26/77, letter dated January 1942.

4   DSIR 26/– [war files Box 59], Ministry of Supply, Zinc Alloy (Mazak) Bomb tail fins, GL 256/1941, GL 2/1944, GL 230/1944; DSIR 26/– [Absorption of Carbon Dioxide from the air of Submarines . . . 1943].

5   DSIR 26/– [war files Box 59], PRO work GL 43/1941, artificial eyes, GL 282/1944, marked 'Confidential', GL 212/1944, 'DSIR Determination of X in Biological Materials GL 266/1944.

6   DSIR 26/338, Miscellaneous.

7   Dr Griffiths carried out work on mustard gas in food. DSIR 26/– [Letters], B A Rose to H Egan 17 November 1982.

8   A G Francis, *Nature*, vol.155, (1945), pp.13–14; R Robertson, 'Sir John Fox', in *Obituary Notices of Fellows of the Royal Society*, vol.5, (1945), pp.141, 153–157. Sir John Fox joined the Laboratory as Jacob Fox. He first appears in references as J J Fox in 1924, remaining Jacob Fox until 1935 in the official list of staff. This changed to J Jacob in 1936 and J J in 1937 when he was the only senior officer not to be given his first name in full. I owe the information on the Carpenter awards to Dr Griffiths.

References to Chapter 18

1   Fox joined the staff of the Laboratory on 1 September 1897 as an attached officer of the Excise department exactly one month before it moved to the new building. The official opening of the Clement's Inn building was on 1 October 1897.

2   A D Mitchell, *Nature*, vol.183, (1959), p.718; E H Nurse *Analyst*, vol.84, (1959), p.260

3   *The Scientific Civil Service : Reorganisation and Recruitment during the Reconstruction Period.* Cmd. 6679, 1945. The salaries recommended were £275–£500 for Scientific Officers (SO, the graduate entry grade), £800–£1100 for Principal Scientific Officers, (PSO) and £1200–£1800 for Senior Principal Scientific Officers (SPSO, the old Superintending Chemist). The Assistant Experimental Officer. (AEO, the lowest experienced staff recruited), £150–£350. In 1947, the date of the Legg Report (see below), the laboratory staff consisted (including vacancies) of the Government Chemist, 3 SPSOs (including the Deputy Government Chemist), 17 PSOs, 25 SSOs, 3 SOs, 13 SEOs, 31 EOs, 41 AEOs, 72 Assistant (Scientific) and 26 various Storekeepers, Porters, Messengers, Cleaners, and an Establishment section of 28 headed by a Higher Executive Officer. These totalled 265. The vacancies were for 2 SSOs, 1 SEO, 2 SOs, 2 AEOs and 1 Assistant.

4   *Report*, 1952, p.1; *Report*, 1953, pp.3–4

5   J B Legg, Department of the Government Chemist: Review. First Report. Objectives, Activities and Organisation, September 1947 (Organisation and Methods Division, HM Treasury); Dr B A Rose,

and A F Weiss, (a PSO and an SSO respectively) at Liverpool (in a memorandum to A M Rheinlander, co-ordinating the responses for submission to Dr Bennett), DSIR 26/– [Legg Report].

6   The descriptions which follow are taken from the brief notes in the Legg Report, amplified where still relevant by the descriptions of each division drawn up in 1954.

7   *Report*, 1952, p.1.

8   Confirmed by Dr R J Mesley, who points out that the Bunsen burners used for distillation were probably a greater fire hazard.

9   *Report*, 1950, p.30 12. Annual Report, 1951, pp.6, 12

10 See *Report* for 1950 *et seq*, particularly the Introductory paragraphs.

11 For the survey of work on this and subsequent pages see particularly *Report*, 1955, and back references there given.

12 Description of gas chromatograph, from document describing functions and work of LGC, drawn up in 1958, section, Coal and Oil Products, division 10, (DSIR 26/– [LGC 1954]). The 'home made' gas chromatograph is described in *Report*, 1956, p.18. The early use of paper chromatography (a filter paper between two glass plates, the sample and eluting liquid being dropped through a hole in the top one), was in the detection of carrot jam. I owe this information to Dr Griffiths.

13 Antibiotics testing: letter from Dr E G Kellett, Officer-in-Charge of Health and Food Standards division (as it was then known), to his Superintendent, Mr J King, 13 June 1957, DSIR 26/– [New Laboratory].

14 Legg (Final) Report, DSIR 26/– [Legg Report, GC(58)13 Final Report, 17 September 1958]

15 R D Haworth, 'Professor G M Bennett' in *Biographical Memoirs of Fellows of the Royal Society*, vol.5, (1960); A D Mitchell, *Nature*, vol.183, (1959), p,718; draft letter of Dr Bennett to the Treasury, (DSIR 26/– [Letters].

Notes and References to Chapter 19

1   DSIR 26/– [Linstead Committee]; The Dispersal of Government Work from London, 1973, Cmnd 5322, p.187; *Department of Scientific and Industrial Research Act* 1956, 4 & 5 Elizabeth 2, *cap.* 58.

2   *Report* 1959, pp.1–2; *Report* 1960, p.2

3   R D Haworth, *Biographical Memoirs of Fellows of the Royal Society*. Vol 5 (1960), p.26.

4   *Report*, 1959, p.2.

5   The number of samples reported in the 1959 *Annual Report* was 680713, but this *Report* in fact covered 20 months, from 1 April 1958

to 31 December 1959, putting it on a calendar year basis for the first time. The number of samples in 1990 was approximately 100,000.

6   *Report*, 1952, p.15; *Report*, 1953, p.14; *Report*, 1954, p.15 (this describes the beginning of the fluoride work); *Report*, 1955, p.28. File DSIR 26/– [store box 19/245] 'Kerr Experiment' gives details of the toothpaste experiment.

7   The beginning of the detergent work is described in *Report*, 1954, p.16.

8   Minutes of first meeting of Agricultural Research Council. Research and Development Co-ordinating Committee on Insecticides, Analytical sub-committee, 9 October 1946 [document held in LGC]; *Report*, 1954, p.10; *Report*, 1956, p.22; *Report*, 1955, p.9; *Report*, 1969, pp.80 *et seq* for a survey of residues then found in wildlife, *Report* 1967, pp.96 *et seq.* for residues in food.

9   *Report*, 1967, p.3.; *Report*, 1968, pp 2, 6, 164

10 *Report*, 1966, p.1.

11 *Report*, 1965, pp 11, 111. The work on automation developed considerably in the Laboratory in later years, and much work on robotics was carried out.

12 *Report*, 1970, p.9.; *Report*, 1976, pp.115–116; *Report* 1980, p.7.

13 *Report*, 1978, pp 57. 58.

14 *Report*, 1972, pp 83–85, 145–148, and subsequent *Reports*; *Report*, 1978, p.57.

15 *Oil in Navigable Waters Acts*, 1955, 3 & 4 Elizabeth 2, *cap.* 25; 1963, 11 & 12 Elizabeth 2, *cap.* 28; *Report*, 1953, pp.17–18; *Report* 1967, p.78; *Report*, 1969, pp 60–61; *Report*, 1972, pp.89– 90; *Report*, 1976, pp.3, 105. Other similar services were set up at the same time, giving advice on pesticides, the carriage of dangerous goods, chemical nomenclature and general analytical advice.

16 *Report*, 1961, p.4; *Report*, 1971, p.85 for description of methods used at that time by the RFTU; *Finance Act* 1960, 8 & 9 Elizabeth 2, *cap.* 44. Schedule 2 paragraph 5 specifies the Government Chemist as an authorised analyst to test for the presence of markers.

17 *Report*, 1966, pp.50, 51.

18 *Report*, 1950, p.22; *Report*, 1955, p.14; K P Oakley and C R Hoskins, 'New Evidence on the Antiquity of Piltdown Man', *Nature*, 1950, vol.165, pp.379–382; C R Hoskins and C F M Fryd, 'Determination of Fluoride in Piltdown and Related Fossils', *Journal of Applied Chemistry*, 1955, vol. 5, pp.85–87; 'Further Contributions to the Solution of the Piltdown Problem, 6: Chemical Changes in Bones, a Note on the Analysis', C F M Fryd, *Bulletin British Museum*, 1955, vol. 2, pp.266–267. Laboratory papers on this subject are DSIR 26/– [Box 67]; M J Glover and G F Phillips, 'Chemical Methods for the Dating of Fossils', *Journal of Applied Chemistry*, 1965, vol. 15, pp.570–576.

19   *Report*, 1964, p.69; *Report*, 1972, p.92.

20   DSIR 26/– [Correspondence on dispersal of samples, Box 63].

Notes and References to Chapter 20

1   *Report*, 1962, p.1.

2   *Report*, 1963, p.3.; *Report*, 1966, p.5.

3   *Report*, 1959, p.1.; Report, 1962, p.4.; *Report*, 1964, p.3. See also *The Laboratory of the Government Chemist at Cornwall House*, by P W Hammond, LGC, 1989.

4   *Report*, 1952, p.4.; *Report*, 1953, p.5.; *Report*, 1966, pp.32, 46–50; *Report*, 1965, pp. 26, 26–28. The latter is a summary of the history of the outstations up to 1965. Testing of dried egg samples, letter of B A Rose, 17 November 1983, DSIR 26/– [Letters LGC History]

5   *Report*, 1975, p.4.; *Report*, 1981, p.13;

6   Sir Burke Trend, *Report of the Committee of Enquiry into the Organisation of Civil Science*, 1963, Cmnd 2171; *Report*, 1965, p.1.

7   *Report*, 1974, p.5. The benches, fume cupboard and steam bath were from the Crown Contracts Laboratory at Clement's Inn, and the desk and stool (both used by the present writer) from the "Butter Room" also known as the "Reference Room".

8   *Report*, 1966, p.1; *Report*, 1962, p.1. By 1966 the division of work between revenue protection, health and safety, and the quality control of government supplies was estimated at 42%, 28% and 19% respectively, the remaining 11% being investigatory and research work including the statutory reference and analytical functions.

9   *Framework for Government Research and Development*, 1972, Cmnd. 4814, p.3

10   *Report*, 1973, pp.1–2; *Report*, 1974, pp.1–2. The 25 subject oriented programmes (from *Report* 1971, pp.144–145) were:

Staff Engaged

Division 1

(1)	Revenue, beverages, alcohol and sugar	52
(2)	Revenue, industrial, alcohol and distillation	10

Division 2

(3)	Tariffs, classification	26
(4)	Revenue, hydrocarbon oil	8
(5)	Revenue, tobacco	22

Division 3

(6)	Procurement for Government, chemical analysis	9
(7)	Industrial, transportation and consumer hazards	9
(8)	Rocks and minerals	4
(9)	Documents, examination and inspection	6
(10)	Poisonous gases	8

Division 4
(11)	Food, chemical analysis	29
(12)	Armed Services Food supplies, procurement and inspection	17

Division 5
(13)	Agriculture, chemical analysis	6
(14)	Pesticides	20

Division 6
(15)	Water, sewage and industrial effluents	18
(16)	Radiochemistry	15

Division 7
(17)	Drugs, chemical analysis	11
(18)	Bacteriology	6
(19)	Dentistry material specification	8

Division 8
(20)	Chemical analysis research	17
(21)	Instrumentation and chemical analysis	18

Central Services
(22)	Technical Services	11
(23)	Library, information and editorial	7
(24)	Administration	40
(25)	Directorate	6
	Total	383

11 *Report* 1972, p.1; *Report* 1973, p.2.

12 *Report*, 1967, pp 170–175 gives a summary of preliminary work on dental materials. For recommendations of the Linstead Committee see chapter 18 above.

13 *Report* 1978, p.11. Structure of directorate, see *Annual Reports*.

14 Dr Egan died in June 1983.

15 For a summary of the career of Mr Williams to his appointment as Government Chemist, see *Chemistry in Britain*, May 1989, pp.449–450.

Notes and References to Chapter 21

1 *Report*, 1973, p.4; *Report*, 1976, p.4; *The Dispersal of Government Work from London*, Cmnd 5322, HMSO, 1973. Dispersal was not a new idea for the Laboratory; removal to a site at Woolwich was considered in 1959, DSIR 26/– [DSIR, Proposed site for Laboratory at Woolwich].

2 *Report*, 1974, p.2.; Hansard.

3 *Report*, 1976, p.4; *Report*, 1977, pp.5–6.

4 *Report*, 1978, pp.10, 133–134; *Report*, 1979, p.10; LGC Dispersal Information 2/79. April 1979.

5 *Report*, 1977, p.6; *Report*, 1978, pp.9–10; LGC DI 2/79 (as above). Elaborate plans for the staffing and operation of the London out-station are preserved in the LGC files.

6 LGC Staff Notice 27/79; *Report*, 1979, p.10.

7 *Report*, 1979, p.10. (and references there); *Report*, 1980, p.8; *Health and Safety at Work, etc, Act*, 1974, 22 & 23 Elizabeth 2, *cap.* 37.

8 'The long-term use of Cornwall House for LGC'. September 1979, in DSIR 26/– [Cornwall House 1961]; *Report*, 1981, p.13.

9 *Report*, 1985, p.3. (the detailed Annual *Reports* from 1982 onwards were not published, copies are held at LGC); *Report*, 1988/89, p.32; Roderic Bunn 'Services for Chemists', *Building Services*, 1989, pp.19–22.

10 *Improving Management in Government : The Next Steps*, HMSO, 1988; *The Laboratory of the Government Chemist Executive Agency: Policy and Resources Framework*, 1989; *Report*, 1988/89, p.1.

11 See third and fourth *Reports* of the Independant Committee on Smoking and Health, HMSO, 1983 and 1988.

12 Legislation under which the Laboratory has statutory responsibilities:–

National Health Service Act 1946
Diseases of Animals Act 1950
Rag Flock and Other Filling Materials Act 1951
Agriculture (Poisonous Substances) Act 1952
Food and Drugs (Scotland) Act 1956
Food and Drugs (Northern Ireland) Act 1958
Factories Act 1961
Farm and Garden Chemicals Act 1967
Medicines Act 1968
Trade Descriptions Act 1968
Agriculture Act 1970
Misuse of Drugs Act 1971
European Communities Act 1972
Poisons Act 1972
National Health Service Reorganisation Act 1973
Customs and Excise Management Act 1979
Customs and Excise Duties (General Reliefs) Act 1979
Alcoholic Liquor Duties Act 1979
Hydrocarbon Oil Duties Act 1979
Tobacco Products Duty Act 1979
Value Added Tax Act 1983
Food Act 1984
Consumer Protection Act 1987
Food Safety Act 1990

In addition work is performed at the request of the Intervention Board for Agricultural Products under the provisions of the *European Communities Act* 1972 and the EC Cosmetics Directive 76/768/EEC.

13   *Report*, 1990; *Report*, 1990/91, pp.5, 7; Methylated Spirits Regula-
     tions. SI 1979/241.

14   These staff are organised (1992) into four large divisions all under the
     control of a divisional head; Forensic and Customs (which includes
     drug examination and most of the Customs and Excise work);
     Analytical Quality (VAM, Reference Materials and instrumental
     work); Environmental Services (pesticides, water chemistry environ-
     mental work generally, tobacco products); Foods. There is also a
     Biotechnology Unit, Sensors Technology and Materials Technology
     which come directly under the Deputy Director for Biosciences and
     Innovation, and Accounting and Finances, Personnel, a Business
     Development Group and a Services Division which come under the
     Deputy Director for Resources.

Index

References to the activities of the Laboratory under its various titles are indexed under those titles except for when it existed as the Inland Revenue Laboratory, from 1849 to 1894. These have been indexed under Excise Laboratory. This is because contemporaries, who have been quoted, seem frequently to have referred to it as such or as the Somerset House Laboratory.

A

Abbott, Dr D C, Deputy Director, 270
Abel, Frederick, 63
acaricide, early, 208
Accum, Frederick, 86
Adam Hilger Ltd., 206, 207
Adams, C A, 211
addiction, drugs of, 189
Admiralty, 43, 137, 141, 143, 181, 228, 260
Admiralty Marshal, 181, 182
Adulteration of Food and Drink Act, (1860), 87
Adulteration of Food and Drugs Act, (1872), 87, 112, 116, 148
adulteration,
    beer, 60, 61–62, 138, 176
    butter, 141, 157
    coffee, 40–43
    food, 86ff
    pepper, 34–35
    snuff, 22, 24
    soap, 33
    tea, 35–36, 114
    tobacco, 5–6, 14, 138
Agencies, 280
Agricultural Analysts, 121
    qualifications for, 142
Agricultural Research Committee, 247

Agriculture (Poisonous Substances) Act, (1952), 235
Air Ministry, Department of Hydrogen Production, 215
Albert Club, 232
alcohol
    denatured, 140, 187, 201–202
    duty free, 140
    methylation of, 140
    proof strength system, change from, 249
    tables, change of basis, 250
Alcoholometry, 204
Alexander, Robert, 79–80
alkaloids, 183
Allanson, Thomas, matriculated at University College, 323
Allen, A H, 94
American Chemical Society, 151, 223
Analysis and Adulteration of Food, The, 95
Analysis of Paints, Pigments and Varnishes, The, 204
Analyst, 88, 94
Analytical methods, nineteenth century, 92
Army Clothing Department, 137
Army Medical Department, 137
arsenic,
    electrolytic method for determination, 186
    in beer, 139, 150–151, 176

artificial silk, analysis, 193
Arundel Street, 37, 74
    date of move to, 317
Ashmore, S A, 345
atomic absorption spectrophotometry, 248
Atomic Weights, 151
Australian wines, 108
aviation fuel, 202

B

Bainbridge, Robert, matriculated at University
    College, 323
Baker, Alfred, head of Customs laboratories, 108
Bankruptcy Building, 196, 346
Bannister, Richard, Deputy Principal, 69, 94, 98,
    156, 158, 181
        attitude towards Thorpe, 125
        gives Cantor Lectures, 328
        work book of, 318
Barlow Report, 222
Barlow, Sir Alan, enquiry into structure of
    Scientific Civil Service, 220–222
Barry Road laboratory, 236
Barry Road, 199
Bate's saccharometer, 5, 57
    accuracy checking, 58
Bateman, Joseph, 97
beer,
        adulteration, 59–60, 61–62, 138, 176
        analysis, 49 ff
        analysis, at Customs laboratories, 109
        arsenic in, 139–140, 150–151, 176
        dilution, 60
        duties, 1
        herb, 139
        legal method for original gravity, 320
        original gravity determination, 51–54, 57
        original gravity tables, 152–153
        original gravity, basis of method, 319
        salt in, 61–62, 94
        samples, 53
        temperance, 60–61, 139

Bell, Dr James, Principal Chemist, 31, 40, 60
        appointed Lime and lemon juice Inspector, 46
        attends international conference, 66
        beer adulteration, difficulty in detecting, 60
        character and career, 97–98
        chemistry prizewinner, 72
        early career, 81–82
        evidence to 1872 Select Committee, 148
        evidence to 1874 Select Committee, 87, 88
        methylated spirit, suggests change to, 67
        opinion of Customs laboratories, 116
        publications by, 329
        refusal to meet Public Analysts officially, 94
        relations with Public Analysts, 94, 157
        reorganisation of staff, 78
        retirement, 96, 121
        succession as Principal, 81
        and tea analysis, 112–113, 116
Bennett, Dr George, Government Chemist
        appointment, 220
        death in office, 239
        early career, 220
        encouragement of instrumental techniques,
            245
        organises toxic gases committee, 236
        produces first post war report, 234
Biotransformation Club, 284
Blagden, Sir Charles, 3
Board of Trade, 46, 137, 192, 222, 228
bone cements, development, 267
bones,
        analysis, 255–256
        fluoride in, 256
        nitrogen in, 256
Bowles, T H, 204
Boys, Charles matriculated at University College,
    323
Brande, Professor William, 19
Bread Act, (1836), 86
British Engineering Standards Association, 202
British Museum (Natural History), 255
British Standards Institution, 202
Broad Street, 36

Brook, Thomas, matriculated at University College, 323
Brown, Dr Horace, 152
Brown, John, Excise Officer, 25–26
Brussels Nomenclature Tariff Classification, 262
Budget Act, (1901), 145
Burgess, W T, 143
butter
    adulteration, 141, 157
    analysis, 90–91, 93–94, 157
Butter and Margarine Act, (1907), 141

C

Cambridge University, 179
Cameron, A, 342
Campbell, Professor, 19
Carpenter awards, 219
Carpenter Report, 195, 222
Carpenter, Dr William, 40
Carta Mercatoria, 1, 103
Central Control Board of liquor Traffic, 181
Cereals Research Station, 232
Chatt, Miss E M, 179, 343
Cheater, Thomas, 151, 172, 174
cheese, analysis, 192
Chemical Manufacturers Association, 121
Chemical News, 88, 89, 90
Chemical Research Laboratory, 200
Chemical Sensors Club, 284
Chemical Society, 163, 223
    Fellowship of, 142
chemistry, use of in Government, 1
chicory, in coffee, 40–42
chromatography, 234–235, 247
Church, Richard, 173, 174–175, 179
cigarettes, 26–27
Civil Service Gazette, 73
Clark, Alan MP, 279
Clarke's hydrometer, 2
Clarke, F W, 151
Clarke, John, 2

Clement's Inn, 127, 197
Coal Controller, 181
cod liver oil, analysis, 191
coffee adulteration, 40–43
coffee and chicory survey, 40
    cost of, 317
coffee and chicory mixtures, confusion, 318
Coffee, Aeneas, Excise Officer, 5
Coleman, Dr Ronald, Government Chemist, appointed, 270
Committee on the Prevention of Pollution of the Sea by Oil, 253
Connah, James, 99, 145, 153, 172
    acting Government Chemist, 185
    attends International Conference, 165
    description, 174
Connor, Rochfort, drawings, 316
controlled release formulations, 267–268
Cooper, Professor John, 19
Cornwall House,
    conversion problems, 259–260
    laboratory, 218, 233
    move to, 258, 260
    situation of, 259
    unsuitability as laboratory, 272, 276
Cornwall, Andrew, matriculated at University College, 323
Court of Reference, 157
Crawly, J P, 167
Crookes, William, 88
Cumbria,
    LGC move to, 272, 275
    LGC move to, cancelled, 275
Curie, Marie, 152
Custom House branch, 145–147, 172, 173
    rooms occupied, 342
Customer Contractor Principle, 264, 266–267
Customs and Excise officers, seconded to Laboratory, 194, 199
Customs and Excise, Board of, 152, 168, 201, 228, 231, 236, 237, 249, 266, 282, 284
Customs and Inland Revenue Act, (1881), 111, 112

Customs and Inland Revenue Act, (1885), 60
Customs Consolidation Act, (1876), 109
Customs laboratories, 103ff, 233
    beer analysis, 109
    closure of one, 107
    conditions in, 175
    costs of setting up, 106, 330
    evidence for early existence, 330
    organisation, 115
    outports, 330
    reduction in staff, 107
    reorganisation, 112
    Reports, 295–296
    tea analysis in, 112, 113
    training of staff, 104
    volume of work, 115
    well organised, 105
Customs outstations, Government Laboratory
    students and, 160–161
Customs Tariff Amendment Act, (1860), 103, 105
Customs, Board of, 16, 33, 42, 83, 107
    set up own laboratories, 103–104

D

Dalton, John, 151
Dangerous Drugs Act, (1931), 189, 235
Dangerous goods, transport of, 140–141
Davis, H W, Deputy Principal, 167
DDT, 247
Dead Sea resources, 202
Dean Bradley House, 199
denatured alcohol, 140
dental materials, development 267
Department of Employment, 266
Department of Health and Social Security, 251
*Department of Scientific and Industrial Research
    Act*, (1956), 243
Department of Scientific and Industrial Research,
    200, 235, 236, 237, 243
    Directorate of Tube Alloys, 217
    dissolution, 263
Department of the Environment, 284

Department of the Government Chemist,
    advisory and committee work, 200–201,
        208–209, 231
    animal testing, 235–236
    artificial silk, work on, 193
    building overcrowded, 196
    buildings used, 232–233
    coffee and tea breaks, 180
    conditions in laboratories, 173, 175
    consultative work, 176–177
    cooperation with Public Analysts, 203
    Dead Sea resources, work on, 202
    dispersal in war, 211
    fire watching, 214
    food work, 190–191
    functions, review of, 236–239
    Geological Survey, work for, 177
    instrumental methods, 229, 245
    invention of methods, 187, 189, 189–190
    lead in paint, work on, 178
    Legg enquiry, 220, 223–226
    Legg Report, reaction to, 226
    Legg Report, results, 226–227
    methods, 187, 193
    new accommodation, need for, 231, 237
    new method of determining arsenic, 187
    new staff, 193–194, 195
    new storey on building, 197
    new types of work, 186, 190
    organisation, 171
    organisation, post 1945, 227–231
    outstations, 171, 199, 225
    outstations, staff at, 194
    overflow accommodation, 196–198
    polonium collectors, preparation, 209–210
    popular attitude towards, 210
    radium recovery, 208
    relocation, suggested, 243
    reorganisation, suggested, 238–239
    research, 204
    research meetings, 204
    samples, numbers, 186, 192, 193, 246
    sea water analysis, 177

social events, 210

staff as expert witnesses, 203

staff leave, 173

staff numbers, 171, 172, 194

staff salaries, 172–173, 194, 195–196, 349

staff seconded, 200

staff structure, 195

staff training, 173–174

staff, unattached officers, 194, 195

staffing during World War 1, 179–180

standard methods, work on, 203

support staff, 171–172

tea analysis, 183

tobacco work, 176

uranium analysis, 217

war damage, 213–214, 217

war work, world war 1, 180–182

war work, world war 2, 211, 214–217, 218–219

war work, hazardous, 218–219

women staff, 179, 179–180

work after 1945, 232, 234

working conditions, 173

Department of Trade and Industry, 265

Departmental Committee on the use of Lead Compounds, 178

Departmental Committee on the Use of Phosphorus, etc, 1899, 148

Deutsche Chemische Gesellschaft, 151

Dictionary of Applied Chemistry, 123

diesel fuel, marking, 254

Dietz, Dr Roy, Deputy Director, 270

Directorate of Tube Alloys, 217

Dobbie, Sir James, Government Chemist

    alcohol denaturant committee, service on, 201

    appointed, 163

    attends International conference, 165

    attitude to standard methods, 165–166

    attitude towards other Government laboratories, 176

    costume, 177, 185

    early career, 163

    reorganisation of work at outset of war, 180

research interests, 183

retirement, 183–185

salary, 172

service on post war committees, 183

war committees, service on, 182

Dobson, Thomas, Secretary to the Board of Inland Revenue, 49, 50, 51, 52, 57, 63

    as 'deputy Principal', 68

    award for work on original gravity, 55

Dodd, Edward, Assistant Secretary to the Board of Excise, 73, 315

    matriculated at University College, 323

Drinkwater, Joseph, 20, 70, 71, 311

    attended University College, 323

drugs analysis, 231

Dudley House laboratory, 232

Dyestuffs (Inport Regulations) Act, (1920), 190

E

Egan Dr Harold, Government Chemist,

    appointed, 270

    appointed Deputy Chief Scientific Officer, 248

    career in LGC, 270

    Cumbria, attitude to move to, 275

    death, 353

egg, dried, analysis, 262

Ellis, B A, 345

emission spectrography, 207–208, 255–256

Ennos, F R, 177

ether, diethyl, drinking, 331

European Economic Community, 249

Evelyn Street laboratory, 236

Excell, George, 115

Excise Act, (1854), 54

Excise Act, (1856), 54, 55

Excise Assistants, 161, 162

Excise chemists, lack of knowledge, 19

Excise Laboratory,

    amalgamation with Customs laboratory, suggested, 121–122

    annual dinners, 99

attitude of Customs towards, 116–117
authorisation of work requests, 83
conditions in, 37, 84
cost, 38
Customs, work for, 103
equipment, 37, 39
expansion of work, 33, 36, 43
expert staff, 43–44, 48, 78, 92
foundation, 11
library, 78–79
library, early costs, 325
made separate department, 68
methods, 48
milk analysis, allowance for decomposition, 93
morale in 1894, 99–100
organisation of, 84
outstations, 27–29
outstations, used for beer analyses, 55
position in 1894, 98–100
procedure, 54–55
refusal to publish standards, 94, 96
relations with Public Analysts, 94, 96
reputation, 88, 90
rooms occupied in Somerset House, 334–335
rooms, 44
routine, 38, 39
spurious wine, analysis of, 107
staff at University College, 22, 70–7, 87
staff at University College, costs, 72, 74
staff at University College, objections to, 73–74
staff, increase in, 82, 83–84
staff leave, 70, 73
staff levels in 1894, 99
staff living costs, 76
staff salaries, 38, 68, 69, 84–85, 325
staff, appointment as referees, 89–90
staffing, 38, 68, 69
standards in food analysis, 93
Store Keeper, 79–80
students, awards, 75
students, good results, 74, 76

students, research papers by, 324
tobacco work for Customs, 107
trained staff, 89
training of staff, 87
water analysis, 44–45
work, increase in, 83
Excise Officers' Manual, 97
Excise Officers, training at Laboratory, 77–78
Excise still, 5
Excise, Board of, 8, 16, 19, 36, 49, 52, 71, 72, 99

F

Factory Inspectorate, 236
Faraday, Michael, analyst to Board of Excise, 312
feeding stuffs, unfit, 190
Fehling's solution, 48
Fertilisers and Feeding Stuffs Act, (1893), 121, 137, 142
Fertilisers and Feeding Stuffs Act, (1926), 203
fertilisers and feeding stuffs, samples, 142–143
Fertilisers Committee, 182
fertilisers, artificial, 182
Finance Act, (1902), 140
Finance Act, (1907), 153
Finance Act, (1914), 154
Finance Act, (1915), 154
Finance Act, (1960), 254
fish oil, analysis, 191
Fitzsimons, R A, 345
Five and six Clements Inn, 197–198, 232
flame photometry, 248
flour analysis, 232, 234, 235
fluoride, in water, 246
Food Act, (1984), 282
food analysis, 43, 93, 156, 183, 231, 232
    historic samples, 254–255
    state of in nineteenth century, 90
Food and Drugs Adulteration Act, (1928), 235
Food and Drugs Act, (1955), 235
Food and Drugs Acts, 187
Food and Its Adulterations, 86
food composition, 235

Food Products Adulteration Committee, 1892–94, 156, 157

food stuffs, need for standards, 92–93, 96

Foods and drugs samples, numbers, 344

Foreign Office, 166, 181

Foreman, Dr J K, Deputy Government Chemist, 269

forged documents, investigation, 164–165

Forsey, Charles, 52, 53, 68

    matriculated at University College, 323

Fownes, Professor George, 50, 71

Fox, Dr John, Government Chemist, 270

    abilities, 206

    appointed, 206

    assistant to Dr Thorpe, 206

    book on paints, 204

    collaborates with Dobbie, 183

    change of name, 349

    character, 219

    death, 219

    early career, 219

    infra red work, 206–207, 215

    retirement postponed, 215

    service on motor spirit committee, 202

    war work, 215

Framework for Government Research and Development, 265

Francis, Arthur, Deputy Government Chemist, 152

    as Acting Government Chemist, 219

Frankland, Professor Edward, 44, 45, 75, 88, 143

fusel oil, analysis, 109–110, 331

    effect on health, 110–111

G

gas chromatography, 235, 247, 248

gas liquid chromatography, 253–254

gas oil, marking, 254

General Medical Council, 165

Geological Survey Museum, 198, 236

Gilpin, George, 3

Gladstone, W E, 116

opinion of Customs laboratories, 333

Glasgow University, 163

Goulbourn, Frederick, Chairman of the Board of Customs, 112, 118

Government Chemist, title, 123

Government Laboratory,

    centre of excellence, 148

    Conversazione at opening, 128, 131

    cooperation with public analysts, 157–158

    Customs branch, work, 145–146

    description, 297–304

    early reports, 137

    fire regulations, 135

    international interest, 133–134

    layout, 131–133, 297–304

    made separate Government Department, 167–168

    meaning of term, 122–123

    new building, 128

    new organisation, 168

    new site, 127–128

    new staff and salaries, 168

    original gravity, work on, 152–153

    outstations, 166

    planned, 122

    Public Analysts, relations with, 156, 157

    recruitment, problems, 166, 167

    reorganisation of staff, 166–167

    reorganisation, 167–168

    rooms occupied in 1897, 335–336

    salaries and organisation, 125–126

    salaries, increase in, 334

    samples, how should be received, 137

    set up, 122

    sour milk, method for analysing, 156

    staff, direct recruitment of, 161–162

    staff in 1930, 346

    staff training, 158–160

    staff, hours of work, 136

    staffing problems, 166

    students, 159–161

    students, bond, 161

    students, examination, 160–161, 305–307

studentships abolished, 167
support staff, 134–136
support staff, salaries, 136
training of staff, 158
volume of work, 137, 138
vote, 137
women staff, 343
volume of work, 140
Graham, Professor Thomas, 16, 19, 20, 35, 40, 62, 71, 72
income from Excise, 319–320
Gresham House, 8, 11
Grubb Parsons, 207
Gunning test, for methanol, 66

H

Halpin, James, 175
Handbook of Hydrometry, 115
Hardman Report, recommends dispersal of LGC, 272, 277
Harkness, William, 69, 84
Hartridge Reversion Spectroscope, 190
Hassall, Arthur, 34, 87, 91
Analytical Committee of, 86
criticism of Excise Chemists, 35
plans for carrying out work of Excise chemists, 326
Heald, J A, 345
Health and Safety at Work Act, (1974), 276
Health and Smoking Now, 252
Hehner, Otto, 156
Helm, Henry James, 69, 81, 84, 107, 110
hemp, samples of, 189
Her Majesty's Stationery Office, 43, 272
herb beer, 139
Herepath, William, 19
Heriot-Watt College, Edinburgh, 177
Hess, Rudolf, 282
high performance liquid chromatography, 248
His Majesty's Stationery Office, 176
Hofmann, Professor Alexander, 53, 62, 74
Hogg, Quintin, MP, 260

Holmes, John, 153
Home Office, 43
Hooker, Dr Joseph, 40
Hooper, Egbert, 151
Hume, Joseph, MP, 14, 17, 18
Hutchinson, George, 63
Hydrometer, Clarke's, 2
Hydrometer, Sikes', 4
abolished, 250
calibration, 230
hydrometers,
abolition, 250
standard, 58, 59

I

Ibbs Report, 280
Imperial College, 74, 163, 167
Import Duties Act, (1932), 192, 234
Import Duties Act, (1935), 228
India Office, 43, 137
Industries, protection of, 186, 190
infra red analysis, 204, 206–207
Inland Revenue Act, (1880), 56, 57, 152
Inland Revenue Laboratory, see Excise Laboratory to 1894, Government Laboratory after this
Inland Revenue outstations,
Government Laboratory students and, 160–161
Inland Revenue, Board of, 27, 40, 41, 46, 53, 55, 67, 77, 94, 99, 111, 121, 135, 140, 151, 152, 162, 236
insecticides, early, 177, 208
Institute of Brewing, 152
Institute of Chemistry, 97
Fellowship of, 142
International Carriage of Dangerous Goods by Road, and Rail, 283
International Committee for Uniform Methods of Sugar Analysis, 235
International Committee on Bird Preservation, 253

International Conference on Methods of Analysis etc., 1910, 165–166

International Union of Pure and Applied Chemistry, 151, 223

Intoxicating Liquor Act, (1872), 60, 61

Irish Office of Works, 137

J

joggery, 176

Johnson, E I, Deputy Government Chemist, 270
    sets up War Office laboratories, 211

Johnstone, James, head of Customs laboratories, 103, 105, 107–108, 110
    distillation apparatus, 329

K

Kay, George, Deputy Principal, 38, 82, 98, 110
    appointed deputy principal, 69
    chemistry prizewinner, 72
    date of joining Laboratory, 323
    death, 69
    work on spirit strength, 155

Keene, James, 106, 108, 115

Kellett, E G, 350

Key Industry Duty, 192

King, John, 345, 350

L

Laboratory of the Government Chemist, see also Excise Laboratory and Government Laboratory
    Agency, 280
    attitude of George Phillips to, 286
    automation, use of, 249, 252
    bone analysis, 255–256
    bone cements, development of, 267
    buildings, poor condition of, 258
    capabilities, 269
    controlled release formulations, development of, 267–268
    Cornwall House, moves to, 258, 260

critical changes, 272

customer orientated, 266

decision to move to Teddington, 277

dental materials development, 267

Department of Industry, becomes part of, 264

Department of Trade and Industry, becomes part of, 263, 264

designing new laboratory, 277

dispersal to Cumbria, 272, 275

environmental work, 283

fluoride work, 246

food samples, analysis of historic, 254–255

formation, 243

historic objects, analysis of, 256

instrumental techniques, use of, 248

Legg enquiry into, 220, 223–226, 236

Liverpool outstation, war work, 261

Ministry of Technology, becomes part of, 263

morale after changes, 268

need to be in central London, 259

oil pollution, work on, 253–254

Open days, 260, 279

organisation of work, 266

outstation move, 260

outstations, 260–261

outstations, closure, 262–263

outstations, research at, 261

pesticide work, 247, 248

Piltdown Man, work on, 256

political issues, greater awareness of, 263

present work, 289

Queen's Award for Technological Development, 267

referee analyst, 282

reorganisation, 248, 263

Requirements Committee, governed by, 264, 265, 266

Rothschild Committee, recommendations of, 265

staff and structure, 1974, 352–353

staff numbers, 269

statutory functions, 267, 354

Steering Committee, governed by, 264, 266

structure, 1992, 355
support work, 282–283
surplus samples, disposal of 173, 256–257
tar and nicotine in tobacco, 251–252
Teddington, decision to move to, 277
Teddington, planning move to, 277–279
Teddington, recommendation to move to, 272
type of work, 245–246
willingness to adapt, 248
Woolwich, suggested dispersal to, 353
Laboratory reports, 293–296
Lancet, 86
Lang, Sir John, 257
lead, in paint, 178
lead, in pottery glaze, 148, 150
League of Nations, 202
Leave, staff, 70, 73, 173
Legg enquiry into the Department of the
    Government Chemist, 220, 223–226, 236
Legg, J B, 221, 223
Legionella pneumophilla, 284
Lemon and lime juice, analysis, 46–47
Lewin, G, 125
Lewis Dr David, Government Chemist
    appointed, 245
    chairs pesticide committee, 247
    character and abilities, 270
    previous career, 245
    retires, 270
Lewis, John, 98
library of Laboratory, foundation, 78–79, 325
lime and lemon juice, analysis, 46–47
Lindley, John, Professor, 16, 40
Linstead Committee, recommendations, 243–245
Linstead Review, 236–239, 258
Linstead, Dr R P, 236
Liquorice in tobacco, 25
Local Government Board, 165, 176
Logie, John, Excise Officer, 5
London School of Economics, 245
London University, 179
Longwell, Dr J R, Deputy Government Chemist,
    223

appointed Deputy, 245
Lord President of the Council, 222

M

Maclean, D, 37
Malt Duty, 56
malt, 49
Manchester University, 179
Manual of Inorganic Chemistry, 123
Manufactured Tobacco Act, (1863), 107
Margarine Act, (1887), 141
margarine analysis, 235
margarine, 141
Marsh, James, 311
matches, phosphorus in, 163–164
McCulloch, John, 20, 68
    matriculated at University College, 323
    stature as analyst, 323
Medical Research Council, 200
    Food Composition Tables, 235
Merchandise Marks Act, (1911), 190
Merchandise Marks Act, (1926), 190, 192
Merchant Shipping Act, (1867), 46
Merchant Shipping Act, (1894), 227
methanol, tests for, 66, 67
methylated spirits, 62–63, 65, 202, 283
    definition of non potability, 322
    purification of, 63–64
    tests for, 66
    use as drink, 64–66
methylation, of alcohol, 140
milk analysis, 90–91, 92–94, 96, 156
    allowance for sourness, 93, 95, 156
Ministry of Agriculture and Fisheries, 200
Ministry of Agriculture, Fisheries and Food, 266
Ministry of Aircraft Production, 217
Ministry of Defence, 260
Ministry of Food, 183, 211, 216, 231, 232, 262
Ministry of Health, 231
Ministry of Labour and National Service, 228
Ministry of Munitions, 181
Ministry of Pensions, 216

Ministry of Supply, 216, 228
    Armaments Research Department, 216
    Controller of Chemical Research, 215
Ministry of Technology, 267
Ministry of Transport, 253
Ministry of Works, 216, 236, 260
molasses, 49
    analysis, 145, 146
Monier-Williams, Dr, 165
More, Andrew, 150
motor spirit, 188, 202
Muir, Pattison, 123
Museum of Practical Geology, 177

N

naphtha,
    addition to spirit, 62–63
    determination of, 140
National Antarctic Expedition, 1901, 143
National Chemical Laboratory, 237, 273
National Health Service Act, (1946), 231, 235
National Mark honeys, 190
National Mark schemes, 190
National Physical Laboratory, 155, 200, 230,
    269, 270, 272, 273
National Whitley Council, 194
Next Steps, The, 280
Nicholls, Dr J R, Deputy Government Chemist,
    192
    President of the Society of Public Analysts,
      223
    overseas visits, 222
nicotine, in tobacco, 251–252
Nitrogen Products Committee, 182
Normal School of Science, 74
Norton, Fletcher, trained at University College,
    323
Nurse, E H, Deputy Government Chemist, 211
    Acting Government Chemist, 239
    retirement, 245
nutritional analysis, 234, 235

O

O'Loghlen, John, Laboratory Book Keeper, 38, 69
Oakley, Dr K P, 256
obscuration, 111, 112
Oding, Charles, matriculated at University
    College, 323
Office of Works, 177, 198
Ogilvie, Robert, 112
Oil in Navigable Waters Act, (1955), 253
Oil in Navigable Waters Act, (1963), 253
oil pollution analysis, 253–254
Oil Spillage Analytical and Identification Service,
    254
Old Age Pensions Act, (1908), 164
Old Broad Street, 8, 11
Oliver, Professor Thomas, 150
opiates, determination of, 189
orange juice analysis, 232, 235
organo- mercurial compounds, 247
organo-chlorine compounds, 247
organo-phosphorous compounds, 248
original gravity of beer, tables, 153
Osborne, Frank, 175
Ottawa Agreement Act, (1931), 191
Oxford University, 179

P

paint, lead in, 178
paper analysis, 320–321
paper chromatography, 350
papers, research, by Excise students, 324
Parsons, A T, 211
pemmican, analysis, 143–144
pepper, adulteration, 34–35
pesticide analysis, 247, 248
pesticides, early, 177–178
Phillips, George, Principal of the Excise
    Laboratory, 8, 270, 285
    adulteration, attitude towards, 18, 19
    appointed lime and lemon juice Inspector, 46
    appointed Principal, 68

award for work on original gravity, 55
character and abilities, 80–81
defence of Laboratory, 88
denatured alcohol, 202
duty free spirit, work on, 62–63
early career, 11
evidence to Select Committee (1844), 14–16
evidence to Select Committee (1855), 17, 87
expenses, 315
fusel oil, devises method for, 110
glass hydrometer, invention of, 58–59
length of service, 325
LGC, possible attitude towards, 286
malted grain, work on, 56
matriculates at Univeristy College, 70
methylated spirit, invention of, 62–63
original gravity determination, 51–54, 57
paper analysis, 320–321
patents, 34, 64
postage stamps, work on, 45–46
retirement and pension, 80
sugar and molasses in brewing, 49–52
tobacco, early methods of analysing, 14–16
tobacco seizures, rewards for, 25
training of staff, 74, 76
wine strength, method to determine, 103
work in own time, 56
Phillips, Richard, analyst to Board of Excise, 16, 19, 20, 313
phosphorus, in matches, 148–149, 163–164
phossy jaw, 148, 164
Pierce, George, matriculated at University College, 323
Piltdown Man, 256
Playfair, Lyon, Professor, 19
Poisons Board, 203
polariscope, 48, 145
polarography, 248
Police Forensic Science Laboratory, 190, 200
polonium collectors, 209–210
Ponting, Clive, 282
Pooley, J, 345
Porch, The, 174, 179

Post Office, 43, 45, 137, 236
Postage stamps, testing, 45–46
pottery glaze, lead in, 148, 150
power alcohol, 201–202
Prescott, Henry, matriculated at University College, 323
Principals, Government Chemists, listed, 290–291
Proceedings of the Society of Public Analysts, 92
proof spirit, definition, 311
proof strength of alcohol, change from, 249
protection of industries, 186
Public Analysts, first appointed, 1860, 87
    relations with Government Laboratory, 156, 157
    rigidity in standards, 328
Public Record Office, 177, 216
Purchase Tax, 215

Q

Qualitative Chemical Analysis, 123
Quantitive Chemical Analysis, 123
Questioned documents work, 164–165

R

radium,
    atomic weight of, 152
    recovery of, 208
Raman effect, 204
Ramsay, M G, 167
Rayleigh, Lord, 155
Rayment, Albert, 173–174
Red List, 284
Redwood, Professor Theophilus, 53, 62
Reports of Principals and Government Chemists, 293–295
Rheinlander, A M, 350
Riche and Bardy test, for methanol, 66, 322
Road Fuel Testing Units, 254
road fuel, marking, 254

Robertson, Sir Robert, Government Chemist,
  appointed Government Chemist, 185
  artificial silk, advice on, 193
  character, 205–206
  early career, 185
  explosives expert, 204
  institutes Laboratory research papers, 203
  new name, methanol, 202
  recruitment of new staff, 193, 195
  retirement, 205
  service on committees, 188, 201, 202–203
Rodway, Thomas, matriculated at University
  College, 323
Rose, B A, 350
Rothschild Committee, 264, 265
Royal Aircraft Establishment, 217
Royal Arsenal, Woolwich, 181, 185, 311
Royal Clarence Yard, Gosport, 260
Royal College of Chemistry, 74, 76
Royal College of Mines, 74
Royal College of Physicians, Report, 251
Royal College of Science, 37, 88, 89, 123, 158,
  163
Royal Commission appointed to enquire into
  Arsenic Poisoning, etc., 150, 151
Royal Commission for Awards to Inventors, 183
Royal Commission on Civil Establishments, 116
Royal Institute of Chemistry, 223
Royal Scottish Museum, 163
Royal Society, 163, 185
  Proceedings, 183
Rutherford, Professor Ernest, 152

S

saccharin, 145
  smuggling, 146
saccharometer, abolition, 250
Saccharometer, Bates', 5, 57
saccharometer, standard, 58
Saccharometer, Tate, 208
Safeguarding of Industries Act, (1921), 190, 192,
  228, 234

salaries, staff, 38, 68, 69, 84–85, 125–126, 136,
  168, 172–173, 194, 195–196, 325, 334, 349
Sale of Food and Drugs Act, (1875), 36, 62, 78,
  89–90, 112, 113, 137, 141, 147, 156, 176, 234,
  282
  clause relating to Laboratory, 327
Sale of Food and Drugs Act, (1899), 17, 45, 157,
  176, 141
Sale of Food and Drugs Acts, 168
samples, disposal of surplus, 173, 256–257
Samuel, W Cobden, Head of Customs laboratory,
  110–111, 115
  deputy to Thorpe, 145
  rebuts Evidence of Bell, 117–118
Science and Art Department, 158, 159
Scott, Captain Robert, 143–144
seals, wax, 177
Select Committee on the Adulteration of Food,
  1855, 86–87
Select Committee on the Adulteration of Food,
  1874, 87
Select Committee on British and Foreign Spirits,
  1890, 110
Select Committee on the Tobacco Trade, 1844, 14
Seubert, K, 151
Sheppard, George, 175, 342
Sheridan, Mr, chief assistant in Customs
  laboratory, 333
Sibthorpe, John, matriculated at University
  College, 323
Sikes' hydrometer, 4, 111, 153
  checking accuracy, 58
  glass, 58
Sikes' Tables, revision, 153
Sikes, Bartholomew, 4
Simmonds, Charles, 150
  standard work on alcohol, 338
Smith, Dr H G, 348
Smoking and Health Now, 251
smoking machine, 252
Snuff, adulteration, 22
  in Ireland, 24
snuff, lime in, 24

snuff, offal, 117

snuff, white, 24

soap, adulteration, 33, 316

Society of Chemical Industry, 223

Society of Public Analysts, 88, 96, 157, 165
food standards, 92, 93–94, 94–95
objections to Excise Laboratory as referees, 89–90
standard method for butter analysis, 203

Somerset House, 36, 37, 74
conditions in laboratories, 37, 44, 84, 127
rooms occupied by Laboratory, 37–38, 39, 44, 74, 83, 334–335

Somerset House Laboratory, see Excise Laboratory

specific gravity bottle, 153
calibration, 173

spectra, absorption, 183

spirit strength, tables, 153–154

spirit, duty free, 62–63
international conference on, 66–67

spirit, measurement of strength, 2–4

Spirits of Wine Act, (1855), 63

St Clement Dane Church, 127

St Katharine's dock, 5

Stamps and Taxes, Board of, 36

standard methods, UK attitude towards, 166

Steering Board, 280

Stevenson, Peter, 52

stoker, extra for early attendance, 336

Strength of Spirit Ascertainment Regulations, (1916), 154

Stronghold, The, 174

Strontium, atomic weight of, 152

Stubbs, George, 151

students, examination papers, 305–308

students, see also under Excise Laboratory and Government Laboratory

sugar analysis, 145, 235

Sugar and Brewing Act, (1847), 53, 54

sugar in tobacco, 16, 19, 20
dispute with trader, 19, 22

sugar, use of in brewing, 49

Supply Reserve Depot, 198
Taunton, 260

Sutcliffe, R, 345

T

Tanner, Arthur, 312

tar, in tobacco, 251–252

Tarrant, Harmon, Excise Officer, 25

Tate saccharometer, 208

Tate, F G H, 204

Taylor, Dr Alfred, 40

tea,
adulteration, 35–36, 114, 316
analysis, 112, 113–114, 117, 146–147, 183
denaturing, 183

Tea Analyst, 113

Tea Inspectors, 113, 114

tea testing, description, 332

Teddington, move to, 272, 277–279

temperance beer, 139

Templeman, H V, 150

Temporary Laboratory Assistants, 194

Terry, John, 80

Thomson, Dr Thomas, 52

Thomson, Robert, 185

Thorpe, Professor Sir Thomas Edward, Principal of the Laboratory, 76
appointed, 123
advice to Antarctic expedition, 143–144
changes training of staff, 158
designs new Laboratory, 128
early career, 123–124
final years, 163
knighted, 163
member of International Committee on Atomic Weights, 151
on cost saving, 141
publishes sour milk method, 339
recruits staff directly, 161–162
reports increase in staff, 172
retirement, 163
serves on committee setting up NPL, 155

speaks to Public Analysts, 157
work on atomic weights, 151–152
work on original gravity of beer, 153
work on spirit strength tables, 153–154
Tilden, Professor Sir William, 158
Tobacco Act, (1840), 6
Tobacco Act, (1842), 6, 8, 30
tin hut, 196
tobacco
    adulteration, 5–6, 14, 138
    adulteration prevented, 32
    analysis, 176
    analysis, new apparatus for, 197
    duty, basis changed, 250–251
    factories, inspection, 24–30
    alkali salts in, 24
    coca leaves in, 32
    early methods of analysing, 14–16
    full analysis of, 20
    gum in, 30
    liquorice in, 25
    method of analysis, 314
    sugar in, 16, 19, 20
    survey of tar and nicotine content, 251–252
    water in, 30–31
Trade, Board of, 83
Transport of dangerous goods, 140–141
Treasury Solicitor, 181, 182
Treasury, 99, 121, 125, 126, 127, 161, 162, 165, 167
    control of Civil Service numbers, 222
    desire to prevent proliferation of new official laboratories, 122, 138, 176, 269
    lays down regulations for samples to Laboratory, 137
    suggested central laboratory, 122
Treatise on Adulteration of Food, 86
Trend enquiry, 263
Trinity House, 137
Tutton, A E H, 148

U

University College of North Wales, 163

University College of Wales, 211
University College, London, 20
    Excise staff at, 22, 68, 69, 70–74, 87, 89
    Excise staff at, costs, 72, 74
    Excise staff at, objections to, 73–74
Ure, Professor Andrew, 8, 19, 35
    hostility to the Excise, 16

V

Validity of Analytical Measurement programme, 284, 285
Voelcker, Dr Augustus, 88

W

Walker, R St J, 264
Wanklyn, Professor James, 45, 87, 88, 89, 91
    on milk analysis, 93
War Office, 141, 143, 181, 183, 222, 228
    analysts seconded to, 211–213
War Trade Department, 181
water,
    analysis, 44–45, 143, 230, 246
    fluoridation, 223
    nitrogen in, 88
    samples, 143
Water Analyst, 143
Water Pollution Research Board, 200
Weiss, A F, 345, 350
Westergaard, Professor, 177
Westminster Hall, work on roof timbers, 177
Wheat Act, (1932), 191
White Phosphorus Matches Prohibition Act, (1908), 163
Williams, Alex, Government Chemist, appointed, 270, 277
    oversees move and change to Agency status, 280
Williamson, Professor Alexander, 71, 72
wine analysis, effect of sugar content on, 111, 112
wine and spirit duties, 1, 103, 106
wine strength, determination, 103

wine, spurious, 107

wines, survey of strength, 108–109

witnesses, cost of consultant, 324

wood naphtha, addition to spirit, 62–63, 140

Wood, John, Chairman of the Board of Excise, 71–72, 73–74

Woodward, James, 167

Works, Commission of, 43

World Health Organisation, 223

Worswick, Dr Richard, Government Chemist, appointed, 280

X

X-ray fluorescence, 256

Y

Yew poisoning, 140

Young, Adam, Assistant Secretary to the Board of Excise, 40, 52, 53, 68, 73, 155

Young, Andrew, matriculated at University College, 323

Printed in the United Kingdom for HMSO
(2014/92) Dd0295334 9/92 C26 G531 10170